Principles of Horticulture:
Level 2

Charles Adams, Mike Early, Jane Brook
and Katherine Bamford

Routledge
Taylor & Francis Group

LONDON AND NEW YORK

First edition published 2015
by Routledge
2 Park Square, Milton Park, Abingdon, Oxon OX14 4RN

and by Routledge
711 Third Avenue, New York, NY 10017

Routledge is an imprint of the Taylor & Francis Group, an informa business

British Library Cataloguing-in-Publication Data
A catalogue record for this book is available from the British Library

Library of Congress Cataloging-in-Publication Data
Adams, C. R. (Charles R.)
Principles of horticulture : basic / C.R. Adams, K.M. Bamford, J.E. Brook, M.P. Early.
pages cm
1. Horticulture. I. Bamford, K. M. (Katherine M.) II. Brook, Jane. III. Early, M. P. (Michael P.)
IV. Title.
SB318.A33 2014
631.5′2–dc23
2014007884

ISBN 978-0-415-85908-0 (pbk)
ISBN 978-1-315-85879-1 (ebk)

Typeset in Univers LT by
Servis Filmsetting Ltd, Stockport, Cheshire

Contents

Preface

Horticulture involves the growing of plants: from the production of flowers, fruit and vegetables outdoors, to the more tender plants under protection. It includes the establishment and maintenance of plants for our enjoyment right through the landscape industry and the provision of sports turf. Those beginning their study of horticulture are often familiar with growing in gardens in order to create an attractive area around them, and to provide leisure facilities such as lawns and sitting out areas. Many embark on the outdoor production of fruit and vegetables in their garden or on an allotment. In some instances the use of greenhouses or other protected areas enables these enthusiasts to extend the growing season, and also to grow tender plants not normally possible in their garden. This background makes an appropriate starting point for an introduction to the principles of horticulture.

There are many techniques involved in horticulture and some familiarity with them is gained through our own experience of gardening. By studying the **principles of horticulture**, you will see how plants grow and develop. In this way, a better understanding of the plant's requirements and its responses to various conditions enables us to grow plants more effectively. The trained horticulturist is able to manipulate plants to suit their own ends, including optimising their growth, fitting them into a pleasing planting scheme or decorative arrangement, or to benefit wildlife.

In the introductory chapter, the many and varied sectors of the horticultural industry serving the gardener are introduced. Gardens do not exist in isolation, so some of the current issues surrounding **sustainable practice** and the importance of **conservation** of our garden plant heritage are explored, including some of the organisations which help to bring this about.

The early chapters introduce **the plant** to the reader. The gardener currently selects from an enormous range of plants that have been collected from all over the world and are adapted to a wide range of climates and habitats. Having acknowledged the work of **plant collectors** in providing such a cornucopia, the challenge of growing the plants they introduced in situations local to us is explored, along with the way these plants relate to each other and to other organisms in the garden through the study of **ecology**. The means of **classifying** and **naming** plants as an important way of communicating accurately with others in gardening and horticulture is established, after which the **internal** and **external features** of the plant and the many and varied ways in which they are adapted to different environments are described. The significance of these features is explained in relation to plant processes such as **photosynthesis, respiration** and **transport**, an understanding of which helps create opportunities for optimising and manipulating plant growth, development and behaviour to the grower's advantage. Along with details of plant reproductive methods including **pollination** and **fertilization**, the nature of **plant propagation** is then described in some depth to illustrate the various means by which plants can be multiplied efficiently.

Growing media include soils and soil substitutes such as composts, aggregate culture and nutrient film technique. Usually the plant's water and mineral requirements are taken up from the growing medium by roots. Active roots need a supply of oxygen, and therefore the root environment must be managed to include aeration as well as to supply water and minerals. The growing medium must also provide anchorage and stability, to avoid soils that 'blow', trees that uproot in shallow soils or tall pot plants that topple in lightweight composts.

The **physical characteristics of soil** are described to help explain how satisfactory root environments can be produced and maintained. **Organic matter, water** and **nutrients** are analysed in detail because they play such an important part in management of productive soils along with **soil pH** which has a major effect on the availability of nutrients. Soil conditions are modified by cultivations, irrigation, drainage and liming, while fertilizers are used to adjust the nutrient status to achieve the type of growth required. **Alternative growing media** and the management of plants grown in pots, troughs, peat bags and other containers are discussed in the context of the restricted rooting volume that makes it very different to growing in soil.

In growing plants for our own needs, we create a new type of community that introduces competition for environmental factors between one plant and another of the same species, between the crop plant and a **weed**, or between the plant and a **pest or disease organism**. These bring about the need to address the challenge of maintaining plant health. It is only by the identification of the competitive organisms (weeds, pests and diseases) and an understanding of their life cycle and biology that the gardener may select the correct approach to keeping them under control. With larger pests recognition is a relatively easy affair, but the smaller insects, mites, nematodes, fungi and bacteria are invisible to the naked eye and, in this situation, the grower must rely on the **symptoms** produced (type of damage) in order to discover the problem. For this reason, the pests are covered under major headings of the organism (most of which are large enough to recognize), while the diseases are described under symptoms. In this section the symptoms of physiological disorders such as frost damage, herbicide damage and mineral deficiencies that may be confused with pest or disease damage are also addressed.

The indexing and key word cross-referencing is to help the reader integrate the subject areas and to pursue related topics without laborious searching. It is hoped that this will enable readers to start their studies at almost any point, although it is recommended that an overview of the subject is gained by reading the early chapters first. Essential definitions are picked out in red boxes alongside appropriate parts of the text. Further details of some of the science associated with the principles of growing have been included in the grey boxes and specialist areas of the horticulture industry are picked out in green boxes. Each chapter concludes with a further reading section on the subjects covered. The companion website is also available at www.routledge.com/cw/adams with extended horticultural information, questions to test your knowledge, syllabus cross-referencing, downloadable tutor and student support materials and all the colour artwork from the text.

This book provides the ideal support for those studying horticulture up to Level Two and is organized to align with the very popular RHS 'Certificate in the Principles of Plant Growth, Propagation and Development' whilst providing the principles underpinning Level 1 and 2 Practical Certificates. In addition it covers the plant science, plant/crop protection and soils units in the Level 2 **Certificate, Subsidiary Diploma, Diploma and Extended Diploma and Work-based Diplomas in Horticulture.** It is also intended to be a comprehensive source of information for the keen gardener, especially for those taking City and Guilds Land-based Services Certificate in Gardening modules.

Charles R. Adams
Jane E. Brook
Michael P. Early

Acknowledgements

We are indebted to the following people without whom this edition would not have been possible:

Katherine Bamford for both the original plant science text and many of the photographs that are found throughout this edition (Figures 3.2–5, 3.8–10, 3.18, 3.23c, 4.8, 4.11–15, 5.1, 5.7, 5.9, 5.11, 7.11, 7.14c, 7.17, 7.19–20, 8.2–5, 8.8, 8.10–12, 8.14–16, 9.1, 11.10, 11.11, 11.13, 16.9, 16.11, 17.22b, 18.1, 18.2, 18.4b, 18.6a, 18.11a 18.11c); Ray Broughton for photographs and PowerPoint topics covering plant propagation and cultivation including those available on the companion website; Carl Dacus, CEO Dublin School of Horticulture, for the illustrations of fern propagation; the two diagrams illustrating weed biology and chemical weed control (Figures 17.5, 17.11) are reproduced after modification with permission of Blackwell Scientific Publications; Dr E.G. Coker who provided the photograph of the apple tree root system that he had excavated (Figure 10.1); Nick Blakemore provided the microscope photographs used through the plant section (Figures 6.4–8, 7.9, 9.4); Syngenta Bioline for biological control pictures (Figures 16.7b, 16.10a, b); Dr David Larner and Ellen Walden for guidance and advice.

Thanks are also due to the following individuals, firms and organizations that have provided photographs, specimens and tables:

Avice Hall (Figure 19.9); Madelaine Hills (Figures 16.7b, 16.10a, b); Howards Nursery, Lincs. (Figure 11.1).

CHAPTER 1
Level 2

Horticulture and gardening

Figure 1.1 Decorative border

This chapter includes the following topics:

- The nature of horticulture and gardening
- The horticultural industry
- The plant
- Ecology and gardening
- Sustainability
- Organic gardening
- Conservation

Principles of Horticulture. 978-0-415-85908-0 © C.R. Adams, K.M. Bamford, J.E. Brook and M.P. Early.
Published by Taylor & Francis.

Figure 1.2 A garden idyll

The nature of horticulture and gardening

Most of us are familiar with plants through our **gardens** and **gardening**. Around us are examples of gardens that are little more than surrounds to houses with the priority being to provide standing room for the car, utility areas and a place to sit out when the weather is good enough; the owners' emphasis is probably on minimizing the workload with much hard surface. For others, the area is an opportunity to provide an attractive view, to enhance the look of the property or to have a safe playing area, but without wanting to be more involved with the plants than necessary. Again the emphasis is likely to be on minimizing their input, with a person brought in considered to be a good solution to dealing with the time required and complications to keep good order. In contrast, gardeners consider their garden to be where they fulfil their wish to work with plants and seek to create their 'paradise on earth' (see Figure 1.2).

We learn quickly that gardening is not a simple process because it is a dynamic situation that we face. Plants and plantings change over time. Over the year there are seasonal changes; and as time passes the plants grow in size. Both these have significant consequences in attaining and maintaining what we wish to achieve. We can choose between rigorously maintaining our planned garden or allow it to evolve. There are many skills and techniques associated with both ways forward: planting and replanting; controlling the size of the plant; allowing plants to spread by seed or vegetative means such as runners or placing plants according to carefully devised plans; but, probably above all, selecting the right plant for the situation in the first place. It is an ongoing job for the gardener to hold at bay the undesirable plants and weeds while protecting their chosen ones from the attack of pests and diseases.

A gardener benefits from knowing about the factors that may improve or harm their plant's growth and development. The main aim of this book is to provide an understanding of how these factors contribute to the ideal performance of the plant in particular circumstances. For many gardeners, their intention is to apply the knowledge to improve their garden or allotment. Many others are seeking to build on what can be learned in the private garden in order for them to widen their interests, or seek professional or semi-professional employment.

What goes on in our gardens gives us all an insight into the wider world of horticulture.

The horticultural industry

Most of us are familiar with the products of the horticultural industry in terms of the fruit, vegetables, plants and flowers that we buy. The orchards and the fields of vegetables and flowers that we see on our travels give us some idea of the area over which **outdoor production** is undertaken (see Figure 1.3). The huge blocks of polythene tunnels or greenhouses indicate that **protected cropping** is being undertaken on a scale very different from the greenhouse in the garden (one of the largest tomato producers in Britain and Ireland has a block of 18 hectares, i.e. it could contain 36 football pitches). Even from the road we are able to note, in general terms, the work being done over the year and the wide range of equipment being used in the fields. Less obvious is what exactly is being done and the technology involved, especially with the protected crops. Protected cropping enables plant material to be supplied outside its normal season and to ensure high quality – for example, chrysanthemums all the year round, tomatoes to a high specification over an extended season, and cucumbers from an area where the climate is not otherwise suitable. Also out of sight is all the work being done with specialist equipment in the packing sheds where so much is undertaken, including the processing and grading as well as the packaging of what we see in the shops.

For many, leisure time will be spent in parks, the stately homes or the great gardens of the country where professionals, and some skilled amateurs, are employed in establishing and maintaining what we are more familiar with on a small scale: borders and lawns. Those of us whose leisure includes sport, playing or watching all too easily take for granted the preparation of the surfaces and surrounds of the field of play. **Turf culture** is a specialist part of horticulture that is concerned with the establishment and maintenance of decorative lawns and sports surfaces for football, cricket, golf and so on.

Parts of the industry also come close to our own gardens. **Garden construction** involves the skills of construction **(hard landscaping)** together with the development of planted areas **(soft landscaping)**. Closely associated with this sector is **grounds maintenance**, the maintenance of trees and woodlands **(arboriculture** and **tree surgery)** and **jobbing gardeners** who do so much to maintain and improve our domestic gardens.

The garden centre is where most gardeners see the work of the industry in more detail. But again, this is only the shop front of a large sector that specializes in producing plants in containers for us to put into our garden. A few have some plant production on site, but stock is usually bought in. The **hardy ornamental nursery stock (HONS)** sector is concerned with supplying not just the garden centres, but also all the other sectors of horticulture including the production of soil-grown or container-grown shrubs and trees (see Figure 1.4) and the young stock of **soft fruit** (strawberries etc.), **cane fruit** (raspberries etc.) and **top fruit** (apples, pears etc.). These plants are supplied in the following ways (see Figure 1.5):

Figure 1.3 Bulb field

Figure 1.4 A Hardy Nursery Stock area

Figure 1.5 a) bare root b) root-balled c) containerized d) container grown

▶ **Bare-rooted** plants are taken from open ground in the dormant period. These are cheaper but only available for a limited period and need to be planted out in the autumn or spring when conditions are suitable. In practice, this is mainly October and March. Roots should be kept moist until planted and covered with wet sacking while waiting. Plants received well before the time for permanent planting out should be 'heeled in' (i.e. temporary planting in a trench to cover the roots).

▶ **Root-balled** plants are grown in open ground, but removed with soil and the root ball secured by sacking (hessian). This natural material does not need to be removed at planting and will break down in the soil. This reduces the problems associated with transplanting larger plants.

▶ **Containerized** plants are also grown in open ground, but transferred to containers. Care needs to be taken to ensure that the root system has established before planting out unless treated as a bare-rooted stock.

▶ **Container-grown** plants, in contrast, are grown in containers from the time they were young plants (rather than transferred to containers from open ground). This makes it possible to plant any time of the year when conditions are suitable. Most plants supplied in garden centres are available in this form.

A specialist area within this sector is the **propagation** of plants. Part of this will be the multiplication of seeds and material to grow on until ready for sale but within this sector are the **plant breeders** who specialize in the creation of new cultivars.

It is essential that care is taken when buying plants. Besides ensuring that the best form of the plants are being purchased and correctly labelled, the plants must be healthy and 'well grown': plants should be compact and bushy (see etiolated p. 68), free from pest or disease and with appropriately coloured leaves (no signs of mineral deficiency see p. 270). The roots of container plants should be examined to ensure that they are visible and white rather than brown (see Figure 1.6). The contents of the container should not be root bound and the growing medium not too wet or dry.

The plant

At the heart of our garden is the plant. In order to discuss them, we need to have an unambiguous means of naming and ideally a way of seeing how they relate to each other (see Chapter 4). While they come in a great variety of shapes and sizes, they do have some fundamental similarities in their life cycles (see Chapter 5) and how they grow (Chapter 6).

Figure 1.6 a) healthy white roots b) poor roots in waterlogged pot

Figure 1.7 Corylus avellana 'Red Majestic' (Corkscrew hazel) with insert showing detail of the contorted twigs

It is important to have a clear idea of what a **healthy plant** is like at all stages of its life. The appearance of abnormalities can then be identified at the earliest opportunity and appropriate action taken. This is straightforward for most plants, but it is rather important to be aware of those plants whose healthy leaves are not normally green, that is, variegated, yellow, silver, purple and so on (see p. 114). Many 'stunted' plants are, often expensive, dwarf forms such as *Betula nana*, *Berberis thunbergeii* 'Atropurpurea Nana'. Some appear 'monstrous' – for example, those with contorted stems such as *Salix babylonica* var. *pekinensis* 'Tortuosa' and the corkscrew hazel *Corylus avellana* 'Red Majestic' (see Figure 1.7).

It should be noted that **physiological disorders** account for many of the symptoms of unhealthy

growth, which include nutrient deficiencies or imbalance (see Chapter 14). Damage may also be attributable to environmental conditions. These disorders are examined in Chapter 19 along with the problem of plant diseases, the symptoms of which may look similar. The problems that plant pests bring and how to deal with them is the theme of Chapter 18.

The incorrect functioning of any one factor may result in undesired plant performance. Factors such as the soil conditions, which affect the underground parts of the plant, are just as important as those such as light, which affect the aerial parts. The nature of soil is dealt with in Chapter 12. Increasingly plants are grown in alternatives to soil such as peat, bark, inert materials and water culture, which are reviewed in Chapter 15.

Ecology and gardening

As well as understanding how individual plants operate and having a knowledge of their requirements, it is important for gardeners to appreciate how they relate to their surroundings and the organisms that live there, whether this is other plants, the animals that live with them and on them, or the non-living parts of their environment. The subject of ecology deals with the interrelationships between all three of these groups, although it is the plants which are the focus of horticulturists. Some of these interactions are explored in Chapter 3.

Although the principles of ecology were developed in relation to the natural world, they can also be applied to artificial situations such as gardens. As in nature, a garden is made up of **population**s of plants and other organisms that form a **community** occupying a particular **habitat**. The community together with its non-living components form an **ecosystem** (p. 36). In terms of plant choice, it is useful to have a knowledge of the ecology of a plant's natural habitat, to try to match this to the environment they experience in the garden. Suitable levels of temperature, water and light, soil type and pH need to be considered on the basis of a plant's origin (Chapter 2). Ecologists have categorized and described a range of global geographical zones in which plants grow in the wild called **biomes** (p. 37) that give a broad idea of a plant's environmental requirements.

Choice of plant spacing directly relates to the ecological concept of **competition** between plants for essential environmental resources such as light, water and nutrients. Gardeners need to choose spacings suitable for individual plant species in, for example, a border, taking into account future growth. They can also space plants to harness the effects of competition, or lack of it, for specific purposes (p. 44).

Plants do not exist in isolation and their **interactions** with other organisms in the community are significant for the well-being of the whole community. Some of these organisms might be detrimental to plant health (e.g. pests and diseases) but many will be beneficial, such as natural pest predators and those which recycle nutrients within the garden ecosystem. One of the most important relationships between all garden organisms is how they feed on each other. Complex **food chains** and **food webs** (p. 38) exist through which energy and matter flow and these need to be kept in balance to support the whole community. Successful garden management takes account of these interactions.

The ecological concept of **succession** (p. 40) describes how plant communities change with time. Gardeners need to manage this change whether it is by controlling weeds on a newly prepared plot or halting succession by interfering with natural progression through, for example, pruning or harvesting.

Finally, paying attention to the ecological processes in gardens and using that knowledge to inform planting

Figure 1.8 A bumble bee feeding on a *Scabeous* flower

and management has an added advantage. Gardens are becoming increasingly recognized as an important **habitat for wildlife** providing suitable habitats for organisms to feed, shelter and breed. Again, an understanding of the ecology of gardens helps us to manage them appropriately to encourage wildlife (see Figure 1.8), which has also been shown to be good for our well-being too.

Beyond the garden gate, the importance of protecting and maintaining ecosystems, both locally and globally, is recognized because they provide humans with many useful '**ecosystem services**'. These include the supply of food, medicines and water, climate and disease control, nutrient cycling and crop pollination, flood control and prevention of soil erosion and also cultural benefits both spiritual and recreational. Gardeners as well as professional horticulturists, whether producing food, providing amenity planting on a large scale or simply growing plants in a domestic garden, can contribute to maintaining these ecosystem services through thoughtful design and management.

Sustainability

There are many ways to define the word '**sustainable**' but essentially this mean that, if an action or process is sustainable, it provides the best for the environment and people, socially and economically, both now and indefinitely into the future. Environmental sustainability means making decisions and taking actions that do not degrade the natural world irreversibly. The end result of most environmentally unsustainable practices is **loss of**

biodiversity, that is, a reduction in the number of habitats and species present in the wild, less genetic variation in the wild and fewer cultivated species (p. 45). Threats to biodiversity include habitat destruction, pollution, introduced species, global warming and overexploitation of natural resources.

When focusing on environmental sustainability, there are many issues that have relevance for gardeners and horticulturists. Some of these are:

▶ peat
▶ invasive non-native species
▶ 'greenhouse gas' emissions
▶ waste
▶ removal of rare species from the wild.

Peat is used in composts (Figure 1.9), and sometimes as a soil conditioner too. It has unique characteristics which make it ideal as a growing medium (see Chapter 15). Peat is extracted from peatlands mainly in the north temperate zone in areas of high rainfall and low temperature and is formed of the incompletely decomposed remains of sedges, mosses, grasses and reeds. Peatlands are important for four main reasons:

▶ They form a unique natural habitat supporting many rare species.
▶ They are an important carbon sink, locking up about a third of the world's soil carbon.
▶ They are an archive containing archaeological and geochemical historical information going back hundreds of years.
▶ They play an important role in the water cycle contributing to flood prevention, water quality and quantity of freshwater.

Peatlands form very slowly at a rate of only 1 mm a year. A 10 m deep layer of peat has taken around 10,000 years to form, so when peat is extracted, the damage is irreparable. Outside the European Union the importance of peatlands is recognized but legislation to protect them varies. Within the EU, most peatlands are now so rare that they are being designated as Special Areas of Conservation and member states are now required to protect them rather than permit them to be mined and then restored. To tackle the problem, targets have been set to phase out peat in bagged composts by 2020 for amateur gardening and by 2030 for professional growers. The main difficulty with reducing peat usage is finding suitable alternatives which are themselves sustainable, consistent, the right cost and deliver similar results to peat-based composts. Acceptable peat alternatives for seed sowing and some groups of plants such as acid-loving plants have been difficult to develop. Some of these alternatives are discussed in Chapter 15.

Figure 1.9 Bags of compost at a garden centre containing peat

Invasive non-native species include many weeds, pests and diseases which have arrived in this country from elsewhere. Their impact is second only to habitat destruction in reducing biodiversity. In the heyday of plant collecting (see Chapter 2), little thought was given to the effects on native wildlife of foreign plants. While most were well behaved, some garden escapees are now causing increasing problems for native plants and habitats and also pose a challenge to forestry, tourism, agriculture and construction. Since they originated in countries far removed from Britain and Ireland, they were not accompanied by their natural diseases and predators that might have kept them in check. Three examples of invasive plant species are:

▶ *Fallopia japonica* (**Japanese knotweed**) was introduced in the nineteenth century by E.A. Bowles as an attractive ornamental (Figure 1.10a). It can grow at an alarming metre or more a month, penetrating tarmac and concrete and is highly resistant to weedkillers. The annual cost to the construction industry for removal is significant (it cost £70 million to remove it from the London Olympic site).
▶ *Rhododendron ponticum* is a familiar weed in woodlands with attractive purple flowers in spring (Figure 1.10b). It is one of nature's most successful plants, spreading readily by seed and rooted branches, and its dense evergreen canopy cuts out the light and produces poisons and a deep leaf litter which suppress the growth of everything beneath it including woodland tree seedlings. It was introduced as an ornamental garden shrub in 1763 and also used as a

Figure 1.10 Three invasive non-native plant species; (a) *Fallopia japonica* (Japanese knotweed) at Myddleton House where E.A. Bowles first planted it in his garden; (b) Clearing *Rhododendron ponticum* from a woodland, note the lack of other plant species in the cleared area; (c) *Impatiens glandulifera* (Himalayan balsam) along a river bank.

rootstock for grafting. It was first recorded in the wild in 1893 and has spread to woodland and moorland, particularly on acid soils. The toxins in its leaves mean that it is undamaged by insects and can also harbour fungal diseases such as *Phytophthera ramorum* and *P. kernoviae* (see p. 44) which can spread to oaks, beech and nursery stock.

▶ ***Impatiens glandulifera* (Himalayan balsam)** can grow to 3 metres tall and is a serious problem along riverbanks (Figure 1.10c). It was introduced in 1839 from Northern India and grown for its attractive pink and white slipper-shaped flowers. Its seed pods can project the seed explosively up to 7 metres away and the seeds are carried downstream to colonize new areas. The dense stands shade out other species and it produces more sugary nectar than any other European plant species, attracting pollinating insects such as bees away from native plants. When it dies back in late autumn, it leaves riverbanks exposed and liable to erosion.

Invasive species are particularly troublesome along and in waterways where it is difficult to use chemical control. Aquatic species such as water fern, Australian swamp stonecrop, floating pennywort, Canadian pondweed and parrots' feather are particularly invasive. Others include *Buddleja*, *Cotoneaster*, some broadleaved bamboos and some exotic honeysuckles. Gardeners can best prevent the spread of these damaging weeds by knowing precisely which plants they are buying, choosing non-invasive species and disposing of plant material carefully. See 'Invasive Plants' on the companion website.

The global trade in plants has also led to the introduction of many new **'alien' pests and diseases** (see Chapters 18 and 19). Ash dieback (*Chalara fraxinea*) is just the latest in a series of fungal diseases affecting our trees.

'Greenhouse gas' emission, largely due to the burning of fossil fuels and conversion of land for agriculture, has been identified as the major cause of global warming. The fruit and vegetables we eat and the plants we grow will all have a **carbon footprint** made up of the

Figure 1.11 Plastics waste at a nursery

Figure 1.12 *Encephalartos ferox*

'greenhouse gas' emissions incurred in their production and supply. Fossil fuels are used directly for cultivation, heating and transport of produce and plants. They are also used indirectly to supply energy for lighting, refrigeration, water treatment and manufacture of materials such as plastics, fertilizers and pesticides. Issues such as the sale of all-year-round produce which involves heating and lighting in winter and long distance transport ('air' and 'road miles') are all hotly debated. Growers can reduce their 'carbon footprint' by reducing their energy use or using alternative 'green' energy. For example, combined heat and power generated from green waste can be used for tomato production in the winter and wood chip burners are used by *Alstromeria* growers to heat their greenhouses in areas where gas is not readily available. Some major greenhouses are located beside power stations to make use of 'waste' heat. A new greenhouse production area generates all its energy from crop waste and exports the remainder to the National Grid. A successful scheme in Suffolk uses 'waste' carbon dioxide produced in an adjacent factory to boost photosynthesis (see p. 113) in its greenhouse-grown tomatoes, thus reducing the amount lost to the atmosphere.

Waste is an issue because of the damage to the environment from landfill and the energy used to collect, dispose of and recycle it. Horticulture uses many types of plastic from trays and pots (see Figure 1.11) to the polythene used to cover polytunnels, fleece to insulate crops and packaging for sale of fruits and vegetables, cut flowers and plants. Growing also generates green wastes such as plant material, including unsaleable produce which is not harvested, and composts.

Removal of rare species from the wild is a result of our demand for new and ever more interesting plants. In particular, overcollection of wild bulbs has been an issue in recent years. Many of these are now

protected to some extent by CITES (Convention on Trade in Endangered Species) which monitors and issues quotas for imports and exports of threatened species. CITES produces a checklist which can be downloaded from its website, most of the plants listed are orchids, cacti, cycads (Figure 1.12), some succulents and some carnivorous plants.

Sustainable gardening practices

Gardeners can help reduce their environmental impact in many ways including the following:

▶ **Cut down on water** use by planting drought-tolerant plants and mulching (see p. 160). Use rainwater to water plants (purification of mains water is energy intensive) or reuse household water where appropriate. Water only when necessary and do this efficiently, that is, thoroughly and less frequently (see Chapter 12).

▶ **Reduce energy use** by, for example, heating and lighting in glasshouses or power tools such as air blowers outdoors. Insulate greenhouses as an alternative to heating. Source garden materials and plants locally if possible to reduce fossil fuels used in transport.

▶ **Limit the use of fertilizers, pesticides** and other garden chemicals and use manures correctly. Only treat pests and diseases when necessary and follow the instructions carefully using only approved products. Incorrect and excessive use might kill beneficial insects and pollute soil, ponds and groundwater.

▶ **Waste management** – good practice can be summarized as **Reduce, Reuse, Recycle**.
  ▷ **Reducing waste** is the most sustainable approach to waste **management – for example**, by avoiding excessive packaging.
  ▷ **Reuse** items such as plastic pots and trays (Figure 1.11), which should be thoroughly

Figure 1.13 A range of rocks sold for garden use

cleaned and sterilized before use (see p. 130), if reducing waste is not possible. Hard landscaping materials such as slabs, rocks, building materials and timber can be reused, which also avoids further damage to habitats (Figure 1.13). Limestone pavement, for example, is a natural rock formation which provides a habitat for many rare species and was previously used for rock gardens. Although it is now legally protected in Britain and Ireland, removal still occurs.

▷ **Recycle** items that cannot be reused, although many horticultural plastics such as seed trays and polystyrene can only be landfilled or incinerated. Garden waste can be composted at home (see Chapter 13), which avoids energy being used for collection and disposal as well as providing a valuable soil improver. Organic waste in landfill produces methane which is 25 times stronger than carbon dioxide as a greenhouse gas.

▶ **Timber** for garden structures should be produced from sustainable sources such as those certified by the Forestry Stewardship Council (FSC).

▶ **Peat alternatives** should be used wherever possible and never use peat as a soil improver or mulch. If possible, choose plants that have been grown in peat-free composts.

▶ **Never collect plants from the wild** and ensure that purchased plants are sustainably sourced.

▶ **Check that imported rare plants** have appropriate CITES certificates.

▶ **Avoid planting invasive species** especially near or in ponds and waterways.

Managing a garden for wildlife will help to conserve biodiversity both in the garden itself and in the wider area (see Chapter 3). See 'Sustainable Gardening' on the companion website.

Organic gardening

There are many different ways of gardening that make use of the principles of horticulture outlined in the following chapters. One particular philosophy of growing that demonstrates the wide range of approaches is that of growing 'organically'. Most of the book deals with the nature of plants and soils common to all growers, but those growing 'organically' will focus on some specific parts of the text and not make use of others such as the use of artificial fertilizers and pesticides. The degree to which the gardener adopts an 'organic' approach can range from the strict adherence to the 'organic' philosophy to those who are just inclined to 'no chemicals' but are less concerned about obtaining organically acceptable plants, propagation material, manures and so on.

Organic gardeners view their activities as an integrated whole and try to establish a sustainable way forward by conserving non-renewable resources and eliminating a reliance on external inputs. The soil is managed with as little disturbance as possible to the balance of organisms present. Organic growers maintain **soil fertility** by the incorporation of animal manures (see p. 163), composted material (see p. 160) and green manuring (p. 165). The intention is to ensure plants receive a steady, balanced release of nutrients through their roots: 'feed the soil, not the plant'. Besides the release of nutrients by decomposition (see p. 158), the stimulated earthworm activity incorporates organic matter improving soil structure which can eliminate the need for cultivations (see 'no dig' methods p. 150). Emphasis is placed on the **balanced nutrition** of the plants because of its role resisting pests and diseases (p. 192) – for example, excess nitrogen feed leads to soft growth that is vulnerable to diseases, strong growing plants can grow away from pest attacks.

The main cause of an imbalance of soil organisms is considered to be the use of **quick-release fertilizers** and **pesticides**. The greater number of soil species (biodiversity) that is encouraged prevents the increase of pests and diseases. Pests and diseases are managed 'organically' by this approach along with the use of resistant cultivars (p. 190), physical barriers, pesticides derived from plant extracts (p. 203), by careful rotation of plant species (p. 193) and by the use of naturally occurring predators and parasites (p. 195–6). Weeds are controlled by using a range of physical and cultural methods including mechanical and heat-producing weed control equipment.

There are many organizations that provide support for those wishing to succeed with an 'organic' approach to gardening. Notably there is 'Organic Growing', formerly the Henry Doubleday Research Association (HDRA), which promotes organic methods of gardening through their advisory and education work. They also have demonstration gardens open to the public.

The European Community Regulations (1991) on the 'organic production of agricultural products' specify the substances that may be used as 'plant-protection products, detergents, fertilizers, or soil conditioners'.

Those intending to sell produce with an organic label need to comply with the standards set by the International Federation of Organic Agricultural Movement (IFOAM). These standards set out the principles and practices of organic systems which, within the economic constraints and technology of a particular time, promote broadly:

▶ the use of management practices which sustain soil health and fertility
▶ the production of high levels of nutritious food
▶ minimal dependence on non-renewable forms of energy and burning of fossil food
▶ the lowest practical levels of environmental pollution
▶ enhancement of the landscape and wild life habitat
▶ high standards of animal welfare and contentment.

Certification is organized nationally with a symbol available to those who meet and continue to meet the 'organic standards' overseen by a recognized organic body with whom, by law, the grower must be registered. In Britain and Ireland, the Department for Environment, Food and Rural Affairs (DEFRA) and the Department of Agriculture, Food and the Marine (An Roinn Talmhaíochta, Bia agus Mara) approve the Organic Control Bodies (CBs) who license individual organic operators. Control Bodies include the 'Organic Farmers and Growers', 'Irish Organic Farmers and Growers Association', 'Scottish Organic Producers' Association', 'Quality Welsh Food Certification Ltd' and the 'Soil Association'. See 'Organic Growing' on the companion website.

Conservation

Conservation is the management of animals, plants and other organisms to ensure their survival as a resource for future generations. It focuses on reducing threats to biodiversity in the wild but is also concerned with conserving cultivated plants and their wild ancestors. These are the gene pool on which future plant breeders can draw for further improvement of plant species.

Figure 1.14 Millenium seedbank at Wakehurst Place

In situ **conservation** involves the creation of natural reserves to protect habitats, and the wild species they contain. Sometimes cultivated plants are also conserved *in situ* – for example, the East of England Apple and Orchards Project is one of numerous local initiatives protecting orchard fruits and their habitats.

Ex situ **conservation** includes whole plant collections in botanic gardens, arboreta, pineta and genebanks where seeds, vegetative material and tissue cultures are maintained – for example, the Millenium Seedbank of the Royal Botanic Gardens, Kew at Wakehurst, Surrey (see Figure 1.14). The botanic gardens are coordinated by the Botanic Gardens Conservation International (BGCI), which is based at the Royal Botanic Gardens, Kew in London and are primarily concerned with the conservation of wild species.

For cultivated species, large national collections include the National Fruit Collection at Brogdale, Kent which holds over 3,500 nut and fruit cultivars in its orchards and the Warwick Genetic Resources Group which collects, conserves, documents and researches into a wide range of vegetable crops and their wild relatives. The Garden Organic Heritage Seed Library conserves old varieties of vegetables which were once commercially available, but which have been dropped from the National List (and so become illegal to sell), operating a seed exchange programme. Plant Heritage (formerly the National Council for the Conservation of Plants and Gardens (NCCPG)) was set up by the Royal Horticultural Society at Wisley in 1978 and is a excellent example of professionals and amateurs working together to conserve stocks of garden plants threatened with extinction. The aim is to ensure the availability of a wider range of plants and to stimulate scientific, taxonomic, horticultural, historical and artistic studies of garden plants. There are over 600 collections of ornamental plants encompassing 400 genera and some 5,000 plants. A third of these

Figure 1.15 Hamamelis x intermedia 'Barmstedt Gold'

are maintained in private gardens, many of which are open to the public.

Others are held in publicly funded institutions such as colleges – for example, *Sarcococca* at Capel Manor College in North London, *Escallonia* at the Duchy College in Cornwall, *Penstemon* and *Philadelphus* at Pershore College and *Papaver orientalis* at the Scottish Agricultural College, Auchincruive. Hillier Gardens in Hampshire alone holds nine genera, including the witch hazels, many of which are featured in the winter garden (see Figure 1.15). Rare plants are identified and classified as 'pink sheet' plants. Plant Heritage's 'Threatened Plants Project' aims to identify and assess for heritage value all named British and Irish plant cultivars and to identify rare and threatened plants in its collections. It develops conservation plans with partner organizations fulfilling global targets for conserving the diversity of plants with cultural and socio-economic value.

Further reading

Clevely, A. (2006) *The Allotment Book*. Collins.

Dowding, C. (2012) *Vegetable Course*. Frances Lincoln.

Brickell, C. (ed.) (2006) *RHS Encyclopedia of Plants and Flowers*. Dorling Kindersley.

Brickell, C. (ed.) (2003) *RHS A–Z Encyclopedia of Garden Plants*. 2 vols, 3rd edn. Dorling Kindersley.

Brookes, J. (2001) *Garden Design*. Dorling Kindersley.

Hessayon, D.G. (n.d.) *Garden Expert Series*. Expert Publications.

Lampkin, N. (1990) *Organic Farming*. Farming Press.

Pears, P. and Strickland, S. (1999) *Organic Gardening*. RHS. Mitchell Beazley.

Thomas, H. and Wooster, S. (2008) *The Complete Planting Design Course*. Mitchell Beazley.

Please visit the companion website for further information:
www.routledge.com/cw/adams

CHAPTER 2
Level 2

Plants of the world

Figure 2.1 *Sequoiadendron giganteum* (Giant redwood): seeds of this species were collected in California by William Lobb in 1853

This chapter includes the following topics:

- Geographical origins of plant culture
- 'Centres of origin'
- World climatic zones
- Britain and Ireland climatic zones
- Plant hunters
- World trade of horticultural plants

Principles of Horticulture. 978-0-415-85908-0 © C.R. Adams, K.M. Bamford, J.E. Brook and M.P. Early. Published by Taylor & Francis. All rights reserved.

Origins

Horticulture and agriculture have a very long history. There is evidence to suggest that areas now known as Iraq were growing forms of flax, lentil and wheat by 7000 BC. It also appears that Chinese farmers were cultivating rice by 6000 BC and that Papua New Guineans had root crops such as yams as early as 5000 BC. Central American cultures are claimed to have had *Phaseolus* beans, marrows, chilli peppers and maize by 4000 BC. South American and Egyptian cultures had ornamental gardens as early as 1500 BC. India, Persia and Greece, among many others, later developed ornamental horticulture and landscape gardening .The fabled 'Hanging Gardens of Babylon', listed as one of the 'wonders of the world', were constructed around 600 BC.

Influences in Europe came first with gardens established by the Romans across their empire up until about AD 300. At about the same time, China (and three centuries later, Japan) was developing its own quite different styles of gardening on the other side of the world. The 'Silk Road' from central China to the eastern Mediterranean was also developing at about this time and was one of the ways that plant species were distributed across the continents. *Rosa* species from China, for example, were seen in Persian gardens over two thousand years ago.

Much later, in southern Europe, the Spanish (fourteen and sixteenth centuries AD), French (thirteenth and eighteenth centuries AD) and Italians (fifteenth and sixteenth centuries AD) developed a wide range of gardening skills. Northwest Europe, including Britain and Ireland, gained a keen interest in horticulture in fifteenth century AD and this passion has grown since then to equal any other part of the world.

Plants have been obtained from other areas of the world by many cultures as awareness of suitably decorative or edible species became known. At the present time, an estimated 8,000 species of plant are grown worldwide.

Centres of origin

Worldwide, distinct '**centres of origin**' can be considered as important sources of crop and ornamental species (Figure 2.2). Each of these centres can be studied in terms of its plant **diversity** (the number of **wild species** of a particular genus found within the area (see also p. 34). Information of this type is taken to indicate whether the area is a **focal**

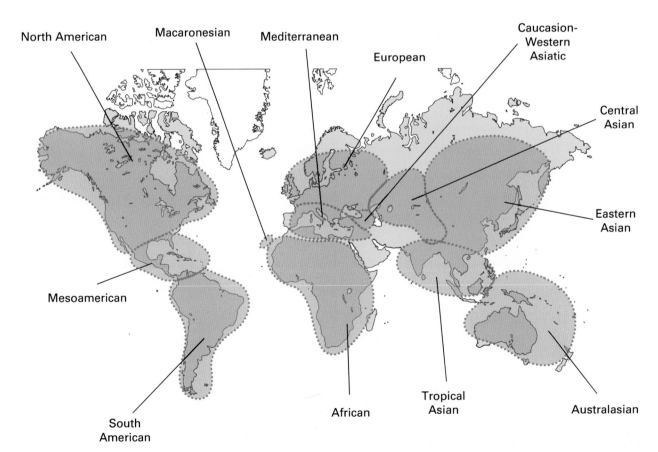

Figure 2.2 The main centres of origin of plants

Figure 2.3 (a) Antirrhinums originate from the Mediterranean area; (b) *Strelitzia reginae* originates from southern Africa

point for that genus. For example, the genus *Fuchsia* has most of its species in the tropical part of the South American Centre, indicating that this genus probably originated in this area.

Second, the 'centre of origin' of plant cultivars is often seen as an indication that a community of people have shown the interest and ability to select and develop **improved plant varieties** and cultivars from a wild ancestor species. For example, the wide range of tomato cultivars seen in the Andes/Central America/Mexico area was developed from wild *Solanum* species by the Mayans, Aztecs and Incas.

Important examples of plant resistance against tomato diseases has recently been identified from different wild South American *Solanum* species:

▶ tomato mosaic virus resistance from *Solanum chilense* (from Peru and Chile)

▶ corky root disease resistance from *Solanum corneliomuelleri* (from Peru)

▶ *Fusarium* wilt resistance from *Solanum habrochaites* (from Ecuador and Peru)

▶ *Alternaria* leaf spot and *Cladosporium* leaf spot resistance from *Solanum peruvianum* (from Peru).

Below, twelve important '**centres of origin**' distributed around the world are briefly described. Emphasis is given here to the **geographical** position of the area. On page 19, a review of **climatic** aspects of plant distribution is given.

1. The **Mediterranean** Centre and surrounding areas have given rise to more than 1,000 ornamental species (about 20% of the total of common garden plants in cultivation). Popular commercial plants in Britain and Ireland native to the Mediterranean Centre include: antirrhinum, daffodil (*Narcissus hybida*), hyacinth (*Hyacinths hybrid*), *Cyclamen*, stock (*Matthiola incana*), iris (*Iris hybrida*), some carnations (*Dianthus*), garden pansy (*Viola x wittrockiana*), horse chestnut (*Aesculus hippocastanum*), hyacinth, laurel (*Laurus nobilis*), oleander (*Nerium oleander*), lily (*Lilium candidum*), fig (*Ficus carica*), cornflower (*Centaurea*), poppies (*Papaver*) and Christmas rose (*Helleborus niger*).

Edible plants originating from this region include asparagus, beetroot, Brussels sprouts, cabbage, cauliflower, celery, fennel, grape, lettuce, onion, parsnip, rhubarb, rosemary and wheat.

2. The **African** Centre is a source of about 600 ornamental plants, or about 13% of the world total. These include African violet (*Saintpaulia ionantha*), castor oil plant (*Ricinus communis*), geraniums (*Pelargonium*), *Strelitzia reginae*, *Streptocarpus hybrida*, *Amaryllis belladonna*, summer hyacinth (*Galtonia candicans*), *Gerbera jamesonii*, *Gazania* spp., some *Erica* spp., mother-in-law's-tongue (*Sansevieria trifasciata*), *Protea* spp., *Gladiolus* spp. and red-hot poker (*Kniphofia* spp.).

Edible crops deriving from this region are black-eyed peas (*Vigna unguiculata*), barley, egg plant, flax, sorghum, coffee, pearl millet (*Pennisetum glaucum*) and okra or lady's fingers (*Abelmoschus esculentus*).

3. The **North American** Centre has given rise to about 600 useful species, representing about 13% of the plants in cultivation worldwide. These include black cherry (*Prunus serotina*), black walnut (*Juglans nigra*), eastern red cedar (*Juniperus virginiana*), box elder (*Acer negundo*), balsam poplar (*Populus balsamifera*), *Ceonothus*, red oak (*Quercus rubra*), aster, *Penstemon*, phlox (*Phlox paniculata*), perennial

Figure 2.4 (a) Marigolds originate from the Mesoamerican centre; (b) Calico vine (*Aristolochia littoralis*) originates from Brazil

lupin (*Lupinus polyphyllus*), *Rudbeckia* spp., *Coreopsis*, *Aquilegia*, evening primrose (*Oenothera*), *Trillium* spp. and *Echinacea purpurea*.

Edible crops deriving from this region are a grape (*Vitis labrusca*), blueberries and cranberries.

4. The **South American** Centre account for about 600 species or 13% of the world's ornamental flora. They include *Aristolochia*, *Canna*, *Dieffenbachia*, *Philodendron*, many species of *Fuchsia*, gloxinia (*Sinningia*), *Gunnera*, *Salvia splendens*, heliotrope (*Heliotropium arborescens*), nasturtium (*Tropaeolum* spp.), orchid (e.g. *Epidendrum*), *Verbena* hybrids (mainly from *Verbena peruviana*) and morning glory (*Ipomoea alba*, *I. purpurea*, *I. tricolor*).

Edible species deriving from this region include potato, *Phaseolus* bean, peanut, pineapple, sweet potato (*Ipomoea batatas*) and tomato.

5. The **Mesoamerican** Centre (Southern Mexico, Central America and West Indies) has about 600 species or 13% of world's ornamental species. These include cacti (e.g. *Opuntia*, *Echinocereus*, *Mammillaria*), orchids (e.g. *Cattleya*, *Odontoglossum*, *Oncidium* and *Vanilla*), begonia (*Begonia imperialis*), *Cosmos bipinnatus*, dahlia, marigolds (*Tagetes erecta*, *T. patula*, *T. tenuifolia*), *Ageratum houstonianum*, *Commelina tuberosa* and *Zinnia violaceae*.

Edible species deriving from this region include maize, capsicum, pepper, marrow, tomato, avocado, vanilla, papaya, sunflower, cassava, tobacco and strawberry.

6. The **Tropical Asian** Centre covers the Indian and Indochinese regions, including the important Indo-Malayan area. Some 450 ornamental species, about 8% of world ornamentals, have been obtained from this area. Not surprisingly, many species are tender

and are grown in Britain and Ireland in greenhouses or indoors. The list includes orchids (e.g. *Cymbidium*, *Phalaenopsis* and *Vanda*), begonias (e.g. *Begonia rex*), India rubber tree (*Ficus elastica*), cockscomb (*Celosia argentea*) and garden balsam (*Impatiens balsaminea*).

Edible species deriving from this region include chickpea (*Cicer arietinum*), mango, orange and banana.

Two interesting historical stories of crop movements relating to this area of the world may be mentioned.

▶ **Breadfruit**. The tropical breadfruit (*Artocarpus altilis*) is a tall dicotyledonous tree species, probably originating in New Guinea and used as a form of staple carbohydrate in those areas. It was spread around the South Eastern Asia area by the Polynesian peoples over 3,000 years ago. **Joseph Banks** (see p. 23), while on the first Captain Cook exploration in the South Pacific in the eighteenth century, noticed the potential of the fruit as food for plantation workers in the West Indies. On his return to Britain and Ireland, Banks encouraged the voyage of the *HMS Bounty* under **Captain Bligh** in 1787 in which the infamous mutiny of the sailors prevented any breadfruit reaching the Caribbean. In 1791, a second attempt by Bligh in the ships *HMS Providence* and *HMS Assistant* was successful in transporting large numbers of young plants from Tahiti in the South Pacific to St Helena, St Vincent and Jamaica in the Atlantic area. The breadfruit was thus spread to the western hemisphere. After all this effort, the fruit did not become as important a part of the diet as was anticipated.

▶ **Spices**. A second example of plants for worldwide use are Indonesian spices. Their origin was the 'Spice Islands' (the islands concerned are more accurately called the Malukus or Maluccas), located in northeast Indonesia, between Sulawesi and New Guinea. They were for many centuries the main area for the production of **pepper**, **cloves** and **nutmeg** (**mace** also comes from the nutmeg plant). These products were traded with countries such as China and India, and also were taken overland via the '**Silk Route**' to cities such as Venice. Sea routes to Europe, established from about 1520, led to increased trade in spices between Indonesia and Europe. The resulting competition for dominance in this trade between Portugal, Holland and Britain was part of the reason for the colonial expansion of these three nations in the tropical South East Asian area.

7. The **European** Centre includes the Atlantic European (this includes Britain and Ireland), Central European and Eastern European areas. About 300 (6%) of ornamental species are from this centre. These include lime (*Tilia platyphyllos*, *T. cordata* and *T. europaea*), Norway Maple (*Acer platanoides*), sycamore (*Acer pseudoplatanus*), birch (*Betula pendula*), hornbeam (*Carpinus betulus*), European beech (*Fagus sylvatica*), European ash (*Fraxinus excelsior*), English oak (*Quercus robur*), Scots pine (*Pinus sylvestris*), European larch (*Larix decidua*) and yew (*Taxus baccata*), horned violet (*Viola cornuta*), sweet violet (*Viola odorata*), snowdrop (*Galanthus nivalis*) and wild daffodil (*Narcissus pseudonarcissus*).

Edible species deriving from this region include, damson, blackberry, raspberry, blackcurrant, parsnip, turnip, mustard, radish and spinach

8. The **Eastern Asiatic** Centre includes China, Japan and the extreme eastern section of Russia. About 250 ornamental species from this area represent 5% of the world total in general cultivation today. About 50% of its imported native species can be grown outside in Britain and Ireland, including *Rhododendron* (*Azalea*) spp., chrysanthemum, clematis, day lily (*Hemerocallis*), *Forsythia*, hollyhock (*Alcea rosea*), peony (*Paeonia lactiflora*), tree peony (*Paeonia suffruticosa*), royal lily (*Lilium regale*), *Sedum kamtschatichum*, wisteria (*Wisteria sinensis*), China aster (*Callistephus chinensis*) and *Kerria japonica*.

Figure 2.5 (a) Snowdrops (*Galanthus nivalis*) originates in the European and West Asia centre; (b) *Forsythia* originates from the Eastern Asiatic centre; (c) *Forsythia x intermedia* (close up)

Figure 2.6 *Phormium tenax* originates in the Australasian Centre

Edible species deriving from this region include apricot, Chinese cabbage, cherry, cucumber, Kiwi fruit (*Actinidia chinensis*), onion, peach, pear, rice and soybean.

9. The **Caucasian-Western Asiatic** Centre is a small region around the Black Sea next to southern Russia providing about 150 ornamental species (3%) for world horticulture, including autumn crocus (*Colchicum speciosum*), *Pyrethrum coccineum* (syn. *Chrysanthemum*) and poppy (*Papaver orientale*).

10. The **Central Asian** Centre includes the mainly mountainous regions of the Central Asian Republics including Afghanistan. This area contributes about 130 species, or 2.5% of ornamental species such as *Malus sieversii*, *Rhododendron arboretum*, *Rhododendron niveum*, *Rhododendron ponticum*, *Tulipa kaufmanniana*, *Tulipa fosteriana*, *Tulipa greigii*, *Allium karatawiense* and *Colchicum luteum*.

11. The **Australasian** Centre, comprising Australia, New Zealand, New Guinea and immediately surrounding islands, has some importance as a source of ornamentals (about 100 species representing about 2% of world horticulture). These include *Schefflera*, *Eucalyptus*, *Casuarina*, *Callistemon*, *Hebe*, *Helichrysum bracteatum*, New Zealand flax (*Phormium tenax*).

Edible species include macadamia nuts (*Macadamia* spp.).

12. The **Macaronesian** Centre covers the Canary Islands and Madeira off the west coast of Africa. Although small in area, its isolated islands provide about 50 species of ornamental species (about 1% of world species), including the dragon tree (*Dracaena draco*), date palm (*Phoenix canariensis*), several species of *Aloe* and Azores bellflower (*Azorina idalii*).

Three early cultures

Over the centuries, there have been many people who have searched in different parts of the world for novel ornamental or edible plants. It is likely that wandering peoples, on their travels, selected and planted desirable plant specimens, especially fruit-bearing shrubs and trees that would be available in strategic places on their journeys if the soils and climate were favourable for the plants' growth. These species gradually became accepted as edible/decorative additions to the horticultural needs of the explorer's home community.

Three centres of social development are described.

▶ **Central America** now covering modern Honduras, Belize, Guatemala, El Salvador and southern Mexico was inhabited by the **Mayan** culture from around 2000 BC and it is thought that they developed horticultural plants such as *Begonia* cultivars and strawberry. This area had fertile volcanic soils, but its occasional severe drought periods may have been involved in the Mayan culture's sudden decline.

▶ The **Chinese** culture that developed from about 2000 BC was unified by the first emperor Chi in 217 BC. His main headquarters at that time was in central China near the present day city of Xian (the site of the 'terracotta warriors') and located close to the Yellow River. This culture has been associated with the development of horticultural plants such as chrysanthemum and cucumber. Xian was the eastern starting point for the 'Silk Road' along which many plant species such as *Paeonia* and *Rosa* cultivars may have been carried to Europe.

▶ The '**Fertile Crescent**' at the eastern end of the Mediterranean that includes modern nations Egypt, Cyprus, Lebanon, Israel, Jordan, Syria, southern Turkey and Iraq is known to have had established settlements as early as 900 BC, with notable water sources such as the rivers Nile, Tigris and Euphrates. The cultures within this area were notable for developing many horticultural plants such as the grape and the hyacinth.

World climatic zones

Seven major climatic zones found around the world that have provided some useful garden and green house species are as follows:

▶ **Polar** – very cold and dry all year, e.g. Greenland. Example British garden species – arctic poppy (*Papaver nudicaule*).

▶ **Temperate** – mild summers and cold winters, e.g. Britain and Ireland. Example British garden species – clustered bellflower (*Campanula glomerata*).

▶ **Arid** – dry, hot all year, e.g. Northern Africa. Example British garden species – many coloured spurge (*Euphorbia polychroma*).

▶ **Equatorial** (tropical) – hot and wet all year, e.g. Amazonian South America. Example British garden species – sweet tobacco (*Nicotiana alata*).

▶ **Subtropical** – hot, humid summers and generally mild to cool winters, e.g. south east China and south east USA. Example British garden species – African lily (*Agapanthus*).

▶ **Mediterranean** – dry hot summers and mild winters, e.g. around the Mediterranean. Example British garden species – sweet pea (*Lathyrus odoratus*).

▶ **Alpine** – very cold all year in mountainous areas, e.g. European Alps. Example British garden species – alpine pink (*Dianthus alpinus*).

Climate in large countries

It should be emphasized that large countries such as China, USA and Brazil (that have historically been sources of many plant species for worldwide distribution) may have several different climatic zones within their border. Ecologically, these climatic zones will relate to different **biomes**, with their distinctive groupings of plants and animals (see p. 37).

These different plant species originating in a single country, but from different zones, may also have different requirements for growing in Britain and Ireland, reflected in their Royal Horticultural Society (RHS) **hardiness rating** (see p. 21).

For example:

▶ **China** encompasses tropical, temperate and alpine zones. Snake bark maple (*Acer davidii*) has a RHS hardiness index of H5, while star-jasmine (*Trachelospermum jasminoides*) is rated H4.

▶ **USA** encompasses polar, temperate, alpine and tropical zones. Creeping juniper (*Juniperus horizontalis*) is rated H7, while blue blossom (*Ceanothus thyrsiflorus*) *is* given an H4 rating.

Figure 2.7 Sweet pea originates in the Mediterranean area

Figure 2.8 Creeping juniper originates in the USA

▶ **Brazil** encompasses tropical (wet, moist and dry zones) and also a temperate zone in the south. Blue passion-flower (*Passiflora caerula*) is rated H4, while paper flower (*Bougainvillea glabra*) has a H1c rating.

Climate in Britain and Ireland

Differences in climatic conditions do occur across the whole of Britain and Ireland. The differences are not as great as those described above for world climatic zones, but are enough (especially during the winter months) to cause serious damage to certain plants growing outdoors. For this reason, plants have, over the years, been classified according to their temperature sensitivity (**hardiness**) using different systems.

An American (**USDA**) system has been used for several years in Britain and Ireland and is described briefly on the companion website: www.routledge.com/cw/adams.

Hardiness

The current Royal Horticultural Society hardiness list gives nine hardiness ratings for plant species based on growers' experiences of species planted in Britain and Ireland. The RHS system can give a fairly accurate indication of a plant's likely survival under Britain and Ireland conditions, although variables such as 'condition and age of the plant' and 'position in the garden' may influence its survival. When the RHS are considering a plant species' eligibility to be awarded an '**RHS Award of Merit**', its hardiness is carefully assessed as part of this process. As a consequence, all 'Award of Merit' species and cultivars are given one of the nine 'H' ratings.

In Table 2.1, several examples of plant species in each 'hardiness category' are included, along with their area of origin. The temperatures referred to are the **lowest winter temperatures tolerated** (not the mean low temperature measured over an extended period).

Figure 2.10 *Lavandula* species (lavender), with an H4 hardiness rating, originates in the Mediterranean area

Figure 2.11 *Clematis armandii*, with an H5 hardiness rating, originates from China

Figure 2.9 *Begonia* a genus with an H2 hardiness rating, originating from South America

Table 2.1 Royal Horticultural Society hardiness categories

Category	Hardiness rating	Species examples
H1a	Tolerate a temperature as low as +15°C. Require a heated greenhouse, or grown as a house plant.	*Begonia rex* (Brazil), *Dieffenbachia seguine* (tropical South America), and *Anthurium andraeanum* (Ecuador)
H1b	Tolerate temperatures as low as 15 to 10°C. Subtropical to tropical plants; need heated glasshouses or grown as a house plant. May be placed in sunny positions outdoors in summer.	*Hibiscus rosa-sinensis* (from southern China) and *Monstera deliciosa* (from Mexico)
H1c	Tolerate temperatures as low as 10 to 5°C. Subtropical plants need a heated glasshouse. Can be grown outdoors in summer throughout most of Britain and Ireland while daytime temperatures are high enough to promote growth.	Most bedding plants such as *Pelargonium zonale* (South Africa) and *Solenostemon scutellarioides* (Malaysia). Also, tomatoes (Central America) and cucumbers (China)
H2	Tolerate temperatures as low as +5 to +1°C. Ornamental species are often grown in frost-free glasshouses. Tolerate lower temperatures, but cannot tolerate any freezing conditions. May survive outdoors in frost-free town locations, or near the coast. Grown outdoors when risk of frost is over.	Originate from warm-temperate and subtropical zones, such as *Canna* spp. (South America), *Begonia x tuberhybrida* (Peru), sweet corn (Central America) and potatoes (Peru)
H3	Tolerate temperatures as low as +1°C to -5°C. Half-hardy, often grown in unheated glasshouse. Survive mild winters, outdoors. Often considered hardy in coastal/mild areas (except in hard winters). May also be hardy elsewhere when protected by a wall.	Warm temperate plants such as *Cosmos bipinnatus* (Mexico), *Osteospermum jacundum* (South Africa) and *Papaver somniferum* (southern Europe). Spring-sown vegetables such as carrots
H4	Tolerate temperatures as low as -5 to -10°C. Hardy in average winters, and throughout most of Britain and Ireland (except for frosty inland valleys, or at altitude, or in central/northerly locations). May suffer foliage damage and stem dieback in harsh winters in cold gardens. Some normally hardy plants may die in long, wet winters in heavy or poorly drained soil. **Plants in pots** are more vulnerable.	Many herbaceous and woody plants such as forget-me-not (*Myosotis sylvatica* from Europe), lavender (*Lavandula x chaytoriae* 'Sawyers' from Mediterranean area). Winter brassicas (from N. Europe) and leeks (from the Mediterranean region)
H5	Tolerate temperature as low as -10 to -15°C. Hardy in cold winters, and in most places throughout Britain and Ireland even in severe winters. They may not withstand open or exposed sites or central/northern locations. Many evergreens suffer foliage damage, and plants in pots will be at increased risk.	Herbaceous and woody plants, such as *Aucuba japonica* (Japan), *Clematis armandii* (China) and *Filipendula rubra* (east and central USA)
H6	Tolerate temperatures as low as -15 to -20°C. Hardy in very cold winters across Britain and Ireland and northern Europe. Many plants grown in containers will be damaged unless given protection.	*Acer palmatum* (Japan), *Chaenomeles x superba* (China), *Clematis tangutica* (China) and *Heuchara cylindrica* (western USA)
H7	Tolerate temperatures colder than –20°C (**very hardy**). Such species are hardy in the severest European continental climates including exposed upland locations in Britain and Ireland.	*Larix deciduas* (northern Europe), *Cornus alba* (northen China), and *Erica carnea* (central Europe)

The new RHS hardiness ratings are listed for a wide range of plants on the companion website: www.routledge.com/cw/adams.

Plant collectors

Many settled cultures throughout human history have improved local wild plant species for their own use and enjoyment. In addition, there have been remarkable **botanists** and **explorers** who travelled widely around different regions of the world to collect promising plant material to grow in their home area. Evidence for such activity in earlier cultures is sparse, but some books written two thousand years or more ago indicate that important species had been introduced by that time. Chinese botanists are recorded as collecting rose species for breeding 5,000 years ago. In Egypt, Queen Hatshepsut is listed as sending collectors around Africa 2,500 years ago in her search for frankincense (*Boswellia carterii*). The Greek philosopher Theophrastus (*c.* 310 BC), a contemporary of Aristotle, recorded exotic species such as bamboo, banana, pepper, cinnamon and cotton brought into Greece from Asia in 323 BC by Alexander the Great's army.

The Chinese emperor Ch'in (*c.* 200 BC) who began the construction of 'the Great Wall' is recorded as having directed the collecting of some 3,000 species of various herbaceous and woody plants for his gardens near Xian. The Roman writer Pliny the Elder (*c.* AD 40) referred to eastern non-indigenous species such as apples, pears, cherries and cucumbers being grown in Rome. Dioscorides (AD 40–90), a doctor in Rome, described over 600 medicinal plants. Some of these such as pepper, ginger and aloe had been brought in from other areas of the world.

In the twelfth century AD, Albertus Magnus, a Bavarian bishop trained in botany, is thought to have introduced many plants to Padua in Italy after travels abroad. In the sixteenth century, Francisco Hernandez, a doctor to King Philip II of Spain, brought back many new plant species from Mexico.

From the end of the sixteenth century, the intensity of plant collecting increased considerably. This was the new age of world travel as well as the renaissance of learning in Europe. Many devoted botanists and explorers, demonstrating exceptional persistence in the face of extreme climates, loneliness, disease and personal danger, were able to find, transport, name and propagate newly found species.

Below is a selected list of some of the better-known collectors.

John Tradescant the Younger (1608–1662). John Tradescant was the son of a famous 'gardener father'. He was born in Kent. Like his father, he was a botanist and gardener, travelling to Virginia, possibly several times from 1628 onwards. He was responsible for importing several genera such as *Taxodium*, *Magnolia*, *Liriodendron*, *Phlox* and *Aster* into Britain and Ireland. On his father's death, Tradescant succeeded him as head gardener to King Charles I.

Figure 2.12 (a) *Liriodendron tulipferae* tree; (b) tulip-shaped flower – this species was brought to Britain and Ireland by John Tradescant the younger; (c) *Gingko biloba*, an unusual tree unrelated to any living plant, originates from China. Its first record by a European was in 1690 by Engelbert Kaempfer

2

Figure 2.13 Joseph Banks was on Captain Cook's first voyage to the South Pacific (source: Wikimedia Commons)

Joseph Banks (1743–1820) was born in London and brought up in Lincolnshire. His interest in natural history took him on a voyage to Newfoundland in 1766. He was invited to join Captain Cook's famous first voyage on *HMS Endeavour* to South America, Tahiti, New Zealand, Australia and Indonesia (1768–1771), and was involved in the collection and painting (by Sidney Parkinson) of about 800 species of plants never previously recorded. These include bottle-brush tree (*Callistemon citrinus*), wintergreen (*Gaultheria mucronata*), tea tree (*Leptospermum scoparium*), New Zealand flax (*Phormium tenax*) and sophara (*Sophara tetraptera*). The Australian flowering shrubs and trees in the genus *Banksia* are named after him. It is interesting to note that on *HMS Endeavour*'s return to Britain in 1771, Banks began compiling his great book **Florilegium**, involving the botanical input of his friend Solander and the skills of several artists and engravers. None of the colour plates were published at that time (in the late eighteenth century). Indeed, it was not until 1905 that about 300 black and white prints were published. Finally in 1982, over 200 years after Banks' return to Great Britain, a limited edition of the complete *Florilegium* was published, including about 700 full colour prints.

Banks was appointed Director of Gardens to King George III, and was largely responsible for establishing the reputation of Kew Gardens. He also encouraged a worldwide search for new species by a large number of plant hunters.

Francis Masson (1741–1805), from Aberdeen, worked for Kew Gardens and, prompted by Joseph Banks, was Kew's first plant hunter. He travelled widely in South Africa, Madeira, West Indies, Portugal and USA, and has the bulb genus *Massonia* named after him. He joined James Cook's second voyage, to South Africa, staying from 1772 until 1775, and sending back to England about 500 plant species. Masson's experiences illustrate the sometimes dangerous life of a plant hunter. His 1778 voyage to Madeira, Canary Islands, the Azores and West Indies involved capture and imprisonment in Grenada by a French ship. The resulting delay led to many of his collected plant species dying. Subsequently, a Caribbean hurricane near St Lucia destroyed many more species. In October 1785 his second voyage to South Africa was unsuccessful largely because of the political strife between British and Dutch settlers. In 1797, Masson's visit to North America was temporarily curtailed with his capture by a French pirate vessel. Relieved to be alive, he was able to proceed to New York in a German ship. During the next seven years, his extensive journeys around eastern USA and Canada led to a rather small collection of 24 new species. He died in Montreal in 1805. Masson's outstanding contribution to horticulture involved the discovery of about 1,700 new plant species (including from South Africa, *Agapanthus inapertus*, *Amaryllis belladonna*, *Zantedeschia aethiopica*, *Strelitzia reginae*, *Protea cynaroides* and *Kniphofia rooperi*; from Canary Islands, *Senecio cruenta*; and from eastern USA, *Trillium grandiflorum*).

Figure 2.14 African lily (*Agapanthus species*), originating in southern Africa, was first collected by Francis Masson

Figure 2.15 (a) *Kerria japonica* was collected by William Kerr in Japan; (b) *Pieris japonica*

John Fraser (1750–1811) was born in Inverness-shire and came to London in 1770 as a draper where he developed an interest in the Chelsea Physic Garden and the Apothecary's Garden. During the next few years, he became involved in plant collecting which led in 1780 to the first of seven visits to North American. At this time, Fraser began the development of his large 'American Nursery' near Sloane Square, London. On a subsequent USA visit in 1785 around South Carolina, Fraser collected *Phlox stolonifera* and *Magnolia fraseri*. In 1789, a mountain trek in the secluded Allegheny mountains of eastern USA led to the discovery of *Rhododendron catawbiense*, an important breeding source for new cultivars. In 1797, Fraser established a brief link with the Russian royal family, selling new and established species of plants. Financial difficulties contributed to Fraser's early death in 1811 at the age of 60. Species named after Fraser include *Abies fraseri* and *Frasera* spp. (gentianworts).

William Kerr was born in Hawick, Scotland (date of birth unknown), and was a gardener at Kew Gardens before becoming, with Joseph Banks' encouragement, a plant collector in China for almost eight years. He also visited Java and Philippines. He is noted for sending the ultra hardy shrub *Kerria japonica*, which bears his name, to Kew Gardens. In addition, he was involved in collecting nearly 250 plants new to Europe. These include *Euonymus japonicus*, *Lilium lancifolium*, *Pieris japonica*, *Nandina domestica*, *Begonia grandis* and *Rosa banksiae*. He died in Colombo, Ceylon in 1812, a short time after becoming the superintendent of the nearby gardens on Slave Island.

David Douglas (1799–1834) was a Scottish botanist, born near Perth, the son of a stonemason. He worked as a gardener in Scone Palace, Perthshire for seven years. There followed a year at a college in Perth, and a short time of employment in Fife. His move to the Botanical Gardens of Glasgow University and the contact with William Hooker (the Garden Director and Professor of Botany) led to him being recommended to the Royal Horticultural Society of London. He was sent, in 1824, on an expedition to Northwest America, where he identified more than nine newly recorded conifers in this harsh high-altitude landscape (*Pseudotsuga menziesii*, *Picea sitchensis*, *Pinus lambertiana*, *Pinus monticola*, *Pinus ponderosa*, *Pinus contorta*, *Pinus radiata*, *Abies grandis and Abies procera*). Most of these species would become important large garden trees and sources of timber in Britain and Ireland. In addition to these trees, Douglas was able to collect shrubs and herbaceous species such as *Ribes sanguineum*, *Gaultheria shallon*, *Lupinus nanus*, *Penstemon douglasii* and *Eschscholzia californica*. Altogether he introduced about 240 species of plants to Britain and Ireland.

In 1830, Douglas visited Hawaii, and then Northwest America for the second time. His second Hawaii visit in 1833 resulted in his early death at 35 years of age. He fell into a pit while climbing a local volcano. He was buried in Hawaii.

William Lobb (1809–1864) was born in Cornwall and employed by an Exeter nursery (Veitch) that had already been responsible for the introduction to England of *Araucaria araucana* (the 'Monkey-Puzzle' tree) from Chile. His brother, Thomas, was sent to

Figure 2.16 *Ribes sanguineum* was introduced to Britain and Ireland by David Douglas

Figure 2.17 Monkey-puzzle tree (*Araucaria araucana*) from Chile was an early introduction to Britain and Ireland

Figure 2.18 (a) A memorial in Cornwall to William Lobb, the collector of giant redwood from California; (b) *Berberis darwinii*, collected by William Lobb (source: Wikimedia Commons)

India and South East Asia to collect plants. William was sent to the Americas.

Though William Lobb did not have botanical training, his personality appeared to Veitch (his employer) to be quite adequate for the difficult task of a plant hunter. At the age of 31, his first journey in 1841 took him to Rio de Janeiro. He had with him a 'peace offering' for the emperor of Brazil; namely the seeds of rhododendron hybrid cultivar 'Cornish Early Red'. So began his impressive plant-hunting career. Lobb soon discovered, among several others, the orchids *Cycnoches pentadactylon*, *Oncidium curtum* and *Begonia coccinea*, and also *Passiflora* spp. All these plants arrived in Britain in acceptable condition.

Nasturtium (*Tropaeolum azureum*), *Abutilon vitifolium* (pale blue) and *Calceolaria alba* (white) were to follow. In 1843, he reached Peru and then Ecuador, arriving finally at Panama City to despatch another consignment for Exeter that included *Oncidium ampliatum* orchid as well as seeds of several *Fuchsia* and *Tropaeolum* species. He was back in England by May 1844. A year later, Lobb went to South America again to collect hardy shrubs and trees. Chile was his main area of interest. He sent home *Berberis darwini*, Chilean firebush (*Embothrium coccineum*), flame nasturtium (*Tropaeolum speciosum*), and seeds of Antarctic beech (*Nothofagus antarctica*) and *Escallonia macrantha*.

His third journey, in 1849, was to look for hardy shrubs and conifers on the west coast of North America. He collected Santa Lucia Fir (*Abies bracteata*), *Lupinus cervinus*, the deciduous *Rhododendron occidentale* and *Aesculus californica* in southern California. To the north, his collection of conifers included large seed quantities of California redwood (*Sequoia sempervirens*). Other new species he recorded included Colorado white fir (*Abies concolor*), Red Fir (*Abies magnifica*), Western red cedar (*Thuja plicata*), California juniper (*Juniperus californica*) and two *Ceanothus* natural hybrids (*C. x lobbianus* and *C. x veitchianus*).

His most memorable experience was in 1853 in the highland forests 200 km east of San Francisco that contained Giant redwood (*Sequoiadendron giganteum* – see Figure 2.1). His decision to collect large quantities of seed and return immediately to England resulted in huge sales of the saplings of these magnificent trees for large gardens and estates, the result of which is still seen in Britain and Ireland today. Lobb returned to California in 1854, but ill health reduced his plant collecting activities. He died in San Francisco in 1864.

Robert Fortune (1812–1880) was born in Berwickshire. After employment in the Royal Botanic Garden Edinburgh, he worked in the Horticultural Society of London's garden. In a first visit to China in 1844, Fortune travelled in the northern areas collecting a number of species of *Azalea*, *Daphne*, *Wistaria*, *Weigela* and *Arundinaria*. On his return to Britain and Ireland in 1846, he became Curator of Chelsea Botanic Garden. In 1848, at the age of 36, he was in China for two and a half years both as a plant hunter and surreptitious 'crop stealer'.

His most notable activity involved the transportation of about 20,000 young tea plants (for the British East India Company) from the southeastern China region of Fujian to the Darjeeling area of north India. The 1842 Treaty of Nanjing between China and Britain allowed Europeans local access only to stipulated ports. But Fortune disguised himself as a Chinese man and was able to successfully transport the young plants, using a number of 'Wardian cases' (a form of mini-greenhouse). Although only a small proportion of the tea plants survived the journey, these plants were partly instrumental in establishing the Indian tea industry. Several Chinese tea growers were persuaded to accompany Fortune to India, and their knowledge was passed on to local growers in India.

Fortune was able to introduce more than 120 plant species from eastern Asia. These included *Buddleja lindleyana*, *Camelia* 'Robert Fortune', *Camellia* 'Captain Hawes', *Chionanthus retusus*, *Clerodendrum bungei*, *Corylopsis pauciflora*, *Daphne genkwa*, *Jasminum*

Figure 2.19 *Jasminum nudifolium*

nudiflorum, *Lonicera standishii*, *Pinus bungeana*, *Rosa fortuniana*, *Rosa* 'Fortune's Double Yellow', *Rhododendron fortunei*, *Dicentra spectabilis*, *Forsythia viridissima* and many chrysanthemum and tree paeony cultivars.

Fortune returned to China in 1853 and 1858, and to Japan in 1860. He also visited Indonesia and Philippines in his travels. He returned to Scotland to farm in 1862. He died in London in 1880.

Jósef Ritter von Rawicz Warszewicz (1812–1866) was born in Wilno, Lithuania (of Polish parents). After working at a local Wilno botanical garden, he moved to the Berlin Botanic Gardens. In 1844, he was recruited for a six-year exploration of Central America (mainly Guatamala) to look for plants of commercial potential. The orchids *Cattleya dowiana* and *Cattleya warscewiczii* were two notable finds. His successful exploits enabled Germany to acquire a reputation for newly introduced plant species. A subsequent exploration in 1851 (mainly to Peru and Bolivia) led to large numbers of species being found .This included *Canna warscewiczii* (used much subsequently in breeding lines) and several new species of Gesneriaceae such as *Sinningia warscewiczii*. Warszewicz is best known, however, for his collection of orchids, including *Cycnoches warscewiczii*, *Catasetum warscewiczii*, *Miltonia warscewiczii*, *Paphiopedilum caudatum*, *Sobralia warscewiczii*, *Brassia warscewiczii*, *Epidendrum warscewiczii*, *Mesospinidium warscewiczii*, *Epidendrum warscewiczii*, *Oncidium warscewiczii*, *Stanhopea warscewiczii* and the genus *Warszewiczella*. In 1853, he accepted the position of Director of Cracow Botanical Gardens until his death in 1866.

Joseph Hooker (1817–1911) was born in Suffolk and spent his childhood in Glasgow. He was the son of Kew Garden's first director, and became a close friend of Charles Darwin. After studying medicine, between 1838 and 1843, he was assistant medical officer on a southern hemisphere expedition on the ships *Erebus* and *Terror* that reached close to Antarctica. He was able to visit Tierra del Fuego, New Zealand, southern Australia, South Africa and several mid-Atlantic islands. There followed a four-year period in Britain, which enabled him to pursue an interest in palaeobotany. Between 1847 and 1851, he travelled extensively on the Indian subcontinent, collecting 25 previously unrecorded species of rhododendron (such as *R. cinnabarinum*, *R. falconeri*, *R.hodgsonii*, *R. thomsonii* and *R. griffithanum*) in the eastern Himalaya region. He also recorded several *Primula* species such as *P. capitata* and *P. sikkimensis*. In total, he collected 700 plant species from the subcontinent.

In 1877, his journey around the USA with the American botanist Asa Gray was notable in establishing a similarity in distribution between some plant genera (such as *Thuja*, *Tsuga*, *Magnolia* and *Catalpa*) in both southeast China and in the far distant southeast USA. Perhaps the most striking example is that of the *Liriodendron* genus (**tulip tree** see Figure 2.12). There are only two known species within this genus. *Liriodendron tulipifera*, a quite common British Isles garden tree, has a native distribution in the eastern states of the USA from Vermont down to Louisiana, especially in the southern Appalachians. *Liriodendron chinense*, however, occurs 10,000 km away in the montane forests of East Asia, from the Yangtze River in China into northern Vietnam. Fossil evidence of *Liriodendron* shows that this plant genus was formerly distributed over the northern hemisphere, and that the present two tulip trees are '**relic species**' (a similar situation to the two widely separated **tapir** mammal species of Malaysia and South America).

Charles Maries (1851–1902) was born in a Warwickshire village, the son of a shoemaker. An early influence was Reverend Henslow, his headmaster, who was to become Professor of Botany at the Royal Horticultural Society. From the age of 18, Maries worked for seven years in his elder brother's nursery in Lytham, Lancashire. He then made the move to join the large London branch of the Veitch horticultural company, where he specialized in Japanese and Chinese plants.

He was sent in 1877 to Japan to obtain seeds of 'new' conifers. In the north of Honshu Island, he collected *Abies mariesii* and *Abies sachalinensis* (previously described but not introduced to the west). On Hokkaido, he collected the shrubs *Acer nikoense*,

Figure 2.20 (a) Charles Maries who collected *Hydrangea macrophylla* in Japan (source: Wikimedia Commons) (b) *Hydrangea macrophylla*

Azalea rollisoni, *Hydrangea macrophylla* 'Mariesii', *Styrax obassia* and *Viburnum plicatum* 'Mariesii', the tree *Abies yessoensis*, the climbers *Actinidia kolomikta* and *Schizophragma hydrangeoides*, the perennial *Platycodon grandiflorus* and the epiphytic fern *Davallia mariesii*.

In 1878, Maries visited China, travelling first to Mount Lushan in Jiangxi Province where he collected *Cryptomeria japonica*, *Hamamelis mollis*, *Larix kaempferi*, *Lilium lancifolium*, *Liriodendron chinenses*, *Pseudolarix amabilis* and *Rhododendron fortunei*. Then began his intended main project in China, to visit the Yangtze valley, but poor communication with the local population led to only a few species such as *Primula obconica* being collected. On a short visit to Japan, he collected the square bamboo (*Chimonobambusa quadrangularis*). He returned to England in 1879.

A year or two later, Maries moved away from plant collecting, taking up a post of garden superintendent in northern India to two maharajahs (one in north Bihar and the other in north Madhya Pradesh). During these 20 years in India, he became an acknowledged expert on mangoes. He died in India in 1902.

His plant collecting legacy is considerable: he discovered over 500 species that were new to Britain and Ireland.

Alice Eastwood (1859–1953) was born in Toronto, Canada, moving to USA at the age of 14. While teaching in Denver, Colorado, she studied botany, which led her to work with the herbarium at the California Academy of Sciences in 1890. She became joint Curator of the Academy in 1892, and then Head of the Department of Botany.

Her plant collecting expeditions in the then remote Big Sur region in California led to the discovery of *Salix eastwoodiae* and *Potentilla hickmanii*. At this time, 1906, Eastwood was able to salvage large number of valuable dried 'type-specimens' from the herbarium that had been set on fire by the San Francisco earthquake of that year. In future years, it became an important aspect of her life's work to build up this herbarium (to reach a total of over 300,000 specimens).

Eastwood was involved in many collecting expeditions and vacations in the western United States, which included visits to Alaska, Arizona, Utah and Idaho.

She had editorial responsibilities in two botanical journals (*Zoe* and *Erythea*) and founded a journal (*Leaflets of Western Botany*) with a fellow botanist John Howell. She also was Director of the San Francisco California Botanical Club during the 1890s. Species such as *Astragulus ceramicus*, *Allium cratericola*, *Aquilegia eastwoodii*, *Ceanothus gloriosus*, many *Lupinus* spp. (including *Lupinus pulcher*), *Oenothera deltoides* and *Penstemon bradburyi* were amongst the many species collected by her. She had eight species named after her. She died in San Francisco in 1953.

Ynes Enriquetta Mexia (1870–1938) was born in Washington DC where her father was a diplomat in the Mexican embassy. She was to become one of the most prolific plant collectors, travelling extensively in Mexico and South America. In 1885, after her school education in Maryland, she moved to Mexico City, where she looked after her father for ten years until his death. She was, in 1904, bereaved by her husband's early death and subsequently a second marriage proved unsuccessful.

By 1921, and after studying for a botany degree in California, at 55 years of age she began her plant-collecting career. She went first to western Mexico and was the first to collect *Mimosa mexiae*. There

followed visits to Brazil in 1926 (*Begonia jaliscana*), back to Mexico in 1927 (*Abutilon jaliscanum*, *Fuchsia decidua*, *Salvia mexicae* and *Sedum grandipetalum*), Argentina, Chile and Alaska in 1928, Brazil in 1929 (*Adiantum giganteum* and *Capsicum microcarpum*), Peru in 1931 (many *Piper* spp.), Ecuador in 1934, Peru and southern Argentina in 1935, and southwest Mexico in 1937. Many of her specimens were sent to herbaria in Harvard and Chicago. *Mexianthus*, named after Ynes Mexia, is a genus in the Asteraceae.

Ynes Mexia is reputed to have collected 150,000 plant specimens in her 17 years of plant hunting. She died in California in 1938 after falling ill on her 1937 collecting trip in Mexico.

George Forrest (1873–1932) was born in Falkirk, Scotland. He worked in a chemist's shop until he was 18, when a small family 'windfall' allowed him to visit and work for several years in Australia. On his return to Scotland in 1902, a chance meeting on a riverbank with Professor Bayley Balfour, Keeper of the Royal Botanic Garden, Edinburgh, changed his life. A year's work in the Botanic Gardens led to a sponsored visit to the southwest China province of Yunnan in 1904, particularly to look for rhododendron species. Skirmishes with politically active Buddhist lamas led to several narrow escapes, but Forrest was to make seven visits to Yunnan province and to the adjacent area of Sichuan province, and then to Burma and eastern Tibet.

On 5 January 1932, while hunting game in Tengchong, the Yunnan town that he often used as a base, he suffered a heart attack, dying instantly. He was buried locally.

In all, this intrepid collector introduced over 1,200 plant species new to science. These included *Acer forrestii*, *Camellia saluenensis*, *Clematis chrysocoma*, *Gentiana sino-ornata*, *Iris forrestii*, *Jasminum polyanthum*, *Mahonia lomariifolia* and *Pieris forrestii*. He also collected many species of *Allium*, *Anemone*, *Aster*, *Berberis*, *Buddleia*, conifers, *Cotoneasters* and *Deutzias*. He is especially remembered for the *Rhododendron* species (*R. forrestii*, *R. sinogrande*, *R. repens*, *R. griersonianum*, *R. intricatum* and *R. giganteum*) and *Primula* species (*P. malacoides*, *P. bulleyana*, *P. nutans*, *P. vialii* and *P. spicata*) that he collected.

Ernest Wilson (1876–1930) was an English plant collector. He was born in Gloucestershire. He began his career working in a nursery near Birmingham before joining the staff at the Birmingham Botanical Gardens. Further experience at the Royal Botanic Gardens, Kew led to the exciting position of 'Chinese plant collector' with the well-known Veitch nursery that had employed William Lobb earlier that century.

Figure 2.21 (a) Ernest Wilson who collected in South East Asia (source: Wikimedia Commons); (b) *Kolkwitzia amabilis* collected by Ernest Wilson (reproduced with permission from 'Cillas')

Wilson's main objective was to collect the dove tree (*Davidia involucrata*). He arrived in China in 1899 and set off for Hubei province where he stayed for two years, collecting almost 400 species that included (in addition to *Davidia*), *Acer griseum* (paper bark maple), *Actinidia deliciosa* (kiwi fruit), *Berberis julianae*, *Clematis armandii*, *Clematis montana*, *Ilex pernyi*, *Jasminum mesnyi* and *Primula pulverulenta*.

On his second China visit in 1903 in the area around Szechuan province, he collected the Regal lily (*Lilium regale*) and was seriously injured by a rock fall. His other collections included the yellow poppy (*Meconopsis integrifolia* – his main target species), beauty Bush (*Kolkwitzia amabilis*), *Rosa willmottiae* and the bush *Sinowilsonia henryi*.

Wilson subsequently collected in China (1907, 1908 and 1910) and Japan for the Arnold Arboretum, Boston, USA. His extended time in Japan (1911–1916) led to the collection of over 50 ornamental cherry cultivars. In 1917–1918, he turned his attention to Korea and Formosa (Taiwan), during which time 50 cultivars of azaleas ('Kurume' azaleas) were sent to America from Japan. Further explorations followed in 1921 to Australia, New Zealand, India and South America, and East Africa. He was appointed Keeper of the Arnold Arboretum in 1927.

Wilson and his wife were killed in the USA in an automobile accident in 1930.

Over 100 plants introduced by Wilson have received the First-Class Certificate or Awards of Merit of the Royal Horticultural Society of London.

Recent plant hunters. In the early twentieth century, Frank Kingdon-Ward collected in the China/Himalayas area, while in the second half of the century, collectors such as Peter Cox, Peter Hutchinson, Roy Lancaster and many others have continued the plant-collecting tradition (see companion website: www.routledge.com/cw/adams, for other plant hunters as well as a summary of the origins of RHS-recommended plant species and, where available, the names of their collector).

World trade in horticultural plants

Home grown

About **93%** of the world's 2.4 billion tonnes of fruit and vegetables are bought and used in their country of production. In this way, costs of transport are reduced, use of local resources and manpower is encouraged and produce arrives at the shop in as fresh a condition as possible. An attitude of 'home grown' is generally accepted by society and by governments as being an environmentally sensible method of production.

Exporting

In spite of the 'home grown' situation mentioned above, there is a large-scale movement of horticultural produce around the world involving diverse types of plant material and each with their individual transport storage requirements (such as temperature, humidity and air-cleansing controls) to ensure healthy arrival at their destination.

Several reasons for this worldwide industry can be given:

▶ **Popular beyond where they can be grown.** Some fruit such as **bananas**, which are popular in temperate climates (and the most popular fruit in the world), are only able to grow in tropical countries (e.g. Ecuador). Acid soil-loving species such as **blueberries** are produced in countries such as USA that have suitable soils.

▶ **Continuity of supply.** Consumers in countries such as Britain and Ireland located in northern temperate regions like to eat fresh vegetables (such as runner beans), and fruit (such as strawberries) **all year round**, even during the winter months when such crops cannot be grown there. Kenya is a major exporter of runner beans during Britain and Ireland winter period. The USA exports strawberries to Britain and Ireland at this time.

▶ **Specialism of the country.** Some countries such as Holland have developed **large-scale industries** of glasshouse food crop and flower production (such as tomatoes and chrysanthemums) and of nursery stock, and is able to compete well for price and quality with produce grown in the other nations.

World production levels

Some examples of **total** world production per year (and the major producers) for vegetables, fruit, and decorative plants are listed below.

Vegetables

Total annual world production of vegetable was estimated to be about 1,800 million tonnes per year in 2009.

Table 2.2 Annual world production of major vegetables (million tonnes)

Vegetable	Tonnes (m)
Potatoes	330
Cassava	230
Tomatoes	150
Onions and garlic	100
Sweet potatoes	100
Legumes	100
Watermelons	100
Brassicas	85
Root vegetables	70
Others (such as carrots, chillies, pumpkins and aubergines	480

Table 2.3 World's nine major producers of vegetable, representing about 65% of world production (millions of tonnes)

Producer	Tonnes (m)
China	680
India	150
Nigeria	85
USA	60
Russia	50
Brazil	45
Indonesia	35
Thailand	35
Turkey	32

Figure 2.22 (a) Potatoes are the world's most popular vegetable; (b) tomatoes are the third most popular; (c) onions are the fourth; (d) brassicas are the eighth

Fruit

Total world production of fruit is estimated to be about 635 million tonnes per year in 2009.

Table 2.4 Annual world production of major fruit (millions of tonnes)

Fruit	Tonnes (m)
Bananas	100
Grapes	70
Oranges	70
Apples	70
Other citrus	50
Dates, figs, olives	45
Stone fruits (plums, etc.)	45
Berries (raspberries, etc.)	10
Other tropical fruits	120
Other temperate fruits	30

Decorative plants

World annual production is estimated to be about 30 billion US dollars. The four sectors of the worldwide decorative plant market are:

Table 2.5 World's major producers of fruit representing about 50% of total production (millions of tonnes)

Producer	Tonnes (m)
China	120
India	70
Brazil	40
USA	30
Indonesia	15
Turkey	15
Nigeria	10
Thailand	10
Russia	5

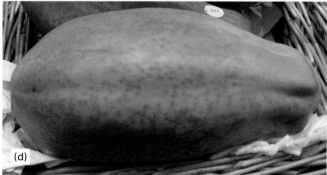

Figure 2.23 (a) Bananas are the world's most popular fruit; (b) grapes are second equal; (c) apples are second equal; (d) paw-paw and (e) rambutan, two popular fruits from the tropics

Figure 2.24 (a) A range of cut flowers; (b) begonia and kalanchoe pot plants

Table 2.6 The 13 major producers of decorative plants (in billions of US dollars) representing about 50% of total world production in 2009

Producer	Tonnes (m)
Holland	10.5
Colombia	1.5
Belgium	1.5
Germany	1.5
Italy	1.5
Denmark	0.7
Ecuador	0.6
Kenya	0.5
USA	0.4
Zimbabwe	0.4
Spain	0.3
Canada	0.3
France	0.2

Table 2.7 Value of decorative plant exports from Holland to 13 countries (in billions of US dollars) representing 85% of the country's sales

Country	Value of exports (bn)
Germany	2.7
Britain and Ireland	1.1
France	1.1
Italy	0.8
Belgium	0.8
Russia	0.8
Poland	0.4
Denmark	0.3
Sweden	0.3
Austria	0.3
Switzerland	0.3
Spain	0.3
USA	0.3

▶ Cut flowers – US $7.3bn (42% of total sales)
▶ Live plants and cuttings – US $7.3bn (42% of total sales)
▶ Bulbs and corms – US $1.5bn (9% of total sales)
▶ Foliage – US $1.2bn (7% of total sales)

Further reading

Cox, P. and Hutchison, P. (2008) *Seeds of Adventure*. Garden Art Press.

International Society for Horticultural Science (ISHS) (2012) *Harvesting the Sun: A Profile of World Horticulture*. Online: www.harvestingthesun.org. ISHS.

Musgrave, T., Gardener, C. and Musgrave, W. (1998) *The Plant Hunters*. Ward Lock.

RHS (n.d.) Hardiness ratings. Online: www.rhs.org. uk/Plants/Plant-trials-and-awards/pdf/2012_RHS-Hardiness-Rating. RHS.

Whittle, T. (1997) *The Plant Hunters*. Lyons Press.

 Please visit the companion website for further information:
www.routledge.com/cw/adams

CHAPTER 3
Level 2

Ecology and garden wildlife

Figure 3.1 Beetles feeding on petals and pollen of *Sedum* flowers. Often overlooked, these and other invertebrates are valuable pollinators as well as providing a food source for other organisms in the garden food web

This chapter includes the following topics:

- Ecology and gardens
- Communities, ecosystems and biomes
- Food chains and webs
- Biomass and energy flow
- Succession
- Interactions between organisms
- Competition
- Gardening for wildlife

Wildlife in the countryside is under pressure as never before from many causes such as changing agricultural practices, use of pesticides, urban spread and introduced species. The value of gardens for wildlife has begun to be appreciated in recent years and they are seen more and more as an important contribution to conservation of our natural heritage. An understanding of the ways in which wildlife can be can be encouraged and supported, while retaining all the desirable qualities of a garden, is useful to know. To know what makes a good garden for wildlife it is useful to understand some ecological principles.

Studies on garden wildlife

Between 1972 and 1991, Jennifer Owens recorded wildlife in her garden in Leicester. She provided a range of heights and growth forms in her planting, grew flowers for insects and plants with fruits for birds. Apart from a few concessions (minimal pruning and clearing, dead heading to prolong flowering and never using pesticides), she managed her garden conventionally and described it as 'a typical suburban garden . . . neat, attractive and productive' with a lawn, some herbaceous borders, a rockery and fruit and vegetables. A huge diversity of organisms was found, some 422 species of plants, 1,602 insects, 121 other invertebrates and 59 vertebrates representing a large proportion of the fauna of Britain and Ireland. Some species were common, many were rare and some were new to science. She commented that 'gardens would seem to be of considerable significance in conservation'.

Since then, studies carried out by the University of Sheffield across the UK (Biodiversity in Urban Gardens in Sheffield or BUGS project) have evaluated urban gardens for wildlife in several UK cities. They too found an astonishing range of plants and animals – for example, 42% of the UK's plants were found in a combined garden area of two football pitches. They concluded surprisingly that small gardens were just as good as large gardens, city centre gardens were as good as suburban gardens and all gardens were good for wildlife whether specifically managed for this or not. Trees were found to be the most important factor and piles of logs were beneficial, while artificial methods of attracting wildlife such as bird boxes and insect hotels varied in their success. They concluded that any reduction in urban gardens would impact on

'biodiversity, conservation, ecosystem services, and the well-being of the human population'.

See companion website for more information on the BUGS project.

An understanding of the basics of **ecology** can help us design and manage gardens in a way that supports all the organisms that live there. In effect, it is the recipe for a healthy, wildlife-friendly garden. Broadly speaking, ecology takes over where the study of individual organisms ends. It investigates the relationships between the organisms themselves and the environment they live in. Ecologists therefore study groups of plants and animals (**populations**) living together in a **community** which, together with their non-living environment, form **ecosystems**. On a global scale, major regional communities of organisms form well-recognized **biomes**.

The term '**ecology**' is formed from the Greek words *oikos*, meaning a house or place, and *logos* meaning knowledge or understanding.

Ecology and gardens

Communities

A **community** is a group of populations in a given area or **habitat**. A **population** is defined as a group of individuals of one species which interbreed – for example, all the weed species *Senecio vulgaris* in a garden, or all the individual hedgehogs within a certain area.

A community is a group of plant and animal species (populations) living within a particular area (**habitat**). 'Wild' communities are defined by either the habitat in which they occur (e.g. a 'lake community') or by a particular plant species which is dominant (e.g. a 'grassland community'). A **microhabitat** is on a much smaller scale (e.g. under a log or against a wall) and its community may differ considerably from that of the larger habitat it occupies. Gardens are themselves a type of habitat and often have many microhabitats which increase the variety (biodiversity) of organisms that live there.

Populations may be restricted to a very specific habitat – for example, *Epilobium palustre* (marsh willow herb) is only found in slightly acidic ponds. However, some populations can occupy a wide range of habitats such as *Rubus fruticosus* (blackberry)

3

Figure 3.2 (a) *Iris pseudacorus* (yellow flag) grows only in waterlogged conditions; (b) *Rubus fruticosus* (blackberry) grows in many different conditions

which is found in heathland, woodland, hedgerows and open fields (Figure 3.2).

In natural communities, a particular environment will often have a specific group of plants growing there which are not found together anywhere else, the plants are said to form **plant associations**. For example, in a chalk habitat *Ophrys apifera* (bee orchid) is often found together with *Centaurea scabiosa* (greater knapweed) and *Poterium sanguisorba* (salad burnet). In the very high rainfall acid bogs of northern Britain and Ireland, *Sphagnum* mosses and *Eriophorum vaginatum* (cotton grass) are found together with *Drosera anglica* (sundew) and *Myrtus gale* (bog myrtle) (Figure 3.3).

Figure 3.3 (a) Bee orchid, found in chalk grassland; (b) cotton grass, found in acid bogs; (c) *Sphagnum* found in acid bogs

Figure 3.4 Marine shoreline, an example of an open plant community

Although gardens are not natural communities, the same applies in that groups of plants often do well together. A hot dry border, for example, will successfully accommodate a range of plant species if it reflects their original habitat. Understanding the habitat to which a particular plant or group of plants is adapted enables us to give it the right conditions in a garden setting, increasing the chance of success. Since many garden plants are introductions collected from many very different habitats worldwide, their origins should be taken into account (see Chapter 2).

Communities can sometimes be '**closed**' in that they receive only minimal contact with outside organisms and materials. A small isolated island community in the middle of a lake would be an example, or the community inside the Wardian cases used by the plant collectors (see Chapter 2), or in a conservatory. However, most communities, are '**open**', that is, organisms and materials can move in and out freely – for example, wind-borne plant seeds can travel between gardens, and birds and other animals migrate widely across their respective ranges and 'belong' to no single garden (Figure 3.4).

The position or role of each species within a habitat is called its **niche**. An organism's niche includes both its physical and its biological environment, as well as how its needs and behaviour vary with time. For a plant, the **physical environment** in which it grows will reflect its requirements such as

▶ light or shade
▶ a particular soil pH
▶ resources such as a good water supply or a high level of a nutrient such as iron.

Its **biological environment** could include any pests, diseases or predators which feed on it. For example,

rosette plants such as daisies originally adapted to survive being grazed by having their growing point at ground level. In lawns the grazing animal has been replaced by a mower, but the daisies still occupy a particular niche in the grassland we call a lawn. An organism's niche may also be described in time. In the case of an annual plant, it flowers in spring and dies at the end of the season, so occupies its niche only part of the year.

A fundamental concept of the niche is that no two species can occupy an identical niche. If they do, sooner or later one will dominate and the other will die out because of competition for resources. However superficially it may appear that organisms occupy the same niche, in reality they rarely do. For example, consider a group of weeds all growing together in the vegetable plot. Although they all grow in the same place, they all have different requirements and behaviours. Some, such as ephemerals, have their main growth period very early in the year and may produce several generations in a single season, surviving as seed throughout the rest of the year. Annual weeds have a single life cycle in a season and produce seed in the autumn, while perennial weeds overwinter, often through the use of underground rhizomes and other perennating organs (see p. 91). These different weed populations therefore vary the time of year when they have greatest need of resources and have their own food supplies when competition is fiercest.

Other plants, which on the face of it grow together, may mine nutrients at different depths in the soil enabling them to survive alongside each other. For example, lettuces with shallow roots can be successfully intercropped with carrots which are deeper rooting. Yet others may need full light for growth whereas their neighbours can tolerate shading – sweetcorn which requires good light intercropped with shade-tolerant spinach would be an example. As in a jigsaw puzzle where each piece has a unique shape and only fits in one place, each species has its own unique niche in its environment and although many plants tolerate a range of conditions, the nearer the gardener gets to providing the ideal conditions for a plant's particular niche, the more likely they are to establish a healthy plant.

Ecosystems

An **ecosystem** is a community of living organsms which, together with their non-living environment, operate as a unit.

An **ecosystem** is composed of all the living organisms (the biotic component) and the non-living environment (the abiotic component) they inhabit, functioning as a unit. Implicit in the term is the idea that all these components react together to form an integrated, balanced and self sustaining system. Non-living factors include the type, structure and pH of soils; climatic conditions such as rainfall, light, wind and temperature and topography including altitude, slope, aspect and degree of exposure. Thus a garden ecosystem in a coastal area in the south west of mainland Britain and Ireland will be quite different compared with a garden ecosystem in the uplands of the Pennines. Understanding the living and non-living components of each ecosystem will help in the successful choice of plants for a particular garden.

Biomes

A **biome** is a group of communities with their own type of climate, vegetation and animal life.

A **biome** is a large geographical area or **global community** of distinctive plants and animals, containing many ecosystems whose communities have adapted to a particular climate. Major biomes include

- ▶ desert
- ▶ grassland
- ▶ tropical
- ▶ temperate forests
- ▶ arctic and alpine tundra.

Types of biome are recognized worldwide based on climate, whether they are land or water based, on geology and soil, or on altitude above sea level. Although each biome will have within it many different habitats, they are characterized by dominant forms of plant life and their associated animals which are adapted to their particular environment. Biomes vary markedly in their capacity for growth, their 'productivity' depending on water and light availability and temperature. For example, a square metre of temperate forest biome may produce ten times the growth of an alpine tundra biome. They also differ markedly in the number of species which live there: the tropical rainforest biome contains more tree species than any other area in the world (see Chapter 2). Sometimes gardeners can seek to recreate a particular biome (e.g. in an alpine house or a tropical conservatory) by providing the appropriate climatic conditions for their plants to flourish. See companion website for more information on biomes.

Alpine tundra biome

Alpine tundra biomes are found in mountainous parts of the world such as the Alps, the Pyrenees, the Andes and the Himalayas usually above 3,000 metres. They form a zone between the tree line and the area of permanent snow and their climate is greatly affected by the altitude. Very low temperatures, averaging 10°C, are typical and the ground often freezes at night and through the winter. Soils are thin, stony and free draining because organic matter decomposes slowly in this environment and the exposed mountainsides are also very windy, so soils are easily blown or washed away. Rainfall can be variable with much precipitation falling as snow. The winter can last from October to May so plants have a short growing season, often only 45–90 days. Because of the thin air, there is less carbon dioxide at this altitude and the sunlight is intense. Plants and animals have had to adapt radically to this severe climate. Most plants are small grasses, sedges, low-growing shrubs and perennials, which grow and reproduce slowly,

often overwintering as bulbs or other perennating organs (see p. 91) and flowering in the spring. They hug the ground often as hummocks or as 'rosette' plants with their growing point below the surface to protect themselves from the cold temperatures and strong dessicating winds. They have adaptations such as hairy leaves to reduce transpiration, succulent leaves to store water and purple pigments to protect them from the concentrated ultra-violet radiation. Many have physiological adaptations to the extreme cold such as production of high levels of sugars in their tissues which act as 'antifreeze', preventing ice formation. Others allow ice crystals to form but limit them to spaces between cells, preventing damage to the cells' membranes and their contents. Mosses and lichens are common.

The horticultural definition of an alpine is a subject of much debate, many plants described as 'alpines' at garden centres have similar growing requirements but do not originate

Figure 3.5 (a) Alpine plants including a gentian species; (b) an alpine environment in the high Andes showing thin rocky soil and small plants, mosses and lichens; (c) lichen; (d) a rosette plant with hairy leaves

from this biome and are more correctly termed 'rock plants'. Alpine enthusiasts aim to provide an environment with good light, free-draining composts and good ventilation to reduce temperature and humidity (alpine plants cannot tolerate wet conditions), often in a specialized alpine house.

Food chains and webs

Charles Darwin is said to have told a story about a village that produced higher yields of hay than the nearby villages because it had more old ladies. Darwin reasoned that the old ladies kept more cats than other people and these cats caught more field mice. Field mice are important predators of wild bees and since these bees were essential for pollination of red clover (and clover improved the yield of hay), the increased number of bees increased the hay yield. This is an example of a 'food chain' which highlights the fact that plants and animals in the ecosystem have important feeding relationships which affect the overall success of the system (Figure 3.6a).

Plants always form the first link in the food chain. They are the **primary producers**, manufacturing their own food through the process of photosynthesis. Organisms which feed on plants are termed **primary consumers**. In turn, organisms which feed on the primary consumers are called **secondary consumers**. A habitat may include a third (tertiary) level and even a fourth (quaternary) level of consumers but rarely

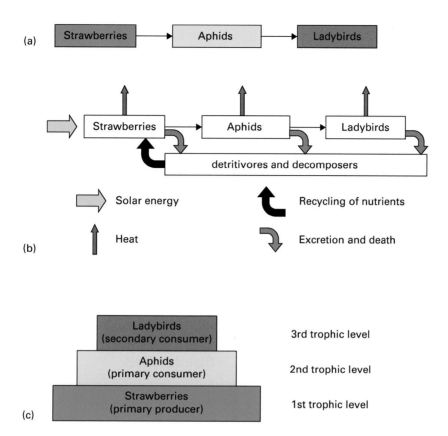

Figure 3.6 Food chains: (a) a simple food chain found in a garden – the strawberry is the primary producer, the aphids the primary consumer and the ladybirds the secondary consumer; (b) flow of biomass and energy through the food chain illustrated in (a) – biomass and energy (heat and chemical energy) are lost from the food chain and nutrients are recycled by detritivores and decomposers; (c) an ecological pyramid representing the loss of biomass and energy at each trophic leveld

more. A garden example could be a rose bush (the primary producer) on which aphids feed (the primary consumer) which in turn are eaten by birds (the secondary consumer). In practice, the rose bush may support several primary consumers including, for example, a fungus causing black spot or caterpillar larvae or slugs some of which themselves may be consumed by a range of secondary consumers, such as hedgehogs. In this way food chains are interconnected, forming complex **food webs** (Figure 3.7). Because of this interdependency, removal of any one of the organisms in a food web can have a profound effect on the whole ecosystem. So in a garden, for example, if we remove the aphids from the rose bush completely, we could also be reducing the numbers of birds.

> A primary **producer** manufactures its own food from simple molecules (e.g. green plants photosynthesizing). A **consumer** feeds on living organisms.

Biomass and energy flow

The weight or volume of living plant and animal material in an ecosystem is called its **biomass**. The flow of biomass in a food chain as organisms feed on each other can be represented as a **pyramid** (Figure 3.6c). At the base is the primary producer (a rose), resting on this is the primary consumer (aphids), and above this is the secondary consumer (birds). Each level in the pyramid is termed a **trophic level**. The pyramid shape reflects the loss of matter from the food chain in that whenever biological material is consumed at each level some matter is lost as waste, or through death of organisms and also through release of gases and water in respiration (see Chapter 9). In our rose example, this means that the total biomass of rose material must always be more than the biomass of aphids which feed on it which, in turn, is always more than the biomass of the birds which feed on the aphids. In this way, if we reduce the rose biomass available in the garden (say by reducing the size of the rose bush or planting fewer

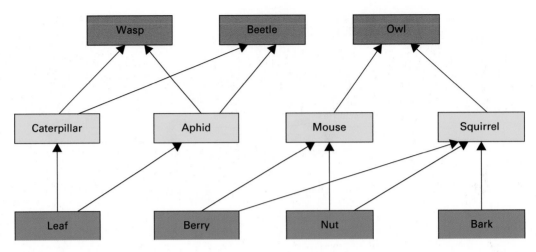

Figure 3.7 A simple woodland food web showing how species interconnect. The bottom level is made up of plants (primary producers). Both primary consumers such as mice and secondary consumers such as predatory beetles may feed on several different food sources. Removal of any species in the food web can have an affect on one or more other species.

rose bushes), this will have an even greater impact on the number of birds because their biomass is comparatively smaller.

In practice, much of the biomass lost as waste and dead organic matter is recycled by a whole group of organisms called **decomposers** and **detritivores** which feed on dead organic matter (see Chapter 13). As well as the flow of biomass, there is also a loss of **energy** through the system (Figure 3.6b). The process of photosynthesis (see Chapter 9) enables the plant to convert sunlight energy into chemical energy, which is stored in the biomass of the plant. As the plant is eaten by primary consumers, approximately 90% of the energy trapped in the leaf is lost either by respiration, or by heat in the consumer's body, by waste which is excreted by the consumer and by death of organisms. Since matter and energy are lost from the food chain at each trophic level, this explains why only a limited number of levels are possible. Eventually a point is reached where there is insufficient biomass and energy to sustain a further level of organisms (which is why large carnivores are comparatively rare).

Gardens are highly productive and can produce a large amount of plant biomass, and the greater the biomass, the more consumers it will support. A densely planted border, for example, is estimated to produce 0.5–2.0 kg of biomass per square metre per year. Therefore in a garden the aim is to provide as rich a variety of organisms as possible, that is, a high biodiversity at every level together with as much plant biomass as can be achieved to maintain a healthy and functioning ecosystem.

Succession

Communities of plants and animals change with time. Within the same habitat, the species composition will change, as will the number of individuals within each species. The process of change is known as 'succession' (Figure 3.8). **Primary succession** starts from uncolonized rock, for example, where plants move in to colonize the area. **Secondary succession** results from distubance of an existing habitat, for example, tree and shrub removal, or burning off vegetation. In this case, existing plants present as seeds in the soil recolonize the area. This kind of succession is common in Britain and Ireland where most land surfaces have been covered by vegetation at one time or another.

Succession involves a characteristic sequence of plant types which each change the environment in some way enabling the next group of plants to become more successful. The first species to establish make up the '**pioneer**' community. For example:

▶ in a cleared woodland this might well be mosses, lichen, ferns and fungi
▶ in the garden these would be ephemeral and annual weeds.

As these pioneer species remove water from and consolidate the soil, they themselves die and contribute to soil formation (see p. 143). Larger herbaceous plants such as bracken, willow herb, foxgloves and tall grasses can then become established and outcompete the pioneer species for light, water and nutrients, taking over. Often these early colonizing species are characterized by being able to:

3

Figure 3.8 Examples of primary succession: (a) a mountain landslide; (b) marram grass on sand dunes

▶ spread rapidly
▶ mature quickly
▶ produce large quantities of seed
▶ extend the period over which they germinate.

In fact, they have typical weed characteristics. Such species are often referred to as **opportunistic** (Figure 3.9).

The third successional stage involves larger plants, shrubs and climbers such as bramble and honeysuckle competing with plants in the previous stage. Eventually, these stages are kept in check or shaded out by trees. This final stage is described as the '**climax**' community. In a cleared garden or newly prepared border or vegetable plot, the pioneer species would be annual and ephemeral weeds which emerge first from the seed bank in the soil while the climax species would be trees (Figure 3.10).

Through the stages of succession there is usually an increase in the number of species found, both plant and animal, meaning an increase in biodiversity. These form increasingly more complex food webs although often in the climax community just a few species dominate. In gardens we constantly intefere with natural succession. If succession is halted at the beginning of the sequence, biodiversity will be low. For example, this could be

▶ through constant cultivation
▶ repeated mowing of a lawn.

Figure 3.9 Sedge, an opportunistic species in succession in damp habitats

Figure 3.10 A mature woodland representing the natural climax community in Britain and Ireland

However, if the garden is managed to allow for further stages in succession to establish, for example

▶ by allowing some weeds to cover the bare soil
▶ by cutting parts of the lawn higher and less frequently

then it will support a greater biodiversity so long as it is not neglected, resulting in a climax vegetation.

> **Succession** is a sequence of changes in the composition of plants and animals in an area over time. The **pioneer** community is the first in the sequence and the **climax** community is found at the end.

Interactions between organisms

Within a garden there is constant interaction between all the organisms in the ecosystem. Sometimes the relationships are beneficial to both partners (**mutualism**). An example is the nitrogen-fixing bacteria which live in the nodules of some plant roots (Figure 3.11). The plant roots provide a home for the bacteria and sugars to utilize, while the bacteria trap nitrogen gas in the atmosphere and convert it to nitrates, a form of nitrogen which the plant can use (see p. 168). Other examples of mutualisms are mycorrhiza (an association between plant roots and fungi) which take the place of root hairs in many plants and lichens (an association between fungi and cyanobacteria) (see Figure 4.14).

In other relationships, the pairing is beneficial to one partner but has no effect on the other – for example, climbing plants and **epiphytic** ferns which

Figure 3.12 Epiphytic ferns growing on a tree branch

live on trees giving them greater access to light for photosyntheis (Figure 3.12).

Predation is a relationship which is harmful to one of the partners and beneficial to the other, often where one partner (the predator) consumes all or part of the other individual (the prey or host). Predation includes **herbivores** which feed on plants (e.g. aphids, vine weevils, slugs, rabbits, deer) (Figure 3.13); **carnivores** which feed on animal tissue (e.g. birds feeding on aphids, nematodes eating bacteria, hedgehogs eating slugs); and **parasites** in which the predator has a very close but detrimental relationship with the host often living inside the host's tissue (e.g. viruses, fungal diseases, some biological controls) (see p. 195).

Some plants such as mistletoe (*Viscum album*) can manufacture their own food through photosynthesis but extract water and mineral nutrients from the host so are only partly parasitic (**hemiparasite**) (Figure 3.14).

Predators and their prey follow closely linked population cycles (predator–prey cycle) which increase and decrease. When there is abundant prey, there is plenty of food and predator numbers increase. This leads to a reduction in the amount of prey which then leads to a reduction in the number of predators as their food supply shrinks. In a balanced ecosystem, the prey population is never completely removed, some prey individuals are able to hide

Figure 3.11 Example of mutualistic relationships – root nodules in a legume (source: Wikimedia Commons)

3

(a)

(b)

Figure 3.15 (a) Larvae of the seven-spot ladybird; (b) rose aphids

Figure 3.13 Muntjac deer – a large herbivore

Figure 3.14 *Viscum album* (mistletoe) – a hemiparasitic plant

in pockets in the garden and survive to reproduce. Once the predator population decreases, the prey population can expand causing the the cycle to start again. In a garden, for example, beneficial insects such as ladybirds (predators) control the numbers of harmful insects such as aphids (the prey)

(Figure 3.15). Understanding predator–prey cycles is important in the use of biological control of plant pests (see Chapters 16 and 18). If broad-spectrum pesticides are used which kill all insects, they can interfere with these natural cycles causing disastrous results. For example, ladybird numbers might be reduced to such a low level they cannot recover as quickly as the aphids, so the latter grow to even greater numbers than before. In this way, the use of chemical pesticides frequently causes minor pests to become serious problems by disturbing the natural controls that keep them in check.

Competition

Organisms compete for shared resources such as light, moisture, nutrients and space, reducing their growth and their potential for reproduction. Competition can be by **exploitation** of the shared resource – for example, when two plants compete for water in a dry environment, the more successful organism is the one which is able to survive on less. It can also be by **interference** where one organism directly affects the uptake of the resource by the other, as in trees and shrubs such as *Rhododendron ponticum* which produce chemicals which prevent seeds germinating beneath their canopies (see Fig. 1.10b). Organisms overcome the effects of competition by moving to new sites through seed and fruit dispersal in the case of plants, or by seeking new territories in the case of some animals.

> **Mutualism** is a relationship between two organisms which is beneficial to both. **Predation** is a relationship which is harmful to one partner (the host) and beneficial to the other (the predator). **Parasitism** is an extreme form of predation where the predator often lives within the host.

The degree of competition between plants as they grow will depend very much on the spacings between them, whether this is in the garden border, in the vegetable plot or in the greenhouse. Competition can be **intraspecific**, that is, between individuals of the same species, and this will be the case in plants grown in a **monoculture**, where only one species is present. For example, if carrots are sown too close together, their roots will be small. Sometimes this can be an advantage, as in carrots grown for canning, closer spacings are used deliberately. Small, even-sized carrots may be specified by supermarkets so spacings can be designed to achieve this. Usually though, close planting will be detrimental. Overcrowded plants may be more susceptible to fungal diseases due to poor air circulation, they may grow leggy with weak stems or seed production may be reduced if shaded and competing for light. Growth may be poor if roots are competing for soil, water and nutrients. Three ways by which gardeners overcome these problems as plants grow is by transplanting seedlings from trays into pots, increasing the spacing of pot plants in greenhouses and hoeing out or 'thinning' a proportion of young vegetable seedlings from a densely sown row. Crops grown in deep bed systems, in which a one metre depth of well-structured and fertilized soil enables deep root penetration, reduce competition by rooting at different depths, allowing plants to be grown closer together.

Competition can also be **interspecific**, between two different species. When designing a mixed border it is important to allow for the future size of the plants and place them accordingly so they do not compete with each other for resources (see 'niche' above).

> **Interspecific competition** is between individuals of different species. **Intraspecific competition** is between individuals of the same species.

Where competition is removed completely growth can take place unchecked. A single plant growing in isolation with no competition is as unusual in horticulture as it is in nature. However, specimen plants such as leeks, marrows and potatoes, lovingly reared by enthusiasts looking for prizes in local shows, grow to enormous sizes when freed from competition. In landscaping, specimen plants are placed away from the influence of others so that they not only stand out and act as a focal point, but also can attain perfection of form.

Gardening for wildlife

When we think of gardening for wildlife it is very easy to just imagine a garden full of flowers and the three 'B's – birds, butterflies and bees – and certainly we want to encourage them. But a healthy garden which is good for wildlife contains much more. It supports a wide range of interdependent organisms, as we have seen above, most of which we never see but on which many more visible organisms depend. Invertebrates (organisms which have no internal skeleton) include not just bees and butterflies but other insects such as flies and lacewings, beetles, spiders, snails and slugs, earthworms and microscopic worms, which all have an important role in food chains. We have 2,400 native moth species, an important food source for bats, but only 60 butterflies which are regularly seen in Britain and Ireland. By providing suitable environments within the garden and managing it appropriately, we can create the right conditions for complex food webs to become established, thereby increasing biodiversity. Fundamentally, it is the primary producers, the plants, which hold the key. If there is a good range of these and plenty of them, the ecosystem's food chains and webs will be well supported.

In addition, the presence of decomposers and detritivores in a healthy soil will cycle organic matter through the ecosystem. To support the widest biodiversity a garden must be designed to provide three factors for all the organisms that live there:

▶ food
▶ shelter
▶ breeding sites.

Some of the ways by which we can increase and maintain biodiversity in a garden are outlined below.

What do we mean by biodiversity?

Biodiversity is generally thought of in terms of numbers of different species in a particular habitat but the definition also has a broader scope. Biodiversity can refer to habitats themselves, a wide range of habitats in a particular location such as a garden will result in greater biodiversity. At the other end of the scale, biodiversity can apply to the range of genetic variability within a species too – for example, the number of cultivars listed for a particular horticultural species such as apple. When we speak about conservation of biodiversity, we therefore mean conserving the genetic variation within a species, the total number of species present and all the habitats they live in.

Garden structure

A good wildlife garden tries to provide the greatest range of habitats within the space available. In nature, woodlands have the highest biodiversity of any terrestrial habitat and borders can be thought of as recreating a woodland edge. The most biodiverse part of a wood is the edge where the light canopy ranges from very shaded under the tree, to open and well lit on the very edge where naturally it would join a path or meadow area. Furthermore, because of the light penetration, the soil moisture content will vary across the edge too. It can be seen that the wood edge provides a wide range of habitats over a short distance. Also, a good vertical structure is important: a woodland has typically at least a three layered structure with an upper canopy of tall trees, a shrub layer beneath and a herbaceous layer at ground level. This can be imitated in a garden where more vertical layers, trees, shrubs, tall perennials, ground cover and bulbs, lead to a better situation for wildlife (Figure 3.16). In addition, such a structure gives a more interesting garden and the opportunity for greater plant choice.

If other habitats can be added this will further increase biodiversity:

Figure 3.16 A multilayered garden structure with trees, shrubs, perennials and ground cover and areas of light and shade

▶ **Trees**. A tree can be thought of as a habitat in its own right with several microhabitats within it. There is a variety of light and shade through the canopy, the leaves, flowers and fruit provide a food source for organisms and it also provides shelter and areas for roosting and nesting birds and overwintering sites for insects. The bark, especially in an old tree, also provides its own microhabitat, with breeding sites and shelter for many insects in its furrows. Old trees also provide support for lichens and epiphytic plants and even small pools of water for many organisms to utilize or live in (Figure 3.17). Rotting wood and leaves are a food source for soil organisms below. Log piles in a garden can provide a similar microhabitat to fallen trees in a woodland. Most conifers support a lower biodiversity than broadleaved trees. Nevertheless, a single tree, however small, has been shown in studies to provide one of the most beneficial habitats in a garden. For example, ornamental *Betula* spp. (birches) potentially host more than 150 species of butterfly and moth.

▶ **Ponds** provide an additional important habitat. As with woods, the pond edge is the most biodiverse part of the pond because it is shallow and warm and light can penetrate easily. A small pond therefore has greater wildlife potential than a large lake where much of the area is deep water and less diverse. In suburban gardens, many small ponds are better than a large pond

Figure 3.17 An ancient tree trunk providing a wealth of microhabitats for many organisms

Figure 3.18 Dragonfly nymphs live underwater then climb up the waterside plants to hatch into adults.

covering the same total area and this also means that the loss of any one pond through development has less impact overall. The area between the highest water level and the lowest (the **draw-down zone**) is the most biodiverse, so sloping pond edges in ponds which are not frequently topped up are the best arrangement. A good range of marginal plants will provide cover for organisms to hide, feed and breed at the water's edge (Figure 3.18).

▶ **Boundaries**. As with borders, a hedge is basically a woodland edge in miniature and provides a good range of habitats for organisms. Hedges can be important for wildlife in three ways:

▷ First, they provide a food source, shelter, breeding and territorial sites for a whole range of organisms from invertebrates to birds and mammals.

▷ Second, hedges act as **wildlife corridors** connecting habitats together and enabling organisms such as hedgehogs and other small mammals to travel further in search of food, shelter or breeding sites and also provide a route for prey to flee from predators. They are commuting highways for bats, butterflies and small mammals between feeding and roosting sites. Where habitats become fragmented (e.g. due to removal of hedgerows or urban development), the populations in them become threatened as they are unable to escape predators (Figure 3.19).

▷ The breeding success of such populations is reduced because there are few individuals and less mixing of genes, so genetic problems such as increased susceptibility to disease or reduced fertility can occur. A collection of suburban gardens can therefore be seen as one large habitat because they are interconnected and they often form wildlife corridors linking larger green areas such as parks or fields to each other.

▷ Third, the hedge itself creates shelter in the garden and protects soils from erosion and moisture loss enabling a wider range of plants to be grown, along with the organisms they support, and a more biodiverse soil community. Although a single species hedge will be better than a fence or wall, a mixed hedge will provide a wider range of food plants and structure.

If hedges are not an option, a sunny wall or fence will support a range of invertebrates and if climbers are grown up them then the habitat is enriched.

▶ **Meadows**. While lawns are an attractive feature in gardens they incorporate only a few plant species, supporting a limited range of organisms. They produce no flowers and are poor cover for invertebrates. Studies show that even a small area of long grass is beneficial for invertebrates for shelter. Incorporation of flowers among the grasses to imitate a meadow will improve the wildlife value through supplying nectar and pollen for insects and seeds for birds. It is important to reduce competition from vigorous grasses and the key to this is reducing soil fertility (see p. 140). Unfertilized, light sandy soils are therefore more suitable than heavy clay soils.

Figure 3.19 (a) Disrupted 'green corridors' in farmland where hedgerows have been removed; (b) farming has also left fragments of disconnected woodland

Yellow rattle

Yellow rattle (*Rhinanthus minor*) is named after its large seeds which rattle in their pods when dry. It was also called 'catch all' by farmers as it was a weed of hay meadows, which reduced yields and was spread in haymaking. The plant is an annual hemiparasite that feeds on the roots of coarse grasses and leguminous plants but can also photosynthesize. It can reduce grass biomass by up to 80% and allows less competitive wild flower species to establish and compete more effectively. Yellow rattle seed is often incorporated into wild flower seed mixes.

Figure 3.20 (a) Yellow rattle; (b) note the species-diverse area in foreground containing yellow rattle compared with the area of coarse grasses behind

▶ **Plant selection**. As primary producers, plants are the foundation of the living component of a garden. They provide food in the form of nectar, pollen, seeds, fruits and green leafy material so should be chosen to supply these useful 'products' over the longest period of time, in particular by including winter-flowering plants such as *Viburnum tinus* and *Mahonia*. *Hedera helix* (ivy) (Figure 3.21) is one of the best plants for wildlife in the garden. It provides evergreen foliage for nesting and shelter for small birds and insects, and it flowers and fruits late in the season when there are few other nectar and pollen sources available.

▶ Cultivated plants with double flowers should be avoided as the male and female flower parts are converted into extra petals unlike the wild forms

Figure 3.21 A mature ivy with flowers and fruits

Figure 3.22 Single (right) and double (left) flowered forms in a paeony. Note the stamens present in the single form

(see petalody p. 102), and so they may be sterile and lacking pollen and nectar (Figure 3.22). Grasses, bamboos and ferns are poor food sources for invertebrates, although they do provide good cover.

Native plants are those which were present at the end of the last Ice Age when mainland Britain and Ireland separated from the rest of Europe. An example is *Quercus robur* (oak).

Naturalized plants are those that have been introduced by humans and have spread and now reproduce in the wild. An example is *Buddleja davidii*.

Exotic plants have been introduced by humans recently and are dependent on human management. An example is *Lavandula* spp (lavender).

Cultivated plants do not exist in the wild anywhere (e.g. plant cultivars such as *Narcissus* 'Tete-a-Tete').

Whether planting should be native or non-native is a subject of much debate. Native plants are often considered the 'best'. This is because they have evolved alongside the organisms that use them over a long period of time so support the greatest number of organisms of which many will be dependent on a particular plant and adapted to it. For example, the oak tree is often cited as supporting over 400 species of invertebrates. However, these will not all be found on a single tree at any one time. Furthermore, most invertebrates are not limited to one tree species, they are generalists rather than specialists. Even within Britain and Ireland, some species are only truly native in some localities. *Pinus sylvestris* (Scots pine), for example, a 'native' tree, is only native to Scotland, that is, records show it was growing there at the end of the last Ice Age. In Southern England it has been planted. Similarly, there is evidence that *Acer platanoides* (sycamore) may have been native in northwest England originally, but elsewhere it is an introduced species and is frequently removed. Even native trees such as *Quercus robur* (oak) may have been planted from European sources sometime in the past, so would be genetically different from local ones. In effect, any tree is useful whether native or not and numerous naturalized and exotic plants such as lavender, buddleja, borage, *Sedum spectabile*, *Cotoneaster* and *Echium* are known to be excellent for wildlife.

In some situations, plants are selected that are local to the site – in this case their **provenance** (see p. 128) is important because it keeps the genetic strains in the locality pure, and this may be necessary. However, most organisms are probably not selective about the plants they use and it is what they can provide that counts rather than what they are. The RHS 'Plants for Bugs Study' is investigating whether the geographical origin of plants makes a difference to the abundance and diversity of garden invertebrates. Three beds have been planted to imitate a garden border each containing plants from a different geographical zone. In total more than 34,000 insects have been recorded in the first two years of the four-year study, which included 13 species of butterfly, 7 bumblebee species, more than 211 ground-dwelling beetle species and 27 species of spider.

Lists of plants suitable for attracting wildlife to the garden are easily obtainable, often they focus on butterflies and bees but it should be remembered that plants and planting should encourage all wildlife in order to provide the greatest variety of organisms to provide a food source for higher trophic levels in the food chain (see Figure 3.23).

Figure 3.23 Some native plants suitable for different habitats in a garden: pond edge, (a) flag iris (*Iris pseudacorus*); damp meadow, (b) snake's head fritillary (*Fritillaria meleagris*), (c) ragged robin (*Lychnis flos-cuculi*), (d) white butterbur (*Petasites albus*); woodland with dappled shade or glade, (e) foxglove (*Digitalis purpurea*), (f) daffodil (*Narcissus pseudo-narcissus*); open ground, (g) teasel (*Dipsacus fullonum*)

Management

Managing a garden to encourage wildlife is not difficult nor should it result in a garden which is overgrown or unpleasant to look at. Many of these management approaches follow that of **organic gardening**: reduced pesticide use, minimal soil disturbance and a focus on good soil health, mixed planting rather than monocultures and a general aim of growing with nature rather than against it. Organic gardening relies on a good range of natural predators for pest control and soil organisms for fertility. Some general principles are:

▶ **Prune to maximize flower and fruit** production and deadhead herbaceous plants to prolong flowering so extending the supply of pollen and nectar.

▶ **Do not use pesticides** because they interfere with the balance of organisms in the garden and disrupt food chains and webs (see above). Pesticides may remove 'beneficial' insects, such as bees or those which feed on 'harmful' species. Even if the pesticide is targeted at a specific pest species, that may be a food source for other organisms. Pests may also become resistant to a pesticide and will no longer be controlled. It is worth tolerating some plant damage in return for improving biodiversity in the garden.

▶ **Minimize interference** with plants, shrubs and trees and carry out operations at a time of year which is least harmful. For example, hedges are an overwintering site for many invertebrates and a nesting site for birds in spring so pruning times should take this into account. In fact, there is evidence that pruning hedges in summer when nesting is over prevents excessive growth compared with pruning at other times of year. Pruning hedges in the winter may disturb overwintering organisms such as small mammals and invertebrates. In the autumn and winter it is important to leave seed heads and dead material for food and shelter rather than have the urge to simply tidy up.

▶ Encourage a good population of soil organisms by allowing organic material such as composted matter, fallen leaves and herbaceous plants to return

Figure 3.24 An insect 'hotel' (source: Arnaud 25, http://commons.wikimedia.org/wiki/File:Insect_hotels_001.JPG?uselang=en-gb. This file is licensed under the Creative Commons Attribution-Share Alike 3.0 Unported licence)

Figure 3.25 Blue tit at a bird feeder

to the soil, recycling nutrients, providing warmth and shelter and protecting the soil from erosion.

▶ There is no need to overcultivate as this can damage soil structure and organisms.

▶ In a small garden where natural vegetation is limited, insect hotels can be provided (Figure 3.24). By using a range of materials, solitary bees, butterflies, predatory insects and other invertebrates can be provided with shelter and refuges especially over winter.

See companion website for information on the RHS wildlife pages.

Garden birds

One of the most visible forms of wildlife in the garden are birds and for many people this is one of the key benefits of a wildlife-friendly garden. Over half the the adults in the UK feed birds in their gardens according to the RSPB. Measures taken to encourage birds with feeders (Figure 3.25) and nest boxes undoubtedly attract birds to the garden, and help to support their numbers. Winter feeding can often lead to earlier laying and increased breeding success and for some species, such as song thrushes, breeding populations in gardens are now better than those of farmlands. Adult birds of other species will feed on supplementary sources of food such as peanuts but their fledglingsrequire a high protein diet of invertebrates such as spiders and worms. Blue tit fledglings, for

example, feed largely on caterpillars. Unless the garden can supply these foods (or the gardener is prepared to provide them artificially) birds may breed earlier in the year but their broods may not survive. It has also been suggested that feeding birds may also increase the spread of disease and may affect natural selection through influencing reproduction and behaviour, enabling birds which are weaker genetically to survive. Therefore, while encouraging birds into the garden undoubtedly has many benefits for both birds and humans, ecological principles remind us that this is no substitute for also providing natural food sources, shelter, roosting, cover and nesting places too.

Further reading

Baines, C. (2000) *How to Make a Wildlife Garden*. Frances Lincoln.

Buczacki, S. (2010) *Garden Natural History*. Collins New Naturalist Library. Collins.

Buczacki, S. (1986) *Ground Rules for Gardeners*. Collins.

Mackenzie, A., Ball, A.S. and Virdee, S.R. (1991) *Instant Notes in Ecology*. Taylor & Francis.

Owen, J. (2010) *Wildlife of a Garden. A Thirty Year Study*. Royal Horticultural Society.

Owen, J. (1991) *The Ecology of a Garden: The First Fifteen Years*. Cambridge University Press.

Ricklefs, R.E. (2000) *The Economy of Nature*. W.H. Freeman.

Thompson, K. (2006) *No Nettles Required*. Eden Project Books.

Townsend, C.R., Begon, M. and Harper, J.L. (2008) *Essentials of Ecology*. Blackwell.

Please visit the companion website for further information:
www.routledge.com/cw/adams

CHAPTER 4
Level 2

Classification and naming of plants

Figure 4.1 A range of plants found in the plant kingdom, including moss growing between stones, a fern and a seed-producing plant (ivy), in their natural habitat

This chapter includes the following topics:

- What is a plant?
- Plant classification – families, genera and species
- Plant names – the binomial system and cultivars
- Gymnosperms and angiosperms
- Monocotyledons and dicotyledons
- Further classifications of plants used in horticulture
- Fungal groups

Principles of Horticulture. 978-0-415-85908-0 © C.R. Adams, K.M. Bamford, J.E. Brook and M.P. Early.
Published by Taylor & Francis.

What is a plant?

The world of living organisms can be grouped in many ways. One such classification is the five kingdom system:

- ▶ Plantae (plants)
- ▶ Animalia (animals)
- ▶ Fungi
- ▶ Prokaryota (bacteria)
- ▶ Protoctista (all other organisms that are not in the other kingdoms, including algae and protozoa).

What distinguishes the plant kingdom from the other four kingdoms? Plants are largely sedentary and live on land (albeit some in aquatic environments such as ponds and rivers). They are multicellular and their cells have cellulose cells walls and a nucleus. Most importantly, they are **autotrophs** (the 'producers' described on p. 38), organisms that are able to convert energy from one form, that is light, into a chemical form stored in organic molecules such as sugar and starch through the process of photosynthesis (see Chapter 9). Animals, among other things, have no cell walls and are **heterotrophs** (the 'consumers' described on p. 38). They rely on eating ready-made organic molecules for their nutrition and feed on plants and other organisms. Fungi do not photosynthesize and although some algae and bacteria do so, the former are aquatic organisms and the latter do not have a nucleus.

The plant kingdom contains a range of plant groups which reflect their evolutionary pathway from simple organisms, such as mosses and liverworts, to the more advanced conifers and flowering plants. Mosses and liverworts are the most primitive and are termed '**non-vascular**' plants as they have no specialized tissue for conduction of water and minerals (Figure 4.2). This, together with the need for water

Figure 4.2 A feathery green moss and a purple liverwort growing on a wet rock. The liverwort structures are approximately 1 cm long

Figure 4.3 Hart's tongue fern (*Asplenium scolopendrium*) a native British fern

for reproduction, limits their size and restricts them to damp shady habitats. Mosses may be a problem weed in some circumstances, particularly in shaded and poorly drained lawns, but may also be used as an ornamental feature in Japanese gardens.

The '**vascular**' plants contain conducting tissue (xylem and phloem) (see Chapter 6) and range from ferns to higher plants such as gymnosperms, which include conifers, and angiosperms (the flowering plants). The '**lower vascular plants**', of which ferns are an example (Figure 4.3), spread by means of spores, in common with the non-vascular plants. They also require water for reproduction so tend to be found in damp places. Ferns provide a wide range of decorative plants in the garden and as houseplants, with many attractive leaf shapes and forms and a variety of sizes. They are especially useful for planting in shade.

In contrast, the '**higher vascular plants**', the angiosperms and the gymnosperms, which are more evolutionarily advanced, produce seed and spread by this means rather than by spores. It is mostly the higher plants which will be examined in this book as they represent by far the biggest and most diverse group of plants used in horticulture.

Plant classification – families, genera and species

Classification involves putting objects or organisms into groups based on characteristics which the members of the group share, it is something we do all the time and it is very useful. For example, the fresh produce section of the supermarket contains a salad section, a vegetable section and a fruit section. The fruit section in turn is subdivided into smaller groupings such as citrus (oranges and lemons), top fruit (apples and pears) and soft fruit (strawberries, blackberries and raspberries).

This enables us to find what we want quickly and know that what we are buying has particular characteristics. In the same way, the plant kingdom is classified into major groups, such as vascular and non-vascular plants described above, which are in turn subdivided into smaller groups. For gardeners, the groupings of plants which are most commonly encountered are **family**, **genus** and **species**.

Plant families were first described comprehensively by the Swedish botanist Linnaeus in the eighteenth century. His classification was originally based on flower structure, although nowadays many other factors such as plant chemistry and genetics are used in addition to their external features. Many of Linnaeus' original family names and groupings still stand. Currently about 240 plant families are recognized. Plant family names always end in –aceae for example, Lamiaceae (the nettle family) or Poaceae (the grass family). The family is an important grouping in horticulture as all plants within a family have certain characteristics in common, so predictions can often be made about other family members. For example, many members of the family Rosaceae (the rose family) are susceptible to the disease fireblight, aiding identification and enabling prediction of the spread of this disease.

Within a family, plants are organized into groups of similar plants called **genera** (sing. genus). A family may contain many genera, such as the Asteraceae (the daisy family) with 1,317 genera including *Lactuca* (lettuce), *Taraxacum* (dandelion) and *Dahlia,* or a few such as Geraniaceae with just five including *Geranium* and *Erodium.*

A genus is made up of groups of similar plants called **species**. A species is a group of individual plants which show the greatest degree of mutual resemblance, and which, most importantly, are able to breed among themselves but not with plants from another species.

Carl Linnaeus – the Father of Classification (1707–1778)

Carl Linnaeus (also known as Carolus Linnaeus or Carl Linné) (Figure 4.4) worked as an assistant and later as professor of botany at Uppsala University in Sweden. He brought together all recorded knowledge of the natural world known at that time and classified it into three kingdoms: minerals, plants and animals. He named some 7,700 plants and 4,400 animals in his lifetime and used an innovative classification system which was first set out in his *Systema Naturae*

in 1735. Each kingdom was subdivided into a hierarchy of classes, orders, genera and species which replaced existing classification systems and is still in use today. For plants, he used the structure of flowers, in particular the male and female parts, as the basis of his classification. This 'sexual system' was a practical and easily learned approach which enabled him and his students to study a large number of species, although the florid way in which he described it caused an uproar at the time!

In addition, Linnaeus is credited with establishing the binomial system of nomenclature. Although the use of binomials for plants was not new, Linnaeus applied them consistently to all plant species alongside the cumbersome many-worded 'phrase names' which were current at the time and thus laid the foundation for a simple and universal naming system which has been adopted ever since. As such, Linnaeus' work *Species Plantarum*, first published in 1753, forms the basis for plant nomenclature right up to the present day.

Figure 4.4 Carl Linnaeus dressed as a Laplander from a painting by Hendrik Hollander in 1853. He explored Lapland in 1732, collecting plants, birds and rocks and used the 600-mile trip to apply his ideas of classification and nomenclature. He described 100 newly identified plant species in his book *Flora Lapponica*

A **genus** is a group of individuals within a family that have characteristics in common. A **species** is a group of individuals within a genus that have characteristics in common and are able to breed among themselves.

Plant names – the binomial system and cultivars

The name given to a plant species is very important and is the key to identification in the field or garden. Botanical plant names are stable and unambiguous; therefore their use avoids confusion. They are an international form of identity used by researchers and gardeners alike, in an internationally understood language. Armed with the botanical name, information on a specific plant can be sourced from books and the internet. A botanical name is required before breeders can legally protect the new plants they have bred and also means that the correct plant can be selected and identified in planting schemes. When dealing with medicinal plants and herbs, poisonous plants can be avoided.

Common names that we use for plants, such as daisy, potato and lettuce, are, of course, acceptable in English, but are not universally used. Common names may vary with location – for example, *Caltha palustris* has 140 names in Germany, 60 in France and 90 local names in Britain and Ireland including marsh marigold, kingcups and Mayblobs (Figure 4.5).

Alternatively, the same common name can describe several different species. Bluebell is the local name for *Campanula rotundifolia* in Scotland, *Hyacinthoides non-scripta* in England, *Wahlenbergia saxicola* in New Zealand, *Clitoria ternata* in West Africa and *Phacelia whitlavia* in the USA, none of which are related (Figure 4.6).

Figure 4.6 Two plants known as bluebells: (a) *Hyacinthoides non-scripta* (English bluebell); (b) *Campanula rotundifolia* (known as bluebell in Scotland and harebell in England)

Figure 4.5 *Caltha palustris*

Common names may be in a variety of languages and scripts and often plants are introduced without a common name (e.g. *Camellia sinensis*) or with one invented by the seller. A scientific method of naming plants therefore enables every plant to be unambiguously identified with an accurate name that is universally recognized.

The name can also provide information about a species, such as its relationship with other species, and can give clues about its origin, its preferred habitat or its characteristics such as its colour, size or form. See 'Plant Names' on the companion website. Linnaeus utilized a naming system which included the name of the genus to which a plant belonged followed by its individual species name written in botanical Latin. This is called a **binomial** after the two named parts. For example, the chrysanthemum used for cut flowers (*Chrysanthemum morifolium*) is in the genus *Chrysanthemum* and is the *morifolium* species; note that the genus name begins with a capital letter, while the species has a small letter. Other examples are *Ilex aquifolium* (holly), *Magnolia stellata* (star-magnolia) and *Ribes sanguineum* (redcurrant). The genus and species names must be written in *italics*, or underlined where this is not possible.

Plants within a species can vary genetically in the wild, giving rise to a number of naturally occurring individuals with distinctive characteristics, much as people vary in their appearance. Where these differ significantly from the original species they may be given an additional name after the species name and are called a subspecies (subsp.), varietas (var.) or forma (f.) depending on the degree of difference (forma being the least different and subspecies the most). These extra names are written in botanical Latin and are italicized. They follow the species name, beginning with a small letter and with the category abbreviated and unitalicized in front of them – for example, *Hydrangea petiolaris* subsp. *anomala, Ceanothus thyrsiflorus* var. *repens, Primula sieboldii* f. *lactiflora*.

In addition, cultivation, selection and breeding by humans have produced variations in species referred to as **cultivated varieties** or **cultivars**, which are distinguished from naturally occurring variants because they have not usually arisen in the wild and must be maintained in cultivation either by specific breeding programmes to produce seed or by vegetative propagation. The cultivar is given a name, often chosen by the plant breeder who produced it, such as *Rhododendron arboreum* 'Tony Schilling' or *Cornus alba* 'Sibirica', and is always a non-Latin (vernacular) name, unitalicized and enclosed in single quotation marks. Cultivar names can also provide information about a plant's characteristics, for example, a dessert apple that is suitable for small gardens, 'Red Devil', produces bright red fruit; a thornless blackberry, 'Loch Ness', was raised in Scotland and shows considerable winter hardiness. This information is useful for gardeners. *Penstemon* and *Pelargonium* genera have the cultivars 'Apple Blossom', which describe well the pale pink and white of the flowers. Cultivar names can also be written, where applicable, after a common name, often for fruits and vegetables – for example, tomato 'Ailsa Craig' and apple 'Bramley's Seedling'.

The general term 'variety' (which is not the same as 'varietas' described above) is often used to refer to any plant type which varies from the original species. As such, it is frequently used interchangeably with 'cultivar'. In this publication, however, the correct term 'cultivar' will be used throughout.

> A **cultivar** is a variation within a species that has usually arisen and has been maintained in cultivation. A horticultural **variety** is a general, non-botanical term for plants that vary from the species.

Examples of plant groupings and how they are named in the family Rosaceae are shown in Figure 4.7.

Plant groups of importance in horticulture

Gymnosperms and angiosperms

Two of the most significant groups in the plant kingdom, the angiosperms and the gymnosperms, are distinguished from all other plants by their reproductive behaviour. These are the seed-bearing plants which spread by dispersal of seeds rather than by spores as in more primitive plants such as mosses and ferns.

Gymnosperms characteristically produce male and female cones which bear only partially enclosed 'naked' seeds, hence the name 'gymnosperm'. By far the largest gymnosperm group is the **conifers**, which include many hundreds of species such as the pines, junipers, spruces and yews (Figure 4.8) and form the vast boreal forests of the northern hemisphere. They are also found in Australia, Papua New Guinea and South America. The Chilean pine or monkey-puzzle tree (*Araucaria araucana*) was once widely planted in Victorian gardens. Conifers have the following characteristics:

▶ primarily perennial, woody trees and shrubs
▶ have male and female cones
▶ male cones produce prodigious amounts of pollen which is spread by the wind
▶ seeds, which are borne on female cones
▶ often found in a limited range of habitats where water is in scarce supply either due to low rainfall or because the ground is frozen for much of the year
▶ frequently display structural adaptations to reduce water loss, for example, needles or scale leaves (Figure 4.9) and branches designed to shed snow
▶ may contain resin in their wood which acts as an antifreeze
▶ mostly evergreen to take full advantage of the short growing season and avoid expending unnecessary energy on producing new leaves each year, although a few genera such as *Taxodium, Metasequoia* and *Larix* are deciduous.

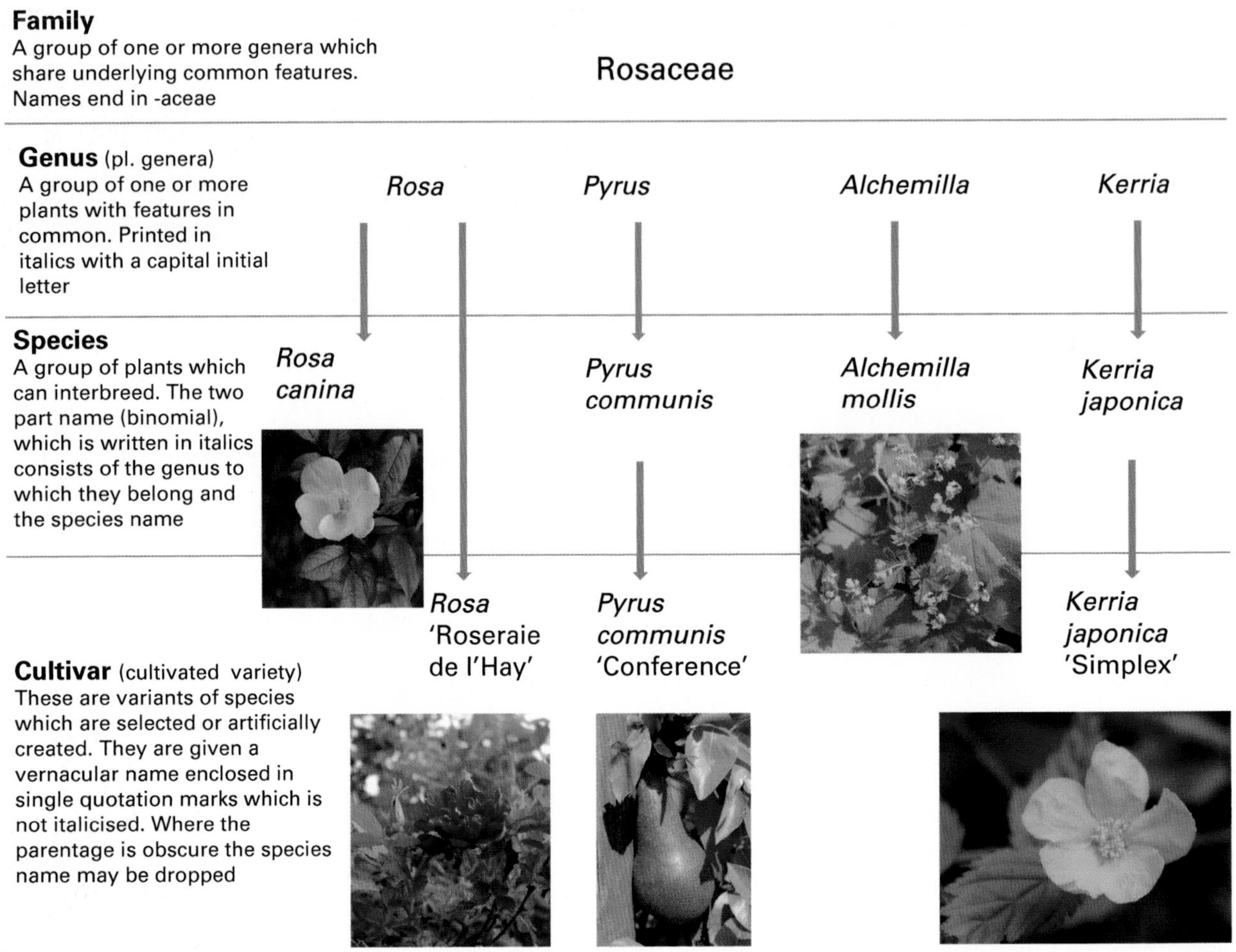

Family
A group of one or more genera which
share underlying common features.
Names end in -aceae

Rosaceae

Genus (pl. genera)
A group of one or more
plants with features in
common. Printed in
italics with a capital initial
letter

Rosa　　　*Pyrus*　　　*Alchemilla*　　　*Kerria*

Species
A group of plants which
can interbreed. The two
part name (binomial),
which is written in italics
consists of the genus to
which they belong and
the species name

*Rosa
canina*　　*Pyrus
communis*　　*Alchemilla
mollis*　　*Kerria
japonica*

*Rosa
'Roseraie
de l'Hay'*

*Pyrus
communis
'Conference'*

*Kerria
japonica
'Simplex'*

Cultivar (cultivated variety)
These are variants of species
which are selected or artificially
created. They are given a
vernacular name enclosed in
single quotation marks which is
not italicised. Where the
parentage is obscure the species
name may be dropped

Figure 4.7 Some plant groups within the family Rosaceae

(a)

Figure 4.8 (a) Yew (*Taxus baccata*) with 'berry'-like arils – a conifer

4

Figure 4.8 (b) conifer cones

Figure 4.9 Conifer leaves: (a) needles in *Pinus* spp; (b) scale leaves in *Thuja*

Figure 4.10 *Ginkgo biloba* leaf in autumn

There are very many important conifers. Some are major sources of wood or wood pulp, but in the garden many are valued because of their interesting plant habits, foliage shapes and colours. The cypress family Cupressaceae, for example, includes fast-growing species (e.g. ×*Cuprocyparis leylandii*), which can be used as windbreaks, and small slow-growing types very useful for rock gardens (e.g. *Juniperus procumbens*). The yews are a highly poisonous group of plants that includes the common yew (*Taxus baccata*) used in ornamental hedges and mazes.

One division of the gymnosperms is represented by a single surviving species, *Ginkgo biloba* the maidenhair tree, which has an unusual slit-leaf shape and distinctive bright yellow colour in autumn (Figure 4.10). It is a survivor from the Carboniferous era and fossils are found dating back 270 million years.

The **angiosperms**, or **flowering** plants, encompass the greatest diversity of plant life with adaptations for the vast majority of global habitats. There are estimated to be some 400,000 species of flowering plant on earth and they represent the most advanced plant life forms. Angiosperms have the following characteristics:

▶ unique in having flowers, usually hermaphrodite, which are pollinated by wind, insects and other agents and, in many cases, these flowers are highly adapted to their specific pollinators
▶ flowers produce seeds inside a protective fruit
▶ life cycles encompass the full range of ephemerals, annuals, biennials and perennials
▶ can be both herbaceous and woody in structure
▶ can be evergreen or deciduous in behaviour
▶ occupy the greatest range of habitats.

Many flowering plants are important in gardens, as crop plants, ornamentals and weeds. Chapters 5 to 10 will focus mainly on the angiosperms.

Monocotyledons and dicotyledons

The angiosperms are split into two main groups generally known as the **monocotyledons** and the **dicotyledons**. The main differences are given in Table 4.1 and are further described in Chapters 6, 7 and 8.

Monocotyledons include some important horticultural families – for example, Poaceae (formerly Graminae) the grasses and bamboos; Alliaceae, the onions; bulbous plants such as Liliaceae, which includes lilies and tulips and Amaryllidaceae which includes daffodils; Orchidaceae, the orchids and some food plants such as Musaceae, the bananas (Figure 4.11).

Dicotyledons have many more families significant to horticulture, including Magnoliaceae, the magnolias; Caprifoliaceae, the honeysuckles; Cactaceae, the cacti; Malvaceae, the mallows; Ranunculaceae, the buttercups; Theaceae, the teas; Lauraceae, the laurels; Betulaceae, the birches; Fagaceae, the beeches; Solanaceae, the potatoes and tomatoes; Nymphaeaceae, the water lilies; and Crassulaceae, the stonecrops (Figure 4.12).

Table 4.1 Differences between monocotyledonous and dicotyledonous plants

Monocotyledons	Dicotyledons
One seed leaf (cotyledon)	Two seed leaves (cotyledons)
Parallel veined leaves, usually alternate and sword-shaped with smooth margins	A variety of leaf vein patterns e.g. reticulate (net) veined and many different shapes and margins
Vascular bundles in stem scattered	Vascular bundles in stem arranged in rings
Vascular tissue (stele) in the root has many arms	Vascular tissue (stele) in the root with up to seven arms
No vascular cambium	Vascular cambium
Fibrous root systems	Both fibrous and tap root (primary) systems
Flower parts usually in threes or multiples thereof, also three seed chambers in fruit	Flower parts usually in fours or fives or multiples thereof, often four or five seed chambers in fruit
Small non-woody herbaceous plants (except palms and bamboos)	Both small and large, woody and herbaceous species with woody stems showing annual rings and bark

Further classifications of plants used in horticulture

Plants can be grouped into other useful categories. A classification based on their life cycle (ephemerals, annuals, biennials and perennials) has long been used by both botanists and growers. Many naturally perennial plants, such as *Pelargonium zonale*, *Lobelia*

4

Figure 4.11 Monocotyledonous angiosperms: (a) a fragrant orchid (*Gymnadenia conpsea*); (b) a bamboo

erinus or *Erysimum cheiri* (wallflower) may be grown as annuals or biennials that are removed and replaced after their first or second season of growth. These may be referred to as 'annuals' and 'biennials' by horticulturists – an example of where botany and horticulture disagree!

A distinction can also be made between the different types of woody plants such as trees and shrubs. Another classification can be made based on a plant's temperature tolerance, separating those plants that are able to withstand a frost (hardy), those that cannot (tender) and those which can withstand a few degrees of frost but may need some winter protection (half-hardy). Plants can be grouped according to their degree of hardiness and various scales are available; the one in Table 4.2 is produced by the Royal Horticultural Society (RHS). It must be remembered that, while withstanding cold conditions is the main factor in a

Figure 4.12 A dicotyledonous angiosperm: rose flower with five petals and a succulent fruit (hips)

Table 4.2 Some commonly used terms that describe the life cycles, structure, leaf retention and hardiness of plants

	Description	Example(s)
Life cycles		
Ephemeral	A plant that has several life cycles in a growing season and can increase in numbers rapidly.	*Senecio vulgaris* (groundsel)
Annual	A plant that completes its life cycle within a growing season.	*Limnanthes douglasii* (poached-egg flower)
Biennial	A plant with a life cycle that spans two growing seasons.	*Digitalis purpurea* (foxglove)
Perennial	A plant living through several growing seasons.	*Quercus robur* (oak), *Vibumum opulus, Acanthus spinosus*
Plant structure		
Herbaceous perennial	A perennial that is non-woody and generally loses its stems and foliage at the end of the growing season. They do not undergo secondary thickening.\n\nA few may be evergreen e.g. *Liriope muscari, Euphorbia characias, Ajuga reptans.*	*Aster* spp. (Michaelmas daisy), *Humulus lupulus* (hop)
Woody perennial	A perennial that maintains a live woody framework of stems at the end of the growing season. They undergo secondary thickening.	Bush fruit, shrubs, trees, climbers (e.g. *Vitis vinifera*, grape)
Shrub	A multistemmed woody perennial plant having side branches emerging from near ground level. Up to 5 m tall.	*Hydrangea macrophylla*
Tree	A large woody perennial unbranched for some distance above ground, on a single stem. Usually more than 5 m tall.	*Aesculus hippocastanum* (horse chestnut)
Leaf retention		
Deciduous	A plant that sheds all its leaves at once, often at the end of the growing season.	*Philadelphus delavayi* (mock orange)
Evergreen	A plant retaining leaves in all seasons.	*Aucuba japonica*
Semi-evergreen	A plant that retains some of its leaves through the year but may shed most leaves under severe weather conditions such as extreme cold or drought.	*Lonicera nitida*
Hardiness*		
Hardy (H4 to H7)	A plant able to survive temperatures below -5°C. Will survive freezing temperatures although some plants may suffer foliage damage or require protection if in pots. Can be divided into further categories depending how far below -5°C the plant can survive.	*Cornus alba* 'Sibirica', *Erica carnea, Magnolia* 'Susan', *Bergenia cordifolia*
Half-hardy (H3)	A plant able to survive freezing temperatures between 1°C and -5°C. Tolerant of a few degrees of frost. Will survive a mild winter but generally requires an unheated glasshouse over winter.	*Pittosporum crassifolium, Petunia* spp., *Plumbago auriculata, Clianthus puniceus*
Tender (H2)	A plant able to survive low temperatures between 1°C and 5°C. Tolerant of low temperatures but will not survive frost. Requires a cool or frost-free glasshouse in winter. Can be grown outside after danger of frost is over.	*Citrus meyeri* 'Meyer'
Heated glasshouse (H1)	A plant requiring temperatures above 5°C to survive. Some can be grown outside in summer when daytime temperatures are high enough or in a sheltered position. Some must be grown under glass or as houseplants all year. Can be divided into further categories depending on the plants requirements for temperatures above 5°C.	*Pelargonium* cvs, *Solenostemon* cvs, *Brugmansia* spp., *Monstera deliciosa, Anthurium andraeanum*

* See Table 2.1. These definitions correspond with current RHS hardiness categories.

plant's hardiness rating, other factors, such as wind and soil conditions (relating to the species origins), will play a part. Table 4.2 brings together these useful terms, providing definitions and some plant examples.

The following terms are derived from the use of plants:

▶ **Bedding**: a fast-growing species, often flowering, used to make a temporary, often formal, display, e.g. *Petunia, Pelargonium, Sedum, Viola x wittrokiana* (pansy).

▶ **Tropical**: a non-native, usually exotic, tender plant used for indoor, or seasonal outdoor, display, e.g. *Canna, Hibiscus.*

▶ **Edging**: low and often, slow-growing species grown in similar groups to create an edge to a path or boundary between planted areas, e.g. *Thymus, Viola*, non-spreading hardy geraniums, and many low-growing, compact perennials.

▶ **Dot plant**: a single plant, usually tall, planted to create a focal point within a bedding scheme, e.g. *Eremurus* (foxtail lily), standard rose, *Phormium, Acer palmatum* (Japanese maple).

▶ **Ground cover**: low growing, usually evergreen plants designed to completely cover the soil, e.g. *Hedera* (ivy), *Vinca* (periwinkle), *Epimedium*.

Fungi

The kingdom Fungi is a very diverse group about which much remains to be studied. It has been estimated that there are 1.5 to 5 million species of which only 5% has been classified to date! Some fungi are single celled (such as yeasts), but others are multicellular, such as the moulds and the more familiar mushrooms and toadstools (Figure 4.13). Most are made up of a mycelium, which is a mass of thread-like filaments (hyphae) which generally remains hidden from view. The mushrooms we see at certain times of year are the spore-producing part of the life cycle. Fungal cell walls are made of chitin not cellulose as in plants.

Fungi obtain their food directly from other living organisms (heterotrophic nutrition), sometimes causing disease (see Chapter 19), or from dead organic matter, so contributing to its beneficial breakdown in the soil (see Chapters 3 and 13). They achieve this by secreting digestive enzymes onto their food source and absorbing the soluble products.

In horticulture, fungi (mushrooms) are also important as a food crop. Mycorrhiza and lichens are examples of a mutualistic relationship between fungi and other organisms (Figure 4.14).

See 'Non-Plant Kingdoms' on the companion website.

(a)

(b)

(c)

Figure 4.14 Lichens – a combination of a fungus and a cyanobacterium: (a) *Xanthoria* spp.; (b) *Parmelia* spp.; (c) *Usnea* spp.

Figure 4.13 Shaggy ink cap, a fungus showing its fruiting bodies

Fungal classification

Fungi are classified into four groups (Figure 4.15):

▶ **Zygomycota** (mitosporic fungi) have simple asexual and sexual spore forms. Damping off, downy mildew and potato blight belong to this group (see p. 253).

▶ **Ascomycota** have chitin cell walls, and show, throughout the group, a wide variety of asexual spore forms. The sexual spores are consistently formed within small sacs (asci), numbers of which may themselves be embedded within flask-shaped structures (perithecia), just visible to the naked eye. Rose black spot (see p. 257), apple canker, powdery mildew and Dutch elm disease belong to this group.

▶ **Basidiomycota** have chitin cell walls, and may produce, within one fungal species (e.g.

cereal rust), as many as five different spore forms involving more than one plant host. The fungi within this group bear sexual spores (basidiospores) from a microscopic club-shaped structure (basidium). Carnation rust, honey fungus and silver leaf diseases belong to this group (see p. 261).

▶ **Deuteromycota** are an artificially derived fourth grouping which is included in the classification of fungi. It includes species of fungi that only very rarely produce a sexual spore stage. As with plants, the sexual structures of fungi form the most reliable basis for classification. But, here, the main basis for naming is the asexual spore and mycelium structure. Grey mould (*Botrytis*) (see p. 258), *Fusarium* patch of turf and *Rhizoctonia* rot are placed within this group.

Figure 4.15 Classification of fungi: (a) damping-off disease – a zygomycota; (b) a courgette with powdery mildew – an ascomycota; (c) pear rust – a basidiomycota; (d) Fusarium wilt – a deuteromycota

Further reading

Allaby, M. (1992) *The Concise Oxford Dictionary of Botany.* Oxford University Press.

Blunt, W. (1971) *Linnaeus The Compleat Naturalist.* Frances Lincoln.

Cubey J. (ed.) (updated annually) *The RHS Plant Finder.* Dorling Kindersley.

Harrison, L. (2012) *RHS Latin for Gardeners.* Quid Publishing.

Hillier (2005) *Gardener's Guide: Plant Names Explained.* David and Charles.

Hodge, G. (2013) *RHS Botany for Gardeners.* Mitchell Beazley.

Ingram, D.S., Vince-Price, D. and Gregory, P.J. (2008) *Science and the Garden.* Blackwell Science.

Johnson, A.T. and Smith, H.A. (2008) *Plant Names Simplified: Their Derivation and Meaning.* Old Pond Publishing.

Stern, K.R., Bidlack, J.E. and Jansky, S.H. (2011) *Stern's Introductory Plant Biology.* Mc Graw Hill.

4

 Please visit the companion website for further information:
www.routledge.com/cw/adams

Plant life cycles

Figure 5.1 Autumn colour in the leaves of *Parthenocissus tricuspidata* (Boston ivy), developing in response to environmental changes

This chapter includes the following topics:

- Growth and development
- The seed – viability, germination requirements and dormancy
- The seedling – hypogeal and epigeal germination, tropisms
- Juvenile growth
- The adult plant
- Senescence and death

Principles of Horticulture. 978-0-415-85908-0 © C.R. Adams, K.M. Bamford, J.E. Brook and M.P. Early. Published by Taylor & Francis. All rights reserved.

Growth and development

Like people, plants follow a series of distinct phases in their lives, described as the plant's 'life cycle'. Unlike humans, the plant life cycle can be either very short, a matter of weeks in ephemerals such as *Senecio vulgaris* (groundsel) or many thousands of years as in *Pinus aristata* (the bristlecone pine), some individuals of which are thought to be more than 5,000 years old. In most cases, a plant's life begins with fertilization and the development of the embryo within a **seed**. Germination of the seed gives rise to a seedling which undergoes a **juvenile** period of growth and development. On reaching **adulthood**, the plant is able to reproduce, then having produced fruits and seeds, a period of **senescence** ending in **death** of the plant ensues (Figure 5.2).

The changes that take place in the structure, form and behaviour of a plant through its life cycle can usefully be described as '**plant development**'. This is in contrast to the term '**plant growth**', which refers to the increase in a plant's weight and size. Plant growth is brought about in two ways:

▶ by new cells being produced in the meristems (see p. 78)

▶ the new cells expanding due to turgor pressure (see p. 120).

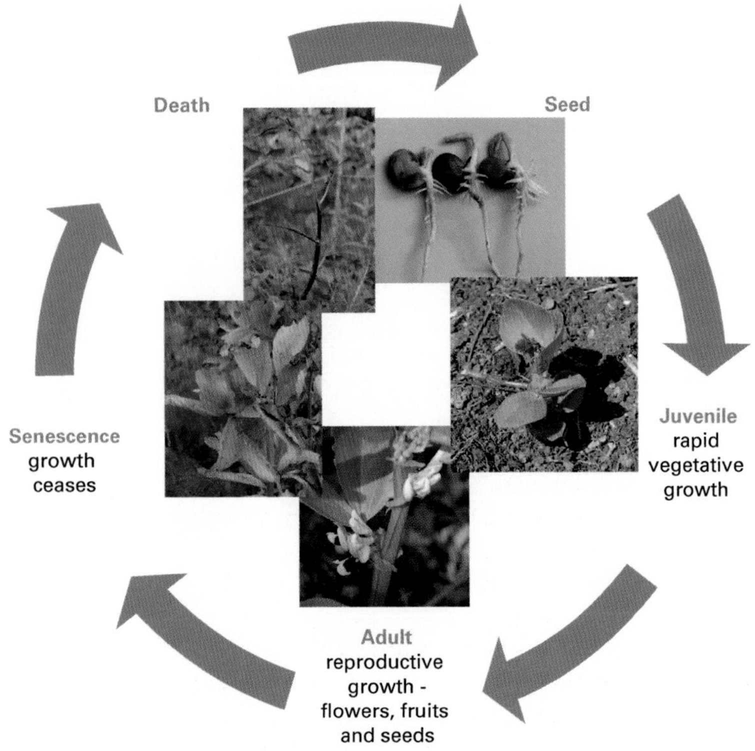

The production of new cells is fuelled by the processes of photosynthesis, respiration and mineral uptake (see Chapters 9 and 10).

The typical life cycle of plants, from seed to death, and some of the horticultural implications for each phase are described in this chapter.

> **Plant growth** is the increase in size of cells, organs or the whole plant due to cell division and cell expansion. **Plant development** describes the changes in structure, form and behavior through its life cycle.

The seed

Following fertilization the seed represents the first stage of the plant's life cycle. Seeds contain and protect the plant embryo, which will grow into the new plant, and a food store which will support growth until the plant is able to photosynthesize and manufacture its own food supply. A seed enables the plant to withstand periods when the environment is not suitable for growth and may have mechanisms which bring about seed dormancy. It is also the means by which plants spread away from the parent plant, and colonize new areas thereby reducing competition for water and nutrients and increasing the success of survival of the species. Seed structure is described in Chapter 8.

Seed viability

A **viable** seed has the potential for germination when the required external conditions are supplied. Its viability is, therefore, an indication of whether the seed is 'alive' or not. Most seeds remain viable until the next growing season but many can remain so for a number of years until conditions are favourable for germination. In general, viability of a batch of seed diminishes with time, its maximum viability period depending largely on the species. For example, celery seed quickly loses viability after the first season, but wheat has been reported to germinate after scores of years. The germination potential of any seed batch will depend on the storage conditions of the seed, which should be cool and dry, slowing down respiration. Some seeds, described as 'orthodox' seeds, can be stored successfully for long periods in these conditions while other 'recalcitrant' seeds lose viability quickly and require more specialist treatment (see p. 129). Cool, dry conditions are achieved in commercial seed stores by means of sensitive control equipment. Packaging of seed for sale takes account of these requirements and often includes a waterproof lining of the packet, which

Figure 5.2 Stages in the life cycle of *Vicia faba* (broad bean)

Death

Seed

Juvenile
rapid
vegetative
growth

Senescence
growth
ceases

Adult
reproductive
growth -
flowers, fruits
and seeds

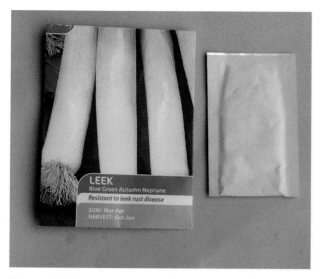

Figure 5.3 A seed packet with a waterproof and airtight insert containing the seed

maintains a constant low water content (and low oxygen levels) in the seeds Figure 5.3 (see respiration p. 116).

> A **viable** seed has the potential for germination when the required external conditions are supplied. A **quiescent** seed is a viable seed which does not germinate because the environmental requirements (water, oxygen, a suitable temperature) are not present. A **dormant** seed is one that does not germinate even though the environmental conditions are suitable.

Seed germination

For seeds to germinate successfully, a number of environmental conditions must be supplied:

▶ water
▶ oxygen
▶ correct temperature
▶ light (some).

> Seed **germination** is defined as the emergence of the young root or radicle through the testa, usually at the micropyle.

▶ **Water** is needed to trigger germination. The water content must be increased from around 10% in the dry seed to 70% or more. Water is initially absorbed (imbibed) into the structure of the seed coat or testa in a way similar to a sponge, softening it and is followed by uptake through a pore in the testa called the micropyle. The cells of the seed take up water by osmosis (see p. 120),

often assuming twice the size of the dry seed and this is a passive process, that is, it does not require energy. A continuous water supply is now required if germination is to proceed at a consistent rate. Eventually the radicle breaks through the seed coat (testa) and emerges followed by the plumule.

▶ **Oxygen** is essential for respiration (see p. 116). On imbibition, respiration rates increase dramatically as the seed's food stores, the cotyledons or endosperm, are broken down to produce energy and building materials for rapid cell division. The growing medium, whether it is outdoor soil or compost in a seed tray, must not be waterlogged, because oxygen would be withheld from the growing embryo.

▶ **Correct temperature** is a very important germination requirement, and is usually specific to a given species or even cultivar. It acts by fundamentally influencing the activity of the enzymes involved in the biochemical processes of respiration, which occur between 0°C and 40°C. However, species adapted to specialized environments may respond to a narrower range of germination temperatures. For example, *Cucumis sativus* (cucumbers) require a minimum temperature of 15°C and *Solanum lycopersicum* (tomatoes) 10°C, whereas *Latuca sativa* (lettuce) germination may be inhibited by temperatures higher than 30°C, and in some cultivars a period of induced dormancy occurs at 25°C. Some species, such as mustard (*Brassica hirta* syn. *Sinapis alba*), will germinate in temperatures just above freezing and up to 40°C, provided they are not allowed to dry out. Typical temperature ranges for various plants are given in Table 5.1. Temperate, winter-flowering and shade-tolerant plants generally germinate at lower temperatures than summer-flowering plants, tropical plants and plants that require full sun. Bulbs, such as tulip, germinate and flower in spring before temperatures become too high, their dormant season being in the summer. Turf grasses used in Britain and Ireland germinate at quite low temperatures.

▶ **Light** is a factor that may influence germination in some species, but most species are indifferent. Seed of *Rhododendron*, *Veronica* and *Phlox* is inhibited in its germination by exposure to light, while that of celery, lettuce, most grasses, conifers and many herbaceous flowering plants is slowed down when light is excluded. When a viable seed fails to germinate because any one of these factors is not suitable (e.g. water is not present or the temperature is too low), the seed is said to be 'quiescent'.

5

Table 5.1 Optimum germination temperature ranges

Seed	Plant type	Temperature range °C
Tulip (*Tulipa*)	Mediterranean bulb	10–15
Busy Lizzie (*Impatiens*)	Summer bedding	20–25
Pansy (*Viola*)	Winter bedding	13–16
Sea holly (*Eryngium*)	Sun-tolerant herbaceous	20–25
Solomon's seal (*Polygonatum*)	Shade-tolerant herbaceous	10–15
Grasses	Turf grasses	10–12
Bottlebrush (*Callistemon*)	Sun-tolerant shrub	20–25
Spotted laurel (*Aucuba*)	Shade-tolerant shrub	15–18
Carrot (*Daucus carota*)	Temperate vegetable	10–30
Tomatoes (*Solanum lycopersicum*)	Tropical vegetable	15–30
Aubergine (*Solanum melongena*)	Tropical vegetable	25–30

Seed dormancy

As soon as the seed **germinates**, the plant is vulnerable to damage from cold or drought. Seeds, therefore, often have **dormancy** mechanisms which prevent germination occurring when poor growing conditions prevail, as in the winter in temperate climates. Dormant seeds are unable to germinate even though water, oxygen and the correct temperature are given to them (unlike quiescent seeds) and they use a variety of mechanisms to delay germination until conditions become more favourable. For example, the seed coat may prevent water and oxygen entering or may be so hard that the embryo is unable to penetrate it, as in many leguminous plants such as *Lathyrus odoratus* (sweet pea). Many seeds (e.g. in *Malus* spp. (apple)) have chemical inhibitors in their seed coat and embryo which prevent germination while others are shed with immature embryos which need a period of time to develop fully before germination can commence (e.g. in *Fraxinus excelsior* (ash)).

In the wild, such dormancy mechanisms are gradually overcome through the abrasive action of the soil and exposure to cold temperature cycles and freezing and thawing which soften and break down the seed coat. Inhibitors are washed out by rain or are broken down chemically, and embryos can mature during a period of dormancy. Artificial methods of breaking dormancy mimic these conditions – for example, by **scarification**, physically damaging the seed coat by nicking or scratching it, as in *Lathyrus sativus* and *Paeonia*, or by **soaking**, as in *Camellia*. **Stratification**, in which soaked seeds are stored in warm or cold temperatures in a moist environment, is also used as in *Lupinus*, *Aconitum* and *Euonymus*.

See 'Seed Dormancy' on the companion website.

The seedling

The emergence of the plumule above the growing medium is usually the first occasion that the seedling is subjected to light. This stimulus prevents rapid extension of the stem so that it becomes thicker and stronger, the leaves unfold and become green in response to light, which enables the seedling to photosynthesize and so support itself. Seedlings which are deprived of light show **etiolated growth** with elongated internodes, few leaves or branches and no chlorophyll. At this stage the seedling is still very susceptible to attack from pests and damping-off diseases. The cotyledons are often the first part of the seed to develop and they may emerge from the testa and remain in the soil, as in *Prunus persica* (peach) and *Vicia faba* (broad bean) (**hypogeal** germination), or be carried with the testa into the air, above the soil where the cotyledons then expand (**epigeal** germination), as in *Solanum lycopersicum* (tomatoes), *Prunus avium* (cherry) and *Phaseolus vulgaris* (French bean) (Figure 5.4)

> **Hypogeal** germination occurs when the cotyledon(s) remain below the ground inside the testa. **Epigeal** germination occurs when the cotyledon(s) emerge above the ground, initially enclosed in the testa.

Cotyledons in epigeal germination may be called '**seed leaves**' when they emerge and they usually look quite different from '**true leaves**'. They turn green and contribute initially to photosynthesis in the seedling but the true leaves very quickly unfold and take over this function (Figure 5.5)

Once the food store in the cotyledons and/or endosperm has been exhausted, the seedling

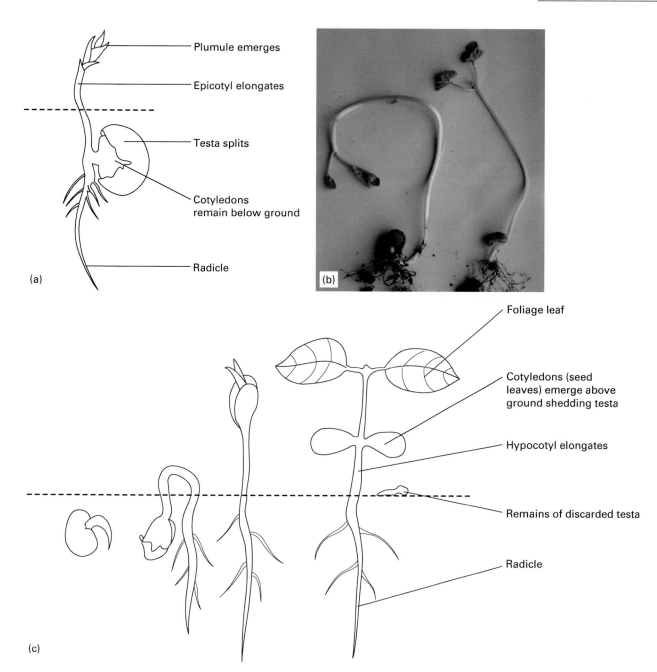

Figure 5.4 (a) Hypogeal germination in *Vicia faba* (broad bean); (b) Broad bean on left, French bean on right; (c) epigeal germination in *Phaseolus vulgaris* (French bean)

must rapidly produce its own food supply and begin to photosynthesize. It must therefore respond to stimuli in its environment to establish the correct direction of growth. Such a response is termed a **tropism**, and is very important in the early survival of the seedling.

> A **tropism** is a directional growth response to an environmental stimulus.

Geotropism is a directional growth response to gravity. The emergence of the radicle from the testa is followed by growth of the root system, which must quickly take up water and minerals to enable the shoot system to develop. A seed germinating near the surface of a growing medium must not put out roots that grow on to the surface and dry out but establish roots that grow downwards to tap water supplies. Conversely, **phototropism** enables the shoot to grow towards a light source that provides the energy for photosynthesis. If the light source is to the side of the plant, a bend takes place in the stem just below the tip as cells in the stem away from the light grow larger than those nearer to the light source.

Figure. 5.5 Germinating tomato seed showing the simpler 'seed leaves' or cotyledons which are produced first, followed by the true leaves

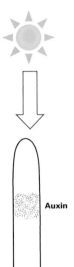

When illuminated from above, auxin produced in the shoot tip is translocated to an area behind the tip where it is evenly distributed across the stem. This brings about cell enlargement and enables the stem to grow up towards the light

A greater concentration of a substance called **auxin** in the shaded part of the stem causes the extended growth (Figure 5.6). Roots display positive geotropic and negative phototropic responses while shoots display the reverse, they are positively phototropic and negatively geotropic.

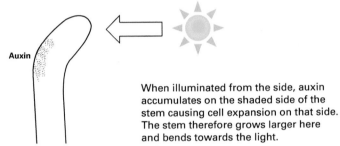

When illuminated from the side, auxin accumulates on the shaded side of the stem causing cell expansion on that side. The stem therefore grows larger here and bends towards the light.

Figure 5.6 Positive phototropism in shoots

Juvenile growth

The early growth stage or **juvenile stage** is a period after germination that is capable of **rapid vegetative growth** and is **non-reproductive**, that is, non-flowering. By putting its energies into growth rather than producing seeds, the plant can establish itself more effectively in competition with others. Juvenile growth can be characterized by certain physical appearances and activities that are different from those found in the later stages or in adult growth. Plant **growth habit** may differ; the juvenile stem of *Hedera helix* (ivy), for example, tends to grow horizontally and is vegetative in nature with adventitious roots for climbing, and its energy is put into internode extension rather than leaves. Adult growth is vertical, with shortened internodes, large leaves and terminal shoots bearing flowers and fruit. Often **leaf shapes** vary, for example the juvenile leaf of ivy is three-lobed while the adult leaf is more oval, as shown in Figure 5.7. The attractive juvenile leaves of *Eucalyptus* vary greatly from the adult leaves (Figure 5.8a). In *Eucalyptus gunnii* they are round and often without petioles (sessile) compared

Figure 5.7 Left: juvenile growth showing adventitious roots and lobed leaf; right: adult growth showing flowers and entire leaf in *Hedera helix* (ivy)

with the adult plant and are much used by florists, while in *Catalpa bignonioides* the juvenile leaves are much larger making an attractive foliage display at the expense of flowers (Figure 5.8b). Other leaf characteristics such as **colour** and **arrangement on the stem** may differ. Differences in juvenile growth

Figure 5.9 Leaf retention in the lower juvenile branches of Fagus sylvatica (beech). The lower branches were produced when the tree was juvenile. Compare these with the upper branches, which grew when the tree reached adulthood and therefore do not show leaf retention

Figure 5.8 (a) *Eucalyptus* leaves. Left: juvenile leaves; right: adult leaves. (b) Juvenile leaves in *Catalpa bignonioides* 'Aurea'

are also common in conifer species, where the complete appearance of the plant is altered by the change in leaf form – for example, in *Chamaecyparis fletcheri* and many *Juniperus* species such as *J. chinensis*. In the genera *Chamaecyparis* and *Thuja*, the juvenile condition can be achieved permanently by repeated vegetative propagation producing plants called **retinospores**, which are used as decorative features in the garden.

> **Juvenile growth** is non-reproductive (vegetative) growth, whereas **adult growth** is reproductive (flowering).

Leaf retention is sometimes a characteristic of juvenility and can be significant in species such

as *Fagus sylvatica* (beech) and *Carpinus betulus* (hornbeam), where the phenomenon is exaggerated (Figure 5.9).

In **propagation**, juvenility is related to rooting success. Softwood cuttings are non-flowering and root easily but as the season progresses and shoots switch from juvenile to flowering growth, rooting becomes more difficult. Rooting hormones may need to be applied to stimulate rooting in semi-ripe and hardwood cuttings for this reason. Since flowering growth often roots less easily, flowers and flower buds must be removed if juvenile material is not available. Adult growth should be removed from stock plants (see p. 133) to leave the more successful juvenile growth for cutting.

Juvenility may also affect **pest and disease resistance**. In Dutch elm disease, juvenile elm trees are resistant but they succumb at around 15–20 years old when they are mature. In some *Citrus* such as lemon, juvenile non-fruiting stems grow vigorously and are thorny for defence against herbivores unlike the

Figure 5.10 *Carpinus betulus* (hornbeam hedge) in winter

adult mature stems which bear fruit, and often these can be found on the same plant.

Pruning for juvenility

If juvenile leaf characteristics are desired rather than flowers, as in *Eucalyptus* spp. and in others such as *Cotinus* spp. and *Liriodendron tulipifera* (tulip tree), juvenility may be maintained by **coppicing**, that is, cutting the plants to a low framework or stool each year in their dormant seasons. Both coppicing and **pollarding** (where trees are cut to a single stem a short distance above the ground) are traditional pruning techniques which have been used for centuries in woodlands and elsewhere to produce wood for many purposes such as fencing or leaves for fodder. When carried out periodically, typically every 15 years in the case of *Corylus avellana* (hazel) for example, trees are maintained in a juvenile or partially juvenile state and this can lengthen their lives considerably.

The retention of leaves is a useful property in beech and hornbeam hedges where annual pruning keeps the plants in juvenile growth and provides an attractive hedge with colourful leaves retained throughout the winter (Figure 5.10). This can create additional protection in windbreaks, although the barrier created tends to be too solid to provide ideal wind protection (see p. 132).

The adult plant

The adult stage is defined by the ability of the plant to **reproduce sexually** and **produce flowers, fruit and seed**. The progression from a vegetative to a flowering plant involves profound physical and chemical changes. This change may simply be genetically programmed with a plant switching to adult growth after a certain number of leaves are produced or when it has reached a certain size. Often, however, an environmental stimulus is required, such as temperature and/or daylength which links flowering to an appropriate season. In this way many plants flower characteristically at particular times of year, for example *Syringa vulgaris* (lilac) in the spring, *Buddleja alternifolia* in early summer, *Hypericum calycinum* in late summer to autumn and *Viburnum x bodnantense* in the winter. In the reproductive phase of the life cycle, flowers are potent 'sinks' for a plant's resources, drawing sugars to them at the expense of vegetative growth. Plant growth therefore slows as all the plant's energies are redirected to producing flowers, fruits and seeds. Adult growth often shows different growth patterns to juvenile growth as in ivy and changes in rootability of cuttings (see above).

In many plants, adult growth is also linked to producing stores of food towards the end of the growing season for overwintering and growth the following spring (**perennation**). Once flowering is over, sugars may be redirected and stored as starch in the stems of woody plants or in modified roots, leaves and stems such as bulbs, rhizomes and tubers in herbaceous plants. These may also be a means of vegetative spread (asexual reproduction) and are useful to the gardener as propagating material (see Chapters 8 and 11).

Pruning in adult plants

Plant pruning is often carried out to reduce the competition within the plant for the available resources. In this way, the plant is encouraged to grow, flower or fruit in the way the horticulturist requires. A reduction in the number of flower buds of, for example, *Chrysanthemum morifolium* (chrysanthemum) will cause the remaining buds to develop into larger flowers; a reduction in fruiting buds of apple trees will produce bigger apples, and a reduction in the branches of soft fruit and ornamental shrubs will allow the plants to grow stronger when planted densely.

To encourage flowering and improve the quantity and quality of blooms, species that flower on the previous year's growth of wood (e.g. *Forsythia*) should be pruned soon after flowering has stopped. Conversely, species that flower later in the year on the present year's wood (e.g. *Buddleja davidii*) should be pruned the following spring, to maximize the growth period for flower production.

As flowers age, they begin to use up a considerable amount of the plant's energy in the production of fruits. In addition, hormones produced by the fruit inhibit flower development and with many species, the maturation of fruits will considerably reduce the plant's ability to continue producing flowers. By removing dead flowers, a pruning technique called dead-heading, the appearance of a garden border can be maintained. An added bonus is that plants that have been dead-headed may continue to flower for many weeks longer than those allowed to retain their dead flowers. Examples of species needing this procedure are seen in bedding plants which flower over several months including *Tagetes erecta* (African marigold); in herbaceous perennials, *Delphinium* and *Lupin*; in small shrubs, *Penstemon fruticosus*; and in climbers, *Lathyrus odoratus* (sweet pea) and *Rosa* 'Pink Perpetue'.

Many species such as *Begonia x semperflorens-cultorum* (wax begonia) and *Impatiens x walleriana* (busy lizzie) used as bedding plants have been specially bred so that flowers do not produce fruits containing viable seed. In such cases, there is not such a great need to dead-head, but this activity will help to prevent unsightly rotting brown petals from spoiling the appearance of foliage and newly produced flowers.

Pruning – some general principles

Pruning affects the shape of the plant, through a property of plant called **apical dominance**. This is where the apical bud inhibits growth of buds further down the stem. By removing the apical bud, lateral shoots are released and develop. The success of such pruning depends very much on the skill of the operator, so a good knowledge of the species habit is required together with an appreciation of the purpose of pruning.

- ▶ **Young plants** should be trained in a way that will reflect the eventual shape of the more mature plant (formative pruning). For example, a young apple tree (called a 'maiden') can be pruned to have one dominant 'leader' shoot, which will give rise to a taller, more slender shape. Alternative pruning strategies will lead to quite different plant shapes. Pruning back all branches in the first few years forms a bush apple. A cordon is a plant in which there is a leader shoot, often trained at 45 degrees to the ground, and where all side shoots are pruned back to one or two buds. Cordon fruit bushes are usually grown against

walls or fences. Similarly, fans and espalier forms can be developed. However, regular trimming of hedges produces a mass of laterals with no single leader making a dense well-shaped hedge.

- ▶ **The pruning cut** should be made just above a bud that points in the required direction (usually to the outside of the plant). In this way, the plant is less likely to acquire too dense growth in its centre. Some plants such as roses and gooseberries are made less susceptible to disease attack by the creation of an open centre to produce a more buoyant (less humid) atmosphere.
- ▶ **Pruning should remove any shoots that are crossing** as they will lead to dense growth and may have damaged bark which could be a point of entry for disease.
- ▶ **Weak shoots should be pruned the hardest** where growth within the plant is uneven, and strong shoots pruned less, since pruning causes a stimulation of growth.
- ▶ **Root pruning** used to be done to restrict over vigorous cultivars, especially in fruit species, but this technique has been largely superseded by the use of dwarfing rootstock grafted on to commercially grown scions. Root pruning is still seen, however, in the growing of bonsai plants.

Senescence and death

The term '**senescence**' in the plant's life cycle refers to the period between adulthood and death of the plant. It is the stage after flowering and fruiting where growth has ceased and a gradual deterioration occurs. This is most obvious in ephemeral, annual and biennial plants which flower and fruit only once before senescence and death.

In perennial plants, the same cycle of seed, juvenile growth, adult growth, senescence and death occurs through the plant's lifetime, as in all plants, but the term 'senescence' is also used to describe the changes that take place through the year and are repeated each season in leaves and fruits. In deciduous trees and shrubs, changes in leaf colour associated with autumn are due to pigments that develop in the leaves and are revealed as the chlorophyll (green pigment) is broken down and absorbed by the plant (Figure 5.11) and waste products accumulate.

Pigments are substances that are capable of absorbing light; they also reflect certain wavelengths

Figure 5.11 Autumn colour in (a) *Vaccinium corymbosum* (blueberry); (b) *Viburnum*; and (c) *Photinia,* showing loss of chlorophyll and emergence of xanthophylls

of light, which determine the colour of the pigment. In the actively growing plant, chlorophyll, which reflects mainly green light, is produced in considerable amounts, and therefore the plant, especially the leaves, appears predominantly green. Other pigments are present – for example, the carotenoids (yellow) and xanthophylls (red) – but usually the quantities are so small as to be masked by the chlorophyll. In some species, such as *Fagus sylvatica* Purpurea Group (copper beech), other pigments predominate, masking chlorophyll. Many colours are displayed in the leaves at this time in such species as *Acer platanoides,* turning gold and red, *Prunus cerasifera* 'Pissardii' with light purple leaves, *Larix decidua* (European larch) with yellow leaves, *Parthenocissus quinquefolia* (Virginia creeper) and *Vitis* spp. with red leaves, beech with brown leaves. Coloured autumn fruits are also prized in the garden such as the berries in *Cotoneaster* and *Pyracantha* and the colouring of stems can become more intense, such as in *Cornus* (dogwoods). These are used in **autumn colour** displays at a time when fewer flowering plants are seen outdoors (Figure 5.12).

In deciduous woody species, the leaves drop in the process of **abscission**, which can be triggered by shortening of day length. To reduce risk of water loss from the remaining leaf scar, a corky layer is formed before the leaf falls.

In many **fruits**, ripening is associated with an increase in respiration and changes in colour, sweetness,

Figure 5.12 Autumn interest in the garden: acers with bright leaves, birch with silver bark and yellow leaves, flowering grasses, leaves and bark of *Eucalyptus*

flavour, texture and scent. This is soon followed by natural progression to a senescent stage in which the fruit deteriorates and the seed ripens before it is shed, although growers aim to harvest fruit before this stage is reached and arrest further development through a range of post-harvest techniques such as the use of reduced temperatures. Eventually, senescence is followed by **death** at the end of the season in annuals or at the end of the plant's life in perennials (or death of the plant organ such as a leaf where this undergoes senescence). All metabolic processes cease and the plant matter is returned to the soil to be taken up by and to sustain future plants.

Further reading

Capon, B. (2005) *Botany for Gardeners.* Timber Press.
Cushnie, J. (2007) *How to Prune.* Kyle Cathie.
Ingram, D.S., Vince-Price, D. and Gregory, P.J. (2008) *Science and the Garden.* Blackwell Science.
Hodge, G. (2013) *RHS Botany for Gardeners.* Mitchell Beazley.

Please visit the companion website for further information:
www.routledge.com/cw/adams

CHAPTER 6

Level 2

Plant cells and tissues

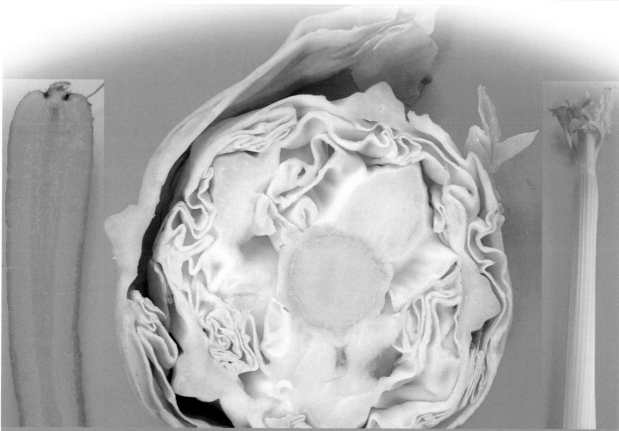

Figure 6.1 Examples of plant tissues, showing packing (parenchyma in the carrot cortex), support (the collenchyma forming celery 'strings') and transport (vascular tissue in the cabbage stem) tissues

This chapter includes the following topics:

- Plant cells and their contents
- Plant tissues
- Stem and root anatomy
- Growth and differentiation of the stem and root

Principles of Horticulture. 978-0-415-85908-0 © C.R. Adams, K.M. Bamford, J.E. Brook and M.P. Early.
Published by Taylor & Francis. All rights reserved.

Plant tissues, and the cells that form them, work together to enable plants to carry out the functions essential for life. As in humans, specialized tissues support the plant body, enable movement of substances around it, manufacture the substances it requires to live and to reproduce and protect it from its environment. An understanding of how these components work and interact with each other enables gardeners to provide successfully for their plants' needs thus optimizing their growth.

The internal structure (anatomy) of the plant is made up of different tissues. Each **tissue** is a collection of specialized cells carrying out one function, such as xylem tissue conducting water and nutrients. An **organ** is made up of a group of tissues carrying out a specific function, such as a leaf producing sugars for the plant.

> A **tissue** is a collection of cells carrying out a specific function.

Plant cells and their contents

Without the use of a microscope, we would not be able to see cells, since they are very small, typically about a twentieth of a millimetre in size. A simple, unspecialized plant cell (Figure 6.2) consists of an outer cellulose cell wall and an inner cell membrane, which enclose the cell contents suspended in the jelly-like cytoplasm.

▶ The **cell wall** is laid down in a mesh of cellulose fibres which allows the wall to stretch as the cell expands. Once the cell is mature, the cell wall loses its elasticity and sets the permanent shape of the cell. In living cells, dissolved substances can pass freely through the open mesh of the cell wall, although larger items such as fungal spores can be excluded. Within the mesh framework are many apertures that, in living cells such as parenchyma, allow for strands of cytoplasm (called **plasmodesmata**) to interconnect between adjacent cells. These strands allow the passage of substances between cells. When a plant wilts, the plasmodesmata normally retain their links with adjacent cells and wilting is reversible. However, in some situations they may break (plasmolysis) and the plant is not able to recover (see p. 120). In all tissues, the cell walls of adjoining cells are held together by calcium pectate (pectin), a glue-like substance which is an important setting ingredient in jam-making. Some types of cell (e.g. xylem vessels) die in order to achieve their usefulness. Here, the wall of cellulose becomes thickened by additional cellulose layers and lignin, which is a strong, waterproof substance, and the

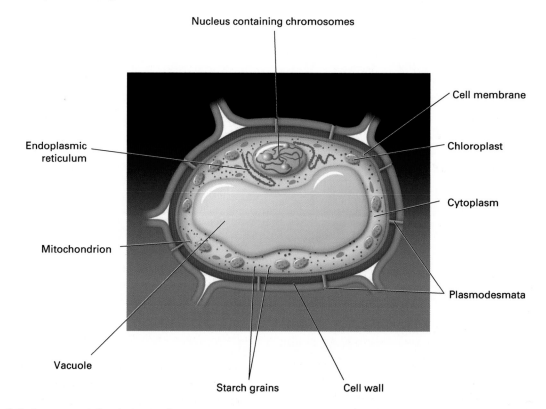

Nucleus containing chromosomes

Cell membrane

Chloroplast

Cytoplasm

Endoplasmic reticulum

Plasmodesmata

Mitochondrion

Vacuole

Starch grains

Cell wall

Figure 6.2 An unspecialized plant cell

cell contents disappear, leaving a central cavity or lumen. Cell walls, in particular those strengthened by lignin, give support and mechanical strength to the cell and ultimately to the plant. As well as acting as a barrier to pathogens, cell walls have substances embedded in them which signal attacks and trigger the cell's defences.

Cells and the microscope

The original inventor of the microscope is hard to establish. Roger Bacon, the English philosopher and Fransciscan friar, studied optical lenses in the 1200s and may have invented the early microscope and telescope. Others credit Zacharias Jansen, a spectacle maker in Holland, with inventing the first microscope in 1590 or Galileo Galilei, the Italian astronomer, mathematician and philosopher, in 1610. It seems likely that the earliest instruments were probably developed in the Netherlands in the early 1600s. In 1665 Robert Hooke, scientist, inventor, architect (and experimenter with flying machines), worked on improving the instrument and examined many natural objects under his microscopes (Figure 6.3). He published *Micrographia*, a book of observations made using microscopes and telescopes which included many illustrations. He coined the term 'cell', referring to the resemblance between the cork cells he studied and the cells which monks lived in. He estimated that a cubic inch of cork would contain about 1,259 million cells! These first microscopes were the familiar optical microscopes which use light to view the subject; these are still widely used today in various forms. They can produce a magnification up to about 2,000x. Modern electron microscopes, however, use beams of electrons to illuminate a subject and produce an image with much greater magnification up to 10,000,000x. Scanning electron microscopes produce images of the surfaces of objects such as the leaf shown in Figure 10.5 which reveal incredible detail. See 'The Size of Things' on the companion website.

▶ The **cell membrane** not only contains the cell contents but also controls the movement of substances entering and leaving the cell. Water uptake takes place through the process of osmosis (see p. 120) across this selectively permeable membrane, which allows water to pass but excludes other substances such as sugars. Some

Figure 6.3 A microscope manufactured by Christopher Cock of London for Robert Hooke (source: Billings Microscope Collection, USA)

mineral nutrients cross the cell membrane through a selective process called active transport, which requires energy (see p. 114) and enables the plant to take in useful substances whilst rejecting others.

▶ The **cytoplasm**, which is largely water, enables substances dissolved in it to move around the cell and take part in chemical reactions within it. It has a network of protein strands which hold the other cell components in place and prevent them sinking to the bottom of the cell due to gravity. Suspended in the cytoplasm are small structures (organelles), each enclosed within a membrane and having specialized functions within the cell.

▶ The **nucleus** coordinates the activities of the cell. The long chromosome strands that fill the nucleus are made up of the complex chemical DNA (deoxyribonucleic acid). In addition to its ability to produce more of itself for the process of cell division, DNA manufactures similar RNA (ribonucleic acid) units, which pass through the nucleus membrane and attach themselves to other organelles. In this way, the nucleus transmits instructions for the assembly of chemicals within the cell. The coded information for these processes is found in genes within the chromosomes.

▶ The **mitochondria** release energy, in a controlled way, by the process of respiration (see p. 116). The meristems of the stem, root and flower, for example, have cells which are packed with mitochondria in order to supply energy for rapid cell division in these areas.

▶ **Chloroplasts**, containing the pigment chlorophyll, are involved in the production of sugar by the process of photosynthesis (see p. 112) and in its short-term storage in the form of starch. Starch grains are often found in the chloroplasts and throughout the cytoplasm in living cells.

77

► The **endoplasmic reticulum** is a complex mesh of membranes that enables transport of chemicals within and between cells and links with the cell membrane at the cell surface. Specialized units made of RNA and protein called ribosomes are commonly located on the endoplasmic reticulum and manufacture proteins, including enzymes which speed up chemical processes.

► The **vacuole** is a sac within the cell, bound by a membrane which contains dilute sugar, nutrients, pigments and waste materials. It may occupy the major volume of the cell and its main functions are storage of waste products and maintaining cell shape through controlling cell turgor (see p. 120), which is important for support in herbaceous plants in particular.

The whole of the living matter of a cell, its membrane, nucleus and cytoplasm, is collectively called **protoplasm**. Plant cells differ from animal cells in having a cell wall, a vacuole and chloroplasts.

Plant tissues

The tissues which make up the structure of a plant can be grouped into five categories according to the functions they perform. **Meristematic** tissues are where new cells are produced by cell division. Meristems are responsible for lengthwise growth at the tips of roots and shoots (**apical meristems**) and, in woody plants, width-ways growth (**lateral meristems**) as the plant increases in size and needs to support itself. Meristematic cells are undifferentiated, that is they are as yet unspecialized for any particular function. They are cuboid in shape with small vacuoles, a large conspicuous nucleus and many mitochondria. **Protective** tissues (e.g. the epidermis) cover the entire plant surface, holding the plant together and protecting it from water loss and damage. **Transport** tissues are the plumbing system of the plant transporting water and nutrients in the xylem, sugars in the phloem and plant hormones in both. The bulk of the plant is made up of **packing** tissue such as parenchyma, whose cells are often unspecialized but may sometimes be adapted for specific functions such as photosynthesis, aeration in plants growing in waterlogged conditions or starch storage. Finally, **supporting** tissues reinforce leaves, stems and roots by having cell walls with extra thickenings of cellulose (collenchyma tissue) or lignin (sclerenchyma tissue) (Figure 6.4).

The term 'tissue' may also be used to describe the different areas within plant organs such as the cortex of roots and stems or the mesophyll of plant leaves.

Stem and root anatomy

When roots and stems are first formed, their tissues are arranged in distinctive ways. In young roots and stems, herbaceous plants and monocotyledons, and in the area behind the root and shoot tips, this anatomy persists while in plants which become woody the structure eventually changes due to secondary thickening.

Dicotyledonous stem tissues

The internal structure of various non-woody dicotyledonous stems, as viewed in cross-section, is shown in Figures 6.4a, b and d and 6.6a. The stem is broadly divided into four areas of tissue, the epidermis, the cortex, the vascular tissues and the pith.

The protective **epidermis** consists of a single layer of cells on the outside of the stem which produces a waterproofing waxy layer of cutin on its surface called the **cuticle**. Pores called **stomata** punctuate the epidermis allowing gases to pass through an otherwise impermeable layer. Opening and closing of the pores is controlled by a pair of modified epidermal cells called guard cells (see Figure 10.5).

The **cortex** of the stem lies beneath the epidermis and is largely made up of **parenchyma tissue** composed of relatively unspecialized parenchyma cells. These cells are thin walled and maintained in an approximately spherical shape by osmotic pressure (see p. 120) with many air spaces between them. The mass of parenchyma cells combines to maintain plant shape. Parenchyma cells also carry out other functions, when required – for example, many of these cells contain chlorophyll (chlorenchyma), giving the stems their green colour (see Figure 7.5), and so are able to photosynthesize. They also release energy, through respiration, for use in the surrounding tissues. In some plants such as the potato they are capable of acting as food stores; the potato tuber is an organ which stores starch (see p. 93). Parenchyma cells can sometimes be triggered to undergo cell division, a useful property when a plant has been damaged. This property has practical significance when plant parts such as cuttings are being propagated, since new cells can be created by the parenchyma to heal wounds and initiate root development. Plants which are adapted to waterlogged situations often have parenchyma tissue

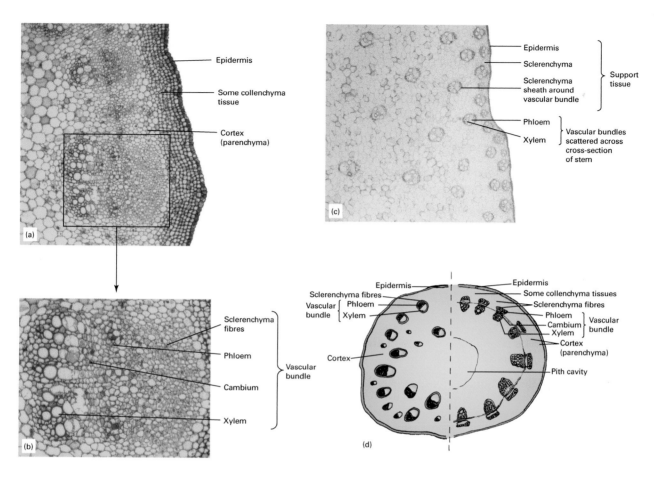

Figure 6.4 Stems: (a, b) Transverse sections of a typical dicotyledonous stem in *Helianthus annuus* (sunflower); (c) a typical monocotyledonous stem in *Zea mays* (maize); (d) a comparison between the stem tissue of monocotyledons (to the left) and dicotyledons (to the right)

with large interconnected air spaces between the cells (**aerenchyma**) to store air for aerobic respiration (see Figure 9.5).

Inside the cortex in dicotyledonous stems is a ring of **vascular bundles**, so named because they contain two vascular tissues that are responsible for transport. The first, **xylem**, contains long, wide, open-ended cells with very thick lignified walls, able to withstand the high pressures of water (with dissolved minerals) which they carry. The second vascular tissue, **phloem**, consists again of long, tube-like cells (**sieve tubes**) and is responsible for the transport of food manufactured in the leaves to the roots, stems or flowers (see p. 112). The sieve tubes, in contrast to xylem, have cellulose cell walls and are living, unlignified cells. They are unusual in not containing a nucleus. The end-walls are only partially broken down to leave sieve-like structures (**sieve-plates**) at intervals along the sieve tubes. Alongside every sieve tube cell there is a small companion cell, which regulates the flow of liquids through the sieve tube. The phloem is found to the outside of the xylem in most species. Also contained within the vascular

bundles of dicotyledons is the **vascular cambium**. This is a **lateral meristem** which contains actively dividing cells producing more xylem and phloem tissue to increase the girth of the stem in woody plants as it grows (**secondary thickening**). The vascular bundles of the stem in dicotyledons are arranged in a ring which gives strength and support to the stem in much the same way as steel rods do in reinforced concrete.

Pith refers to the central zone of the stem, which is mainly made up of parenchyma cells. It may sometimes break down to give a hollow stem.

Collenchyma and **sclerenchyma cells**, forming collenchyma and sclerenchyma tissues, are often found to the inside of the epidermis and around the vascular bundles and are responsible for **support** in the young plant. Both tissues have cells with specially thickened walls. When a cell is first formed, it has a wall composed mainly of **cellulose** fibres. In collenchyma cells the amount of cellulose is increased, providing extra strength, but otherwise the cells remain relatively unspecialized. Characteristically, there are no air spaces between the cells. In sclerenchyma cells, the thickness of the wall is increased by the addition of a substance called **lignin**, which is tough and waterproof and causes the living contents of the cell to disappear. These cells, which are long and tapering and interlock for additional strength, consist only of cell walls with a central cavity.

Dicotyledonous root tissues

The internal structure of a non-woody dicotyledonous root is shown in cross section in Figures 6.5 and 6.6b.

The **epidermis** is comparable with the epidermis of the stem; it is a single layer of cells which has a protective as well as an absorptive function. Unlike the stem it lacks a cuticle since reducing water loss is unnecessary in the root. Inside the epidermis is the parenchymatous **cortex**. The main function of this tissue is respiration to produce energy for growth of the root and for the absorption of mineral nutrients. The cortex can also be used for the storage of **starch** where the root is an overwintering organ.

The cortex is often quite extensive and water must move across it to reach the transporting vascular tissue that is in the centre of the root. This central region, called the **stele**, is separated from the cortex by a single layer of cells, the **endodermis**, which has the function of controlling the passage of water and nutrients into the stele (see p. 122). Water passes through the endodermis to the **xylem** tissue, which transports the water and dissolved minerals up to the stem and leaves. The arrangement of the xylem tissue varies between species, but often appears in transverse section as a star with several 'arms'. Since support is unnecessary in roots surrounded by soil, this arrangement can maximize water uptake. The root also has a **vascular cambium** and in dicotyledons undergoes secondary thickening to increase its girth. As in the stem, **phloem** tissue is present for transporting sugars from the leaves to provide energy for the living cells of the root.

A distinct area in the root inside the endodermis, the **pericycle**, has cells which are able to divide and produce lateral roots, which push through to the main root surface from deep within the structure. As roots age they become thickened with waxy substances and the uptake of water becomes restricted.

The structures of a dicotyledonous stem and root are compared in Figure 6.6.

Monocotyledonous stem and root tissues

These have the same functions as those of a dicotyledon, therefore the cell types and tissues are similar. However, the arrangement of the tissues does differ: monocotyledon stems have their vascular bundles scattered throughout the stem and do not have a clearly defined cortex and pith whereas in dicotyledons they are arranged in a ring between the cortex and the pith tissue (Figure 6.4c and d). Monocotyledonous roots often have stele with multiple arms (Figure 6.7) rather than the relatively few found in dicotyledons.

A major distinction is the absence of a vascular cambium in monocotyledons which therefore do not undergo secondary thickening and show limited increase in stem diameter, generally not becoming woody. In monocotyledons such as palms and bamboos, a 'woody' stem does develop but this comes about through a different mechanism to that in dicotyledons. Mostly, the stem relies on extensive sclerenchyma tissue for support that, in the maize stem shown in Figure 6.4(c), is found as a sheath around each of the scattered vascular bundles.

Growth and differentiation of the stem and root

Plant increase in size (**growth**) (see p. 112) and cell and tissue specialization (**differentiation**) take place in a well-ordered sequence in particular areas such as the tips of roots and shoots.

> **Differentiation** is the change that takes place in a cell, tissue or organ enabling it to perform a specific function.

Growth of stems, for example, is initiated in the **apical** or terminal **bud** at the end of the stem (the apex) (Figure 6.8a). Deep inside the apical bud lies a tiny mass of small cells, each cell with a conspicuous nucleus but no cell vacuole. These make up the **apical meristem** where cells divide frequently to produce the tissues which will give rise to the epidermis, the vascular bundles and the parenchyma, collenchyma and sclerenchyma tissues of the cortex and pith. In addition to its role in tissue formation, the apical meristem gives rise to small leaves (bud scales) which collectively protect the meristem. These scales and the meristem together form the **bud**. Mutation or damage to the sensitive meristem region by aphids, fungi, bacteria or herbicides can result in distorted growth such as fasciation where shoots and other plant parts become flattened or fused together (see p. 272). Buds located lower down the stem in the angle of the leaf are called axillary buds, each with their own apical meristem which often give rise to side branches (see p. 85). Leaf tissues similarly develop from the

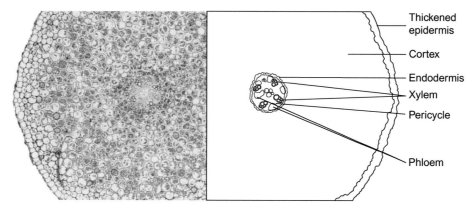

Figure 6.5 Transverse section of a *Ranunculus* (a dicotyledon) root showing thickened outer region, large area of cortex and central vascular region or stele, enclosed in an endodermis

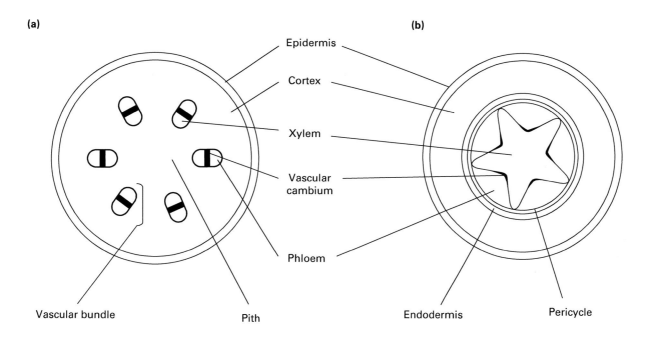

Figure 6.6 Transverse section of a young dicotyledonous (a) stem and (b) root

apex and form specialized tissues to carry out the process of photosynthesis.

In roots, an apical meristem is also found at the root tip which gives rise to all the root tissues. Just behind the root tip (Figure 6.8b) single epidermal cells become hugely elongated to form many thousands of **root hairs** which increase the surface area for water uptake (see p. 122). The root apex itself is protected by a **root cap** (Figure 6.9) which exudes a gel enabling the root tip to grow through the soil more easily and whose cells are continuously worn away and replaced. Further back from the root tip, **lateral roots** develop from the pericycle.

Three distinct zones can be identified in the tips of roots and shoots where growth and differentiation

take place (Figure 6.8). The overall increase in length of the root and stem is brought about by both the increased number of new cells formed by **cell division** in the apical meristems (the **zone of cell division**) and the increase in their size through **cell expansion** in the region just behind the meristem (the **zone of cell elongation**). Cell expansion is caused by water pressure (turgor pressure) within the cell pushing outwards on the cell membrane. In the early stages the cell wall is stretchy, enabling the cell to expand, but later on when the cell has differentiated to its final shape, the cell wall loses its elasticity and the cell can expand no further. It is important therefore to make sure that plants are well watered during periods of growth otherwise cells will not reach their maximum size, resulting in an overall smaller plant.

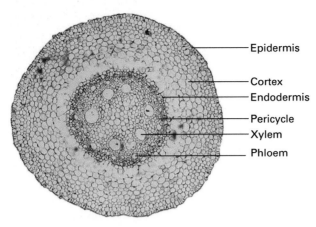

Figure 6.7 Transverse section of a monocotyledonous root in *Zea mays* (sweetcorn)

Epidermis
Cortex
Endodermis
Pericycle
Xylem
Phloem

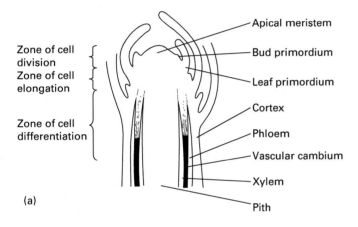

Zone of cell division
Zone of cell elongation
Zone of cell differentiation

Apical meristem
Bud primordium
Leaf primordium
Cortex
Phloem
Vascular cambium
Xylem
Pith

(a)

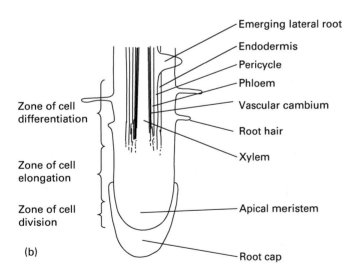

Zone of cell differentiation
Zone of cell elongation
Zone of cell division

Emerging lateral root
Endodermis
Pericycle
Phloem
Vascular cambium
Root hair
Xylem
Apical meristem
Root cap

(b)

Figure 6.8 Longitudinal sections through a dicotyledonous stem and root tip showing zones of cell division, growth and differentiation: (a) stem; (b) root

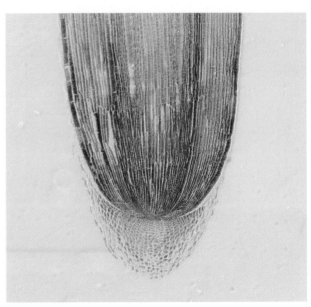

Figure 6.9 A root tip showing the protective root cap

Initially the new cells are unspecialized or undifferentiated, but once expanded, cells change their structure and chemistry to adopt their final role. Their cell walls become rigid and the connections between cells (plasmodesmata) form. The exact shape and chemical composition of the wall is different for each type of tissue cell, since each has a particular function to perform. This takes place in the **zone of differentiation** where the epidermis, cortex, vascular tissues and pith become distinct.

Leaf structure is described in Chapter 9.

See 'Plants and their Structure' on the companion website for further images.

Further reading

Capon, B. (2005) *Botany for Gardeners*. Timber Press.

Clegg, C.J. (2003) *Green Plants: The Inside Story*. III Advanced Biology Series. Hodder Murray.

Clegg, C.J. and Cox, G. (1978) *Anatomy and Actvities of Plants*. John Murray.

Hodge, G. (2013) *RHS Botany for Gardeners*. Mitchell Beazley.

Ingram, D.S., Vince-Price, D. and Gregory, P.J. (2008) *Science and the Garden*. Blackwell Science.

Lack, A.J. and Evans, D.E. (2005) *Instant Notes in Plant Biology*. Taylor & Francis.

Sugden, A. (1992) *Longman Botany Handbook*. Blackwell.

Please visit the companion website for further information:
www.routledge.com/cw/adams

External features of plants

Figure 7.1 Poppy flower emerging from a flower bud

This chapter includes the following topics:

- External features of roots, stems, buds and leaves
- Adaptations of roots, stems and leaves

Principles of Horticulture. 978-0-415-85908-0 © C.R. Adams, K.M. Bamford, J.E. Brook and M.P. Early. Published by Taylor & Francis. All rights reserved.

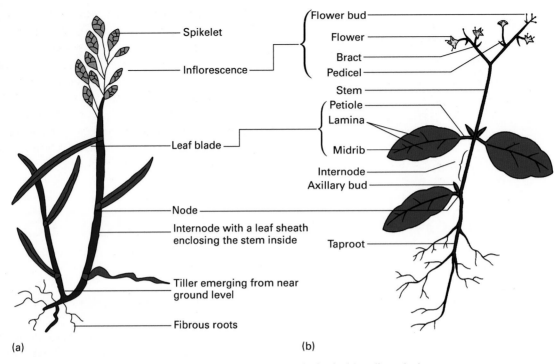

Figure 7.2. Generalized plant forms: (a) a grass (a monocotyledon); (b) a dicotyledon

While the internal structure of a plant organ can give us an idea of its function and how it is designed to carry out that function, it is the external appearance of plant organs, the plant's morphology, with which we are most familiar. Most plant species at first sight appear very similar, since all four organs, the root, stem, leaf and flower, are present in approximately the same form and have the same major functions. However, it is the differences in appearance of these plant parts that enables us to distinguish between them. An appreciation of how plant parts vary in their size, shape, colour or other characteristics helps us to correctly identify plants. See 'Plant Characteristics' on the companion website. Furthermore, knowledge of the terminology involved in describing plant parts is invaluable in using plant keys as an aid to identification. These are systematic lists of plant characters which help to identify them through a process of elimination. See 'Plant Identification Tools' on the companion website. Finally, studying plant form and appearance gives us a better understanding of how they are adapted to their particular habitats.

Some basic terms used to describe the various plant organs are shown for a generalized dicotyledon, and for a grass as an example of a monocotyledon in Figure 7.2.

More detailed descriptions of the parts of flowers, fruits and seeds are given in Chapter 8. The internal structures of roots and stems are described in Chapter 6, and leaf structure, as an organ of photosynthesis, is described in Chapter 9.

External features of roots, stems, buds and leaves

Roots

The main functions of the root system are to:

▶ take up water from the growing medium
▶ take up mineral nutrients from the growing medium
▶ anchor the plant in that medium.

To achieve maximum water and mineral uptake, roots must have as large a surface area as possible. The **root hairs**, which can be seen just behind the root tip (see Figure 6.8), greatly increase the root surface area with as many 200–400 hairs per square millimetre. The loss of root hairs during transplanting can check plant growth considerably, and the hairs can be points of entry of diseases such as club root (see p. 256). Root hairs are replaced frequently as the root grows, a single rye plant producing more than a million a day!

Two main types of root system are produced (Figure 7.2). A **taproot** (primary root) is a single large root which grows directly from the radicle (see p. 106) in the embryo (Figure 7.3). It has many smaller **lateral roots** (secondary roots) growing out from it at intervals.

Taproot systems are a distinctive feature of dicotyledons (e.g. chrysanthemums, brassicas, carrots). In contrast, a **fibrous** root system consists of many roots with no dominant root (Figure 7.14c). It is characteristic of monocotyledonous plants such as

Figure 7.3 Germinating *Vicia faba* (broad bean) seed showing the radicle developing into a tap root with laterals

grasses but can also be found in dicotyledons such as *Senecio vulgaris* (groundsel).

Adventitious roots do not derive from the radicle of the plant embryo. They tend to grow in unusual places such as on the stem or other organs. Most fibrous root systems are made up of adventitious roots which grow from the bottom of the stem, with the primary root failing to develop or dying away.

> A **taproot (primary root)** is a single large root which will have many **lateral (secondary)** roots growing out from it at intervals. **Primary** roots originate from the radicle of the embryo. A **fibrous root system** consists of many roots growing from the base of the stem with no dominant root. **Adventitious** roots grow in unusual places and do not originate from the radicle of the embryo.

Stems

The stem's main functions are to:

▶ physically support the leaves in the optimum position for photosynthesis
▶ physically support the flowers in the optimum position for pollination
▶ transport water, minerals and food between roots, leaves and flowers.

The leaf joins the stem at the **node** and has in its angle (axil) with the stem an **axillary bud**, which may grow out to produce a lateral shoot or in some instances flowers. The distance between one node and the next is termed the **internode**. Stems of herbaceous plants need to maintain a high water content to provide turgor pressure to support the plant (see p. 120). Winter stems in deciduous woody plants often show a **leaf scar** where the leaf was attached and a **bud scale scar** (girdle scar) where last year's **apical bud** was positioned and this can be useful in determining which part of the stem is current, one-year-old or two-year-old wood when pruning (Figure 7.4).

Young stems may be green and carry out photosynthesis (Figure 7.5). In common with other plant organs, stems are enclosed by the **epidermis** which contains **stomata**, pores which allows gas exchange between the air and the living tissues inside the stem (see p. 80). When a stem becomes woody, the epidermis gives way to a waterproof and gas-tight **bark** layer. The stomata are then replaced by breathing pores called **lenticels**. Lenticels may be useful in identification; in *Prunus* species they form very distinctive horizontal lines on the smooth bark (Figure 7.6).

The colour and texture of plant stems are decorative features which can also be an aid to identification. For example, *Cornus alba* 'Sibirica' (a dogwood) has bright red stems in winter (Figure 7.7), many *Salix* spp. (willows) have bright yellow or green stems even when woody. Stem colour is best in young stems so these shrubs are often pruned to the ground each year in winter to produce new stems the following season (coppicing).

Many trees have highly decorative bark, which bring winter interest to the garden. These include many *Eucalyptus* spp., *Betula utilis* var. *jaquemontii* and other white stemmed birches, *Prunus serrula* var. *tibetica* (Tibetan cherry) with shiny red bark, *Acer griseum* with peeling cinnamon-coloured bark and the snakebark maples *Acer davidii*, *A. pensylvanicum* and *A. capillipes* (Figures 5.12 and 7.8).

Buds

A bud is a condensed stem which is very short and has small leaves attached, both enclosing and protecting it (Figure 7.9). Buds are found at the apex of the shoot and in the axils of leaves with foliage buds containing numerous folded leaves and flower buds containing the immature flower.

On the outside of the bud, the leaves are often thicker and darker forming **bud scales** to resist drying and damage from animals and disease. They may contain chemical inhibitors which delay bud break until the

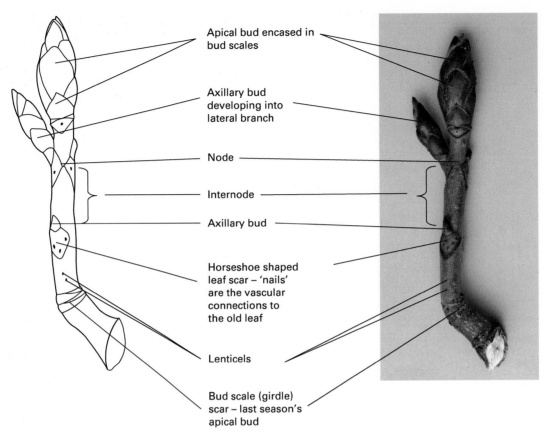

Apical bud encased in
bud scales

Axillary bud
developing into
lateral branch

Node

Internode

Axillary bud

Horseshoe shaped
leaf scar – 'nails'
are the vascular
connections to
the old leaf

Lenticels

Bud scale (girdle)
scar – last season's
apical bud

Figure 7.4 External features of a woody stem in *Aesculus hippocastanum* (horse chestnut)

spring. The *Aesculus hippocastanum* (horse chestnut) buds shown in Figure 7.4, for example, exude a sticky substance to deter insects. A **terminal (apical) bud** is present at the tip of a main stem or branch and contains a meristem from which lengthwise vegetative growth or, less commonly, a flower will emerge. Where leaves join the stem, **axillary buds** may grow into lateral shoots or flowers or may remain dormant.

Bud characteristics may be useful in identifying plants – for example, the native ash *Fraxinus excelsior* has black buds, those of beech (*Fagus sylvatica*) are long and pointed and *Magnolia* buds are hairy (Figure 7.10a–c). Flowering buds tend to be much larger and plumper than the vegetative buds (Figure 7.10d), which give rise to new shoots and leaves and this is useful when pruning. In spur pruning of apples, for example, the vegetative buds on the ends of lateral shoots are easily identified and removed encouraging development of flower buds.

Leaves

The main function of the leaf is to carry out **photosynthesis**.

The **leaf** consists of the leaf blade (**lamina**) and stalk (**petiole**) (Figure 7.2), its shape and arrangement on

the stem depending on the water and light energy supply in the species' habitat. Sessile leaves lack a petiole, while in peltate leaves such as *Nasturtium*, the petiole attaches to the centre of the lamina rather than the base (Figure 7.11d). Many monocotyledonous leaves also lack a petiole.

Leaf shape and structure (Figure 7.11) are useful indicators when attempting to identify a plant and descriptions often include specific terms. A few are described below but many more are used by botanists:

▶ **Simple** leaves have a continuous leaf blade, with an axillary bud at the base of the petiole (Figure 7.11a–f). There are a multitude of leaf shapes, e.g. lanceolate, ovate, obovate, lobed, orbicular and oval. Linear leaves are characteristic of monocotyledons (Figure 7.11a).

▶ **Compound** leaves, notably compound palmate and compound pinnate, have separate leaflets each with an individual base on one leaf stalk, but the only axillary bud is at the base of the main leaf stalk where it attaches to the stem (Figure 7.11g and h).

▶ **Margins** of leaves can be described, for example, as entire (unbroken or smooth) (Figure 7.11e), sinuous, serrate, dentate or crenate.

Figure 7.5 Photosynthetic stems

Figure 7.7 Red stem colour in dogwood (*Cornus alba* 'Sibirica')

Figure 7.6 Lenticels in the bark of *Prunus* (cherry) tree

▶ **Leaf vein arrangement** is another important feature of leaves. Parallel veination is a characteristic of monocotyledons (Fig. 7.11i) whereas dicotyledons have a wide variety of veinations such as the pinnate arrangement shown in Figure 7.11j and the palmate veination of *Geranium* leaves (Figure 7.11f).

In cultivated plants, leaf size and shape, colour (Figure 7.11k) and variegation (see Figure 9.3) are also important identifying features since many cultivars have been bred to differ from the original species in these respects.

In some plants, leafy structures called **stipules** may be found at the base of, or attached to, the petiole as in *Rosa* spp. (Figure 7.12)

The arrangement of leaves on the stem is also an important identifying feature (Figure 7.13). For example, *Salix alba* has leaves arranged alternately along the stem whereas *Cornus alba* leaves are attached opposite each other. Leaves may also be attached in whorls around the stem as in *Lilium*.

The major morphological differences between monocotyledonous and dicotyledonous roots, leaves and stems are summarized in Chapter 4.

Leaves in the garden

The novice gardener may easily overlook the contribution that the shape, texture, venation, colour and size of leaves can make to the general appearance of a garden. Flowers are the most striking feature, but they are often short-lived and it is foliage which gives more permanent interest.

Considering leaf shape and size, the long linear leaves of *Phormium tenax* (New Zealand flax) contrast with the large palmate leaves of *Gunnera manicata*, while on a smaller scale, the shade-loving hostas, with their lanceolate leaves, mix well with the pinnate-leaved *Dryopteris filix-*

Figure 7.8 Trees with ornamental bark: (a) *Acer pensylvanicum* 'Erythrocladum'; (b) *Prunus serrula* var. *tibetica*; (c) *Eucalyptus coccifera*

mas (male fern). Leaf texture is also important. Most species have quite smooth-textured leaves but *Verbascum olympicum, Stachys byzantina* (lamb's tongue) and the alpine *Leontopodium alpinum* (edelweiss) all have woolly textures. In contrast *Ilex aquifolium* (holly) and *Pieris japonica* have striking glossy leaves.

A wide variety of leaf colour tones are available to the gardener. The conifer *Juniperus chinensis* (Chinese juniper), shrubs of the *Ceanothus* genus and *Helleborus niger* (Christmas rose) are examples of dark-leaved plants. Plants with light-coloured leaves include the tree *Robinia pseudoacacia* 'Frisia' (false acacia), the climber *Humulus lupulus* 'Aureus' (golden hop) and the creeping herbaceous perennial *Lysimachia nummularia* 'Aurea' (creeping jenny). Plants with unusually coloured foliage include the small tree *Prunus* 'Shirofugen' (bronze–red), the subshrub *Senecio maritima* (silver–grey) and the shade perennial *Ajuga reptans* 'Atropurpurea' (bronze–purple). In autumn, the leaves of several tree, shrub and climber species change from green to a striking orange–red colour. *Acer japonicum* (Japanese maple), *Euonymus alatus* (winged spindle), and *Parthenocissus tricuspidata* (Boston ivy) are examples.

Variegation gives a novel appearance to the plant. *Aucuba japonica* (laurel), *Euonymus fortunei* and *Hedera helix* (ivy) all have good variegated forms.

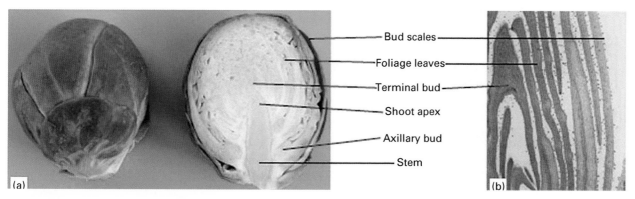

Figure 7.9 Structure of a bud: (a) Brussels sprout; (b) magnified image

Figure 7.10 Bud characteristics: (a) ash; (b) beech; (c) *Magnolia*; (d) *Camellia*, a large flower bud next to a narrow vegetative bud

Adaptations

The features of a typical plant are described above, but there are also many variations on the basic form of the stem, root and leaf. Some adaptations enable plants to spread vegetatively (asexually) (see p. 101) and gardeners can take advantage of this for propagating them (see Chapter 11). Other adaptations enable plants to live in extreme environments, protect them from herbivores or enable them to reach the light for photosynthesis or access air and nutrients more efficiently. Some adapted organs store food in the form of starch over the winter, giving plants a head start when growth resumes in the spring (**perennation**).

> A **perennating organ** is an organ that stores food, enabling the plant to survive unfavourable conditions. Examples are bulbs, corms, tubers and rhizomes.

Figure 7.11 Leaf shape: (a) linear, e.g. *Agapanthus*; (b) lanceolate, e.g. *Viburnum tinus*; (c) oval, e.g. *Garrya elliptica*; (d) peltate, e.g. *Nasturtium*; (e) hastate, e.g. *Zantedischia*; (f) lobed, e.g. *Geranium*; (g) palmately compound, e.g. *Lupin*; (h) pinnately compound, e.g. *Rosa*. Leaf veination: (i) parallel veins in a monocotyledonous leaf; (j) pinnate veins in a dicotyledonous leaf. Leaf colour: (k) a green *Helleborus*, yellow *Berberis* and purple *Ajuga*

Root adaptations

Adapted **adventitious roots** are found in many tropical plants for support including **buttress roots** (Figure 7.14a) which are plank-like outgrowths of the stem supporting tall forest trees growing on shallow soils (e.g. in *Ficus* spp.), **stilt roots** dropping down from branches to the ground (e.g. in mangroves, Figure 7.14b) and **prop roots** at the base of tall stems

(e.g. in *Zea mays* (sweetcorn), and the tropical tree *Pandanus utilis*, Figure 7.14c and d).

Some trees such as *Alnus* spp. (alder) and *Taxodium distichum* (swamp cypress) are particularly adapted to grow in waterlogged soils and produce 'breathing' roots or 'knees' (**pneumatophores**) covered in many lenticels which act as snorkels enabling them to obtain oxygen for respiration from the air above (Figure 7.14e).

Figure 7.12 Stipules of *Rosa* spp. along the petiole

(a)

(b)

Figure 7.13 Leaf arrangements: (a) alternate in hazel (*Corylus avellana*); (b) opposite in *Acer* spp.

In *Hydrangea anomala* subsp. *petiolaris* and in *Hedera helix* (ivy), **adventitious climbing roots** develop along the stem attaching it to vertical surfaces such as tree trunks and walls to increase light for photosynthesis and raise the flowers up for better pollination opportunities (see Figure 5.7). Ivy is not parasitic and obtains all its water and nutrients from its own roots in the ground, its adventitious climbing roots are there just for support. Contrary to popular belief it does not kill trees or damage their bark and, where it grows into the crown, this is most likely where the tree is already in decline or is diseased. Similarly ivy rarely damages walls if the brickwork is sound. Ivy is a valuable plant for wildlife in the garden (see p. 47). On buildings, ivy can insulate in the winter and cool the building in the summer. Unlike ivy, *Viscum album* (mistletoe) is a hemiparasite (see p. 47) which obtains some of its nutrition from a host tree, commonly apple, poplar or lime, to supplement its own photosynthesis. It produces **haustorial roots** which penetrate the vascular system of the host and tap into water and sugars. Some plant roots associate with other organisms such as bacteria (e.g in **nitrogen-fixing roots** and fungi in **mycorrhizal roots**) to improve their uptake of water and nutrients.

Epiphytes are plants which are physically attached to aerial parts of other plants for support, enabling them to reach more light for photosynthesis, such as some ferns, bromeliads and orchids. Epiphytic orchids have specialized aerial roots called **velamen roots**. These have a multilayered epidermis with an outer velamen layer composed of dead cells giving the root a silvery appearance. The velamen may absorb water from the air and reduce water loss from the root tissue and it seems that it is able to take up nutrients dissolved in moisture in the air. Unusually in roots, the underlying tissue is green and photosynthesizes (Figure 7.14f).

Root tubers develop near the base of the plant, often from adventitious roots, as in *Dahlia* or *Ophrys* spp. (bee orchids). Root tubers such as the *Dahlia* tuber (Figure 7.14g) can be distinguished from stem tubers in having lateral roots and no nodes. These, together with **swollen taproots** (e.g. *Daucus carota* (carrot) (Figure 6.1) and *Taraxacum officinale* (dandelion)) are all perennating organs.

Stem adaptations

Many stems are adapted for climbing. In *Phaseolus coccineus* (runner bean), *Lonicera* spp. (honeysuckle) and *Wisteria* spp., **twining stems** wind around other

Figure 7.14 Some root adaptations: (a) buttress roots; (b) stilt roots of mangrove; (c) prop roots of maize; (d) prop roots in *Pandanus utilis*; (e) pneumatophores of swamp cypress; (f) velamen roots of orchid; (g) root tuber of a *Dahlia*

upright structures for support either clockwise (e.g. *Wisteria sinensis*) or anticlockwise (e.g. *Wisteria floribunda*). Such twining stems can become large and woody, which are then called lianes (Figure 7.15a and b).

Ruscus aculeatus (butcher's broom) is a native plant which is adapted to dry, shady woodlands. Its stems are leaf-like **cladodes** carrying out the functions of the leaves which are themselves reduced to a protective spine at the cladode tip. The small buds, flowers and berries are borne in the centre of the cladode indicating that it is a stem (Figure 7.15c).

Thorns, which are modified branches growing from axillary buds (and hence have a vascular connection to the stem), can have a protective function (e.g. in *Crataegus* spp. (hawthorn) and *Pyracantha*) discouraging herbivores (Figure 7.15d). **Prickles** are specialized outgrowths of the stem epidermis (so are easily rubbed off), which not only protect but also assist the plant in scrambling over other vegetation, as in many roses (Figure 7.15e).

Many stem adaptations important to horticulture are also perennating organs storing starch such as some **rhizomes**, which are stems growing horizontally

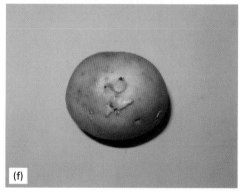

Figure 7.15 Some stem adaptations: (a) twining stems of a mature *Wisteriasinensis*; (b) Strangler fig (*Ficus* spp.) uses a tree for support which it may eventually kill, leaving a hollow core; (c) cladodes of butcher's broom; (d) thorn of *Pyracantha*; (e) prickles on the stem of *Rosa sericea* subsp. *omeiensis* f. *pteracantha*, grown as an ornamental for its large red thorns; (f) stem tuber of potato *solanum tuberosum* with new shoots growing from buds on the stem; (g) stolon of bramble *Rubus fruticosus*

usually underneath and sometimes just above the ground. Nodes and internodes can be seen clearly along the stem together with adventitious roots which anchor it in the soil and take up water and mineral nutrients. As the stem branches, new rhizomes are formed from lateral buds, each with a shoot at its tip. This eventually enables the plant to spread. Rhizomes are found in many *Iris* spp. such as *I. germanica* and its cvs., including the bearded *Iris* in Figure 7.16. **Corms**, found in *Gladiolus, Crocosmia* and *Crocus* for example, are compressed underground shoots in which the stem is swollen

with starch (Figure 7.18). They have dry scale leaves on their outer surface underneath which nodes, internodes and axillary buds can be seen. Each year a new corm forms on top of the old one. As well as small adventitious roots, rhizomes and corms have specialized, thickened contractile roots which help to pull them to an appropriate level in the soil. **Stem tubers** are typified by *Solanum tuberosum* (potato) (Figure 7.15f). The tubers of potato can be distinguished from root tubers by having vestigial nodes and axillary buds (the 'eyes') which grow into shoots. They also turn green (and poisonous) on

7

93

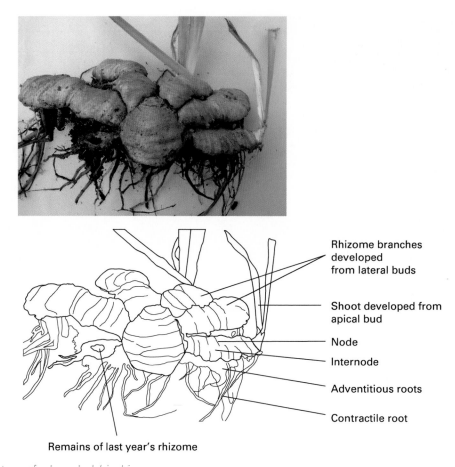

Figure 7.16 Structure of a bearded *Iris* rhizome

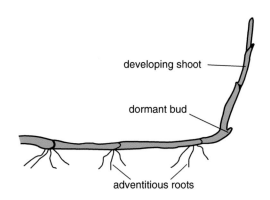

Figure 7.17 Couch grass (*Elymus repens*) rhizome

exposure to light since stems can photosynthesize but roots cannot, hence the need to 'earth up' potatoes to exclude light.

Anemone blanda, A. coronaria De Caen Group, *Begonia* x *tuberhybrida* (tuberous begonia), *Gloxinia* and *Cyclamen* tubers develop from the hypocotyl and are often classed as stem tubers.

Rhizomes and stem tubers can also be methods of vegetative (asexual) spread. Others include the **runners** of *Fragaria* spp. (strawberry), horizontal stems which grow just above the ground and root at nodes along the stem or at stem tips producing plantlets. These are often found in rosette plants and grow from buds at the base of the plant. The term **stolon** is often used interchangeably with runner but it also includes plants such as *Rubus fruticosus* (bramble) (Figure 7.15g) which have long arching branches which root where their tips touch the ground. Non-perennating rhizomes can be a nuisance when trying to eradicate perennial weeds such as *Calystegia sepium* and *Convolvulus arvensis* (hedge and field bindweed) and *Elymus repens* (couch grass) (Fig. 7.17) as their rhizomes can penetrate up to 5 m deep and they can regenerate a new plant from buds on even the tiniest section.

Suckers are stems that grow from adventitious buds (buds which do not derive from the plumule (see p. 106), which are found on the roots of many plants, such as *Rhus typhina* (stag's horn sumach). These can be useful in propagating plants vegetatively but can also be a nuisance if their spread is difficult to control.

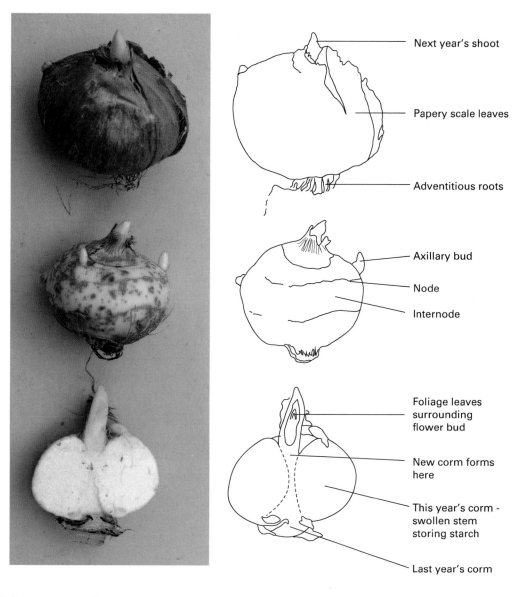

Next year's shoot

Papery scale leaves

Adventitious roots

Axillary bud

Node

Internode

Foliage leaves surrounding flower bud

New corm forms here

This year's corm - swollen stem storing starch

Last year's corm

Figure 7.18 Structure of a *Crocus* corm

Tendrils

Many plants use modified plant parts called tendrils which wrap themselves around a support to climb towards the light. Sometimes the tendril itself will coil like a spring, pulling the plant even closer to the support. Tendrils can be formed from various plant organs and often it is difficult to establish their origin. In members of the grape family (Vitaceae), for example, which includes *Parthenocissus* represented by *P. tricuspidata* (Boston ivy) and *P. quinqefolia* (Virginia creeper), *Vitis vinifera* (grapevine) and *Cissus rhombifolia* (grape ivy), there is much debate about whether the tendrils are formed from a leaf, or a stem tip or even a group of

Figure 7.19 Tendrils of passion flower

flower buds. *Parthenocissus* have an additional mechanism in the form of sucker pads on the ends of the tendrils which secrete a sticky substance enabling them to attach tightly to smooth surfaces such as walls. In *Passiflora* spp. (passion flowers), a group of axillary buds is found at the base of each petiole, one of which forms a tendril, one or two flower buds and another a vegetative shoot making it difficult to decide what plant part the tendril actually develops from (Figure 7.19).

Leaf adaptations

While remaining essentially the organ of photosynthesis, the leaf takes on other functions in some species. The most notable of these are modifications for climbing which enable the plant to compete effectively with other plants for light, therefore increasing its photosynthetic capacity. Leaf **tendrils** may be formed from slender extensions of the leaf, and are of three types. In *Clematis* spp., the leaf petiole curls round the stems of other plants or garden structures to support the climber (Figure 7.20a), while *Lathyrus odoratus* (sweet pea) holds on with tendrils modified from the end-leaflets of the

compound leaf (Figure 7.20b). The monocotyledonous climber *Smilax china* has tendrils provided by modified stipules (found at the base of the petiole). In insectivorous *Nepenthes* spp. (a type of pitcher plant), a tendril develops from the leaf and then the tip of the tendril forms the pitcher (Figure 7.20c).

In *Galium aparine* (cleavers), both the leaf and stipules, borne in a whorl, bear **prickles** that allow the weed to sprawl over other plant species acting as grappling hooks. Leaves may also have a protective function, notably against predation, as in the aloes and, as shown in Figure 7.20d, where some hairs on aubergine leaves have become tiny **spines**. On a larger scale, the spines of *Berberis* spp. are also adaptations of leaves (Figure 7.20e). In many desert species, the leaves are transformed into spines, which have the additional advantage of reducing water loss, and the function of photosynthesis is taken over by the stem (Figure 7.20f).

In a **bulb** (e.g. *Narcissus*, Figure 7.21) the outer papery scale leaves enclose succulent, light-coloured scale leaves containing all the food and moisture necessary for the bulb's emergence making this a perennating organ. The scales are packed densely together around the terminal bud towards the base of the stem,

Figure 7.20 Some leaf adaptations: (a) petiole tendrils of *Clematis*; (b) leaflet tendrils of sweet pea; (c) pitcher developed from a leaf tendril in *Nepenthes*; (d) spines on an aubergine leaf; (e) spines of *Berberis*; (f) spines of the cactus *Ferrocactus emoryi* subsp. *emoryi*

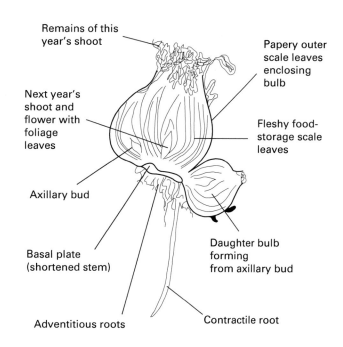

Remains of this year's shoot

Papery outer scale leaves enclosing bulb

Next year's shoot and flower with foliage leaves

Fleshy food-storage scale leaves

Axillary bud

Basal plate (shortened stem)

Daughter bulb forming from axillary bud

Adventitious roots

Contractile root

Figure 7.21 Structure of a *Narcissus* bulb

Figure 7.22 White bracts of flowering dogwood. The insignificant flowers are in the centre of the bracts

minimizing the risks from extremes of climate, or pests such as eelworms and mice. Bulbs have both adventitious and contractile roots (see above) attached to a basal plate. Daughter bulbs form in the axils of the leaves which eventually detach, bringing about vegetative spread.

In many plants such as *Hydrangea* spp., the houseplant *Euphorbia pulcherrima* (poinsettia) and flowering *Cornus* spp. (dogwoods) (Figure 7.22), coloured **bracts** take the place of petals in attracting pollinating insects.

Phyllodes, found in *Acacia* spp., are flattened leaf-like petioles which replace and carry out the functions of the leaves. The leaves of many monocots are believed to have originated from phyllodes. *Sedum*, in common with many plants adapted to arid conditions, have succulent **fleshy leaves** which are able to store water, while in the houseplant *Bryophyllum daigremontianum* the succulent leaf bears **adventitious buds** which are able to drop to the ground below and develop into young plantlets.

Further information on plant adaptations and their use in propagation are found in Chapter 11 and on the companion website.

Further reading

Capon, B. (2005) *Botany for Gardeners*. Timber Press.

Hickey, M. and King, C. (1997) *Common Families of Flowering Plants*. Cambridge University Press.

Hodge, G. (2013) *RHS Botany for Gardeners*. Mitchell Beazley.

Jepson, M. (1938) *Biological Drawings*. John Murray.

McMillan Browse, P. (1979) *Plant Propagation*. Mitchell Beazley.

Stern, K. (2001) *Introductory Plant Biology*. 8th edn. McGraw Hill

7

 Please visit the companion website for further information:
www.routledge.com/cw/adams

CHAPTER 8

Level 2

Plant reproduction

Figure 8.1 Succulent fruits of *Crataegus monogyna* (hawthorn)

This chapter includes the following topics:

- Flower structure
- Characteristics of wind- and insect-pollinated flowers
- Seeds, fruits and their dispersal

Principles of Horticulture. 978-0-415-85908-0 © C.R. Adams, K.M. Bamford, J.E. Brook and M.P. Early. Published by Taylor & Francis. All rights reserved.

Figure 8.2 Range of flowers as organs of sexual reproduction having similar basic structure, but varying appearance having adapted for successful pollination or by plant breeding: (a) *Iris chrysographes* 'Kew Black'; (b) *Eryngium giganteum* ('Miss Willmott's ghost'); (c) *Trollius chinensis* 'Golden Queen'; (d) *Rosa* 'L.D. Braithwaite'; (e) *Hemerocallis* 'Rajah'; (f) *Aquilegia fragrans*; (g) *Oenothera* 'Apricot Delight'; (h) *Helenium* 'Wyndley'; (i) *Helleborus x hybridus*; (j) *Nepeta nervosa*; (k) *Primula vialii*

The flowering plant represents the pinnacle of evolution in the plant world and there is no doubt that in the ornamental garden it is the contribution of flowers which is often at the forefront of plant selection (Figure 8.2).

Flowers are the organs of **sexual reproduction** in flowering plants. Their structure forms the basis, in large part, of plant classification and reflects the different pollination mechanisms used and, in the case of animal-pollinated flowers, the need to advertise their wares! They are also where fertilization, the fusion of the male sex cell in pollen and the female sex cell in the ovule, takes place. Sexual reproduction leads to the development of seeds and fruits, the means by which plants spread. It also brings about mixing of the genes contributed by each 'parent' so that the offspring will be similar but not identical to the parent plants and each other. This range of variation enables plants to withstand changes in environmental conditions as there will always be some individuals which are likely to survive.

In contrast, plants can also reproduce **asexually** through many natural means such as runners, bulbs, or by layering or suckering, enabling plant dispersal

(see Chapters 7 and 11). The resulting plants are genetically identical to the parent plant (clones). It is an alternative to sexual reproduction, useful if pollination is poor or seed production fails, but does not provide the genetic variation which is necessary for long-term survival and adaptation of the species. It is, however, very useful in vegetative propagation where the aim is to maintain the desirable characteristics of the parent plant (see Chapter 11).

> **Sexual reproduction** is the formation of new individuals through fusion of male and female sex cells (gametes). It results in variable offspring. **Asexual reproduction** is the formation of new individuals without fusion of gametes resulting in genetically identical offspring.

Flower structure

However varied flowers appear, they all have the same basic structure with the flower parts arranged in four whorls (Figure 8.3). These are the:

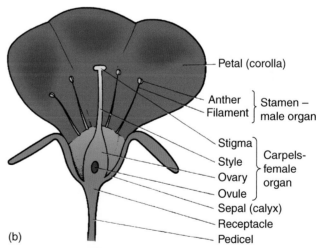

Figure 8.3 Flower structure: (a) flower of (*Glaucium corniculatum*); (b) diagram of a typical dicotyledonous flower to show structures involved in the process of sexual reproduction; (c) long section of a pumpkin flower showing anthers (with yellow pollen), lobed stigma in the centre and developing ovary with immature seeds on the right

▶ Calyx.

▶ Corolla.

▶ Androecium (stamens).

▶ Gynoecium (carpels).

▶ The **calyx** or ring of **sepals** which initially enclose and protect the flower bud. The sepals are often green and can therefore photosynthesize. In some plants (e.g. *Fuchsia*), the sepals may be coloured to attract animal pollinators, while in wind-pollinated plants they may be reduced in size.

▶ The **corolla** or ring of **petals** may be small and insignificant in wind-pollinated flowers, (e.g. many tree species), or large and colourful in insect-pollinated species (Figure 8.4). **Nectaries** may develop at the base of the petals. These have a secretory function, producing substances such as nectar which attract pollinating organisms.

The colours and size of petals can be improved in cultivated plants by breeding, and may also involve the multiplication of the petals or **petalody**, when fewer male and/or female organs are produced, e.g. many 'double-flowered' *Campanula* species and *Gardenia augusta* (Figures 3.22 and 8.5).

▶ The **androecium**, the male organ, consists of **stamens**, which bear **anthers** that produce and discharge **pollen grains**, borne on a **filament** (Figure 8.3). The pollen contains the male sex cells or gametes.

▶ The **gynoecium**, the female organ, is positioned in the centre of the flower and consists of an **ovary** enclosing one or more **ovules** which contain the ovum (the female sex cell or gamete). The **style** leads from the ovary to a **stigma** at its top where pollen is captured. The basic unit of the gynoecium is the **carpel**, made up of a stigma, style and ovary. More evolutionarily primitive flowers such as *Ranunculus* (buttercup) have many separate carpels but in most flowers the carpels are fused to form one large ovary, style and stigma (Figure 8.3).

The flower parts are positioned on the **receptacle**, which is at the tip of the **pedicel** (flower stalk). Associated with the flower head or **inflorescence** are leaf-like structures called **bracts**, which can sometimes assume the function of insect attraction – for example, in *Euphorbia pulcherrima* (poinsettia), *Hydrangea* spp. and some *Cornus* species (see Fig. 7.22). See 'Inflorescences' on the companion website.

In many monocotyledons such as tulips and lilies, the outer two layers of the flower have a similar appearance, making the sepals and petals indistinguishable (**tepals**) (Figure 8.6).

A distinguishing feature of monocotyledons and dicotyledons is the number of flower parts – for

Figure. 8.4 Dicotyledonous *Papaver* (poppy) flower with five petals and a bee seeking pollen and nectar

Figure 8.5 Petalody in a *Gardenia* flower

example, sepals, petals and carpels. In monocots these are in multiples of three, whereas in dicotyledons, flower parts are in multiples of four or five.

The flowers of most species have both male and female organs (**hermaphrodite**), but some have separate male and female flowers on the same plant (**monoecious**) (Figure 8.7), such as water lilies (*Nymphaea*) and members of the cucumber genus (*Cucumis*), and many trees such as walnut (*Juglans*), alder (*Alnus*) and birch (*Betula*), whereas others produce male and female flowers on different plants (**dioecious**), such as holly (*Ilex*), willows (*Salix*) and *Skimmia japonica*.

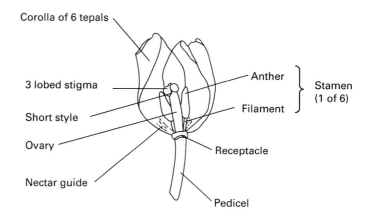

Figure 8.6 (a) Flower of *Tulipa*, a monocotyledon; (b) diagram showing the typical structure of a monocotyledonous flower

Most conifers are monoecious with male and female cones (Figure 8.9).

> A plant possessing flowers with both male and female organs is **hermaphrodite**. Species with separate male and female flowers on the same plant are **monoecious**. Species which produce male and female flowers on different plants are **dioecious**.

Characteristics of wind- and insect-pollinated flowers

Pollination is the transfer of pollen, containing the male sex cell, from anther to stigma either of the same flower or different flowers and is the essential first step in the process leading to fertilization. In **cross-pollination**, pollen transfer is between different plants, while in **self-pollination** it is within or between flowers on the same plant. The commonest natural agents of pollination are **wind** and **insects**.

> **Pollination** is the transfer of pollen from stamen to stigma of a flower or flowers. **Fertilization** is the fusion of a male sex cell (the male gamete) from a pollen grain with a female sex cell (the female gamete) in the ovule to produce an embryo.

Wind-pollinated flowers

The characteristics of **wind-pollinated flowers** (Figure 8.8) are their small size, their green appearance (their petals are reduced in size or absent and

Figure 8.7 Monoecy in courgette *Cucurbita pepo*: (a) female flower showing ovary between the petals and the pedicel and branched stigma within flower; (b) male flower lacking an ovary with stamens and pollen within flower

lack colour), their absence of nectaries and scent production, and their production of large amounts of small, smooth, light pollen which is intercepted by large feathery stigmas. They also often have proportionally larger stigmas and flexible stamens that protrude from the flower to maximize the chances of dispersing and intercepting pollen grains in the air.

Figure 8.8 Wind-pollinated species with small inconspicuous flowers: (a) *Stipa calamagrostis*; (b) *Cyperus chira*; (c) *Luzula nivea*; (d) male flowers of *Fagus sylvatica* (beech) with prominent anthers; (e) female willow (*Salix caprea*) catkins with prominent stigmas

The most common examples of wind-pollinated plants are the grasses, and trees with catkins such as some *Salix* species (willow), *Betula* (birch), *Corylus* (hazel), *Fagus* (beech) and *Quercus* (oak). The conifers also use wind pollination to disperse copious amounts of pollen from the small male cones (Figure 8.9).

Insect-pollinated flowers

The characteristics of insect-pollinated flowers (Figure 8.10) are brightly coloured petals (often with scent production) to attract insects and the presence of nectaries to entice insects with sugary food. Insects such as bees and flies collect the pollen on their bodies as they fly in and out and carry it to other flowers. Petals may have nectar guides, coloured lines to point the way to the nectaries which may only be visible to insects. Some flowers are designed to favour certain insects – for example, in *Antirrhinum majus* (snapdragon) and *Trifolium repens* (clover) the flower physically prevents entry of smaller non-pollinating insects and opens only when heavy bees land on it. Other plant species, such as *Arum italicum* (arum lily), trap pollinating insects for a period of time to give the best chance of successful fertilization. The stigmas and stamens of insect-pollinated flowers tend to be short and sturdy and the pollen they produce is larger, heavier and produced in smaller quantities since pollination success is more likely than in wind-pollinated flowers. See 'Pollination by Bees' on the companion website.

The seed

Pollination may be followed by fertilization, which is the fusion of the male sex cells (male gametes) in the pollen and the female sex cell or ovum (female gamete) in the ovule to produce a new embryo plant contained in a seed. The seed itself is formed from the ovule of the flower and enclosed in the fruit which is formed from the flower's ovary.

The seed (Figure 8.11), resulting from sexual reproduction, creates a new generation of plants that bear characteristics of both parents. The plant must survive often through conditions that would be damaging to a growing vegetative organism, so

Figure 8.9 Male cones of *Cedrus atlantica* Glauca Group (blue cedar)

Figure 8.10 Insect pollinated flowers are brightly coloured and sometimes have guidelines in the petals to guide insects to nectar: (a) *Hemerocallis* (day lily); (b) *Digitalis stewartii*; (c) *Verbascum* 'Cotswold'

the seed is a means of protecting against extreme conditions of temperature and moisture, and is thus often the **overwintering stage**. The seed, together with the fruit, may also enable the embryo to be dispersed away from the parent plant and may have dormancy mechanisms which prevent germination until conditions are favourable.

> A **seed** is the structure that develops from the ovule after fertilization. A **fruit** is formed from the ovary wall usually following fertilization and encloses the seed.

Seed structure

The basic structure of a dicotyledonous seed is shown in Figure 8.12. The main features of the seed are:

▶ **The embryo**. In order to survive, the seed must contain a small immature plant protected by a seed coat. The embryo consists of a **radicle**, which will develop into the root of the seedling to take

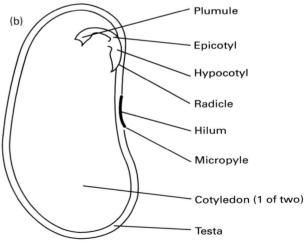

Figure 8.11 Seeds: a range of species. Top: runner bean; left to right: leek, artichoke, tomato, lettuce, Brussels sprout, cucumber, carrot, beetroot

Figure 8.12 Dicotyledonous seed structure: (a) germinating *Phaseolus coccineus* (runner bean) seed showing developing radicle; (b) long section of a *Phaseolus vulgaris* (French bean) seed

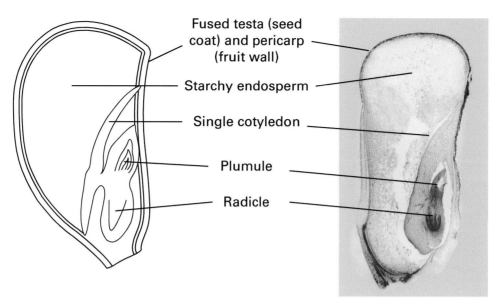

Figure 8.13 Structure of a monocotyledonous seed *Zea mays* (sweetcorn). The kernel is actually a fruit enclosing the seed within

up water and nutrients, and a **plumule**, which develops into the shoot system, bearing leaves for photosynthesizing and flowers for seed and fruit production. The region between the cotyledons and the radicle is the **hypocotyl** while the short length of stem between the cotyledons and the shoot is termed the **epicotyl**. A single **cotyledon** will be found in monocotyledons, while two are present as part of the embryo of dicotyledons. The cotyledons may occupy a large part of the seed, such as in *Vicia faba* (broad bean), and act as the food store for the embryo. In other seeds, such as *Zea mays* (sweetcorn), the single cotyledon remains small and the food store is provided by another tissue called the **endosperm** which is not part of the embryo (Figure 8.13).

▶ The **testa**, also known as the seed coat, is formed from the outer layers of the ovule after fertilization. It is waterproof and airtight and may contain germination inhibitors which enable seeds to stay dormant over winter.

▶ The **micropyle**: this is a weakness in the testa where water uptake occurs triggering germination.

▶ The **hilum**: this is a scar on the testa where the seed was attached to the fruit.

Food storage in seeds

In some species, such as **grasses** and ***Ricinus communis*** (castor oil plant), the food of the **seed** is found in a different tissue from the cotyledons, which is called the endosperm. Plant food in either cotyledons or endosperm

is often stored as the carbohydrate starch, formed from sugars as the seed matures – for example, in peas and beans. Other seeds, such as sunflowers, contain high proportions of fats and oils, and proteins are often present in varying proportions. These substances store energy in a very concentrated form, which is released through the process of respiration when the seed germinates, fuelling rapid growth (see Chapter 9). The seed is also a rich store of nutrients, such as phosphate, which it requires for seedling growth (see p. 168). This explains why seeds are such a useful foodstuff for humans too.

The fruit

The development of a fruit involves either the expansion of the ovary into a juicy **succulent** structure, or the tissues becoming hard and **dry**. Fruits provide a means of protection and often a means of dispersal for the seeds they contain and may also contribute to delayed germination through dormancy. Some dry fruits split to release their seeds (described as **dehiscent**) while others rely on the fruit coat being broken down to release the seeds (described as **indehiscent**) (Figure 8.14). Some methods of dispersal include:

▶ **Explosive** or self-dispersed: the fruit splits open propelling the seeds into the air, e.g. *Cytisus* (broom), *Lupinus* (lupin), *Lathyrus odoratus* (sweet pea), *Erysimum* (wallflower), *Lunaria annua* (honesty), *Cardamine hirsuta* (hairy bittercress) and *Geranium*.

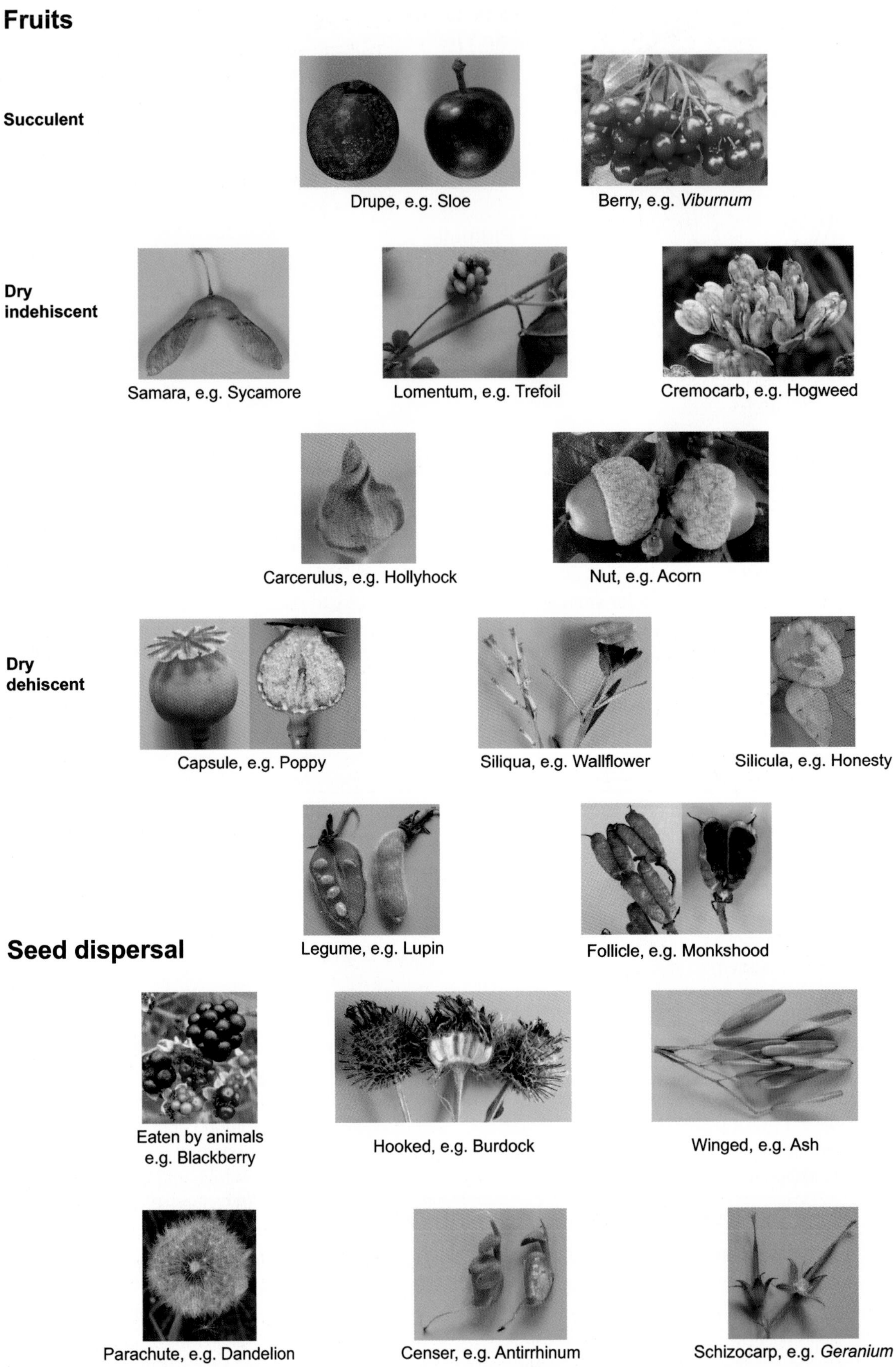

Fruits

Succulent

Drupe, e.g. Sloe

Berry, e.g. *Viburnum*

Dry indehiscent

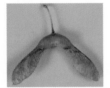
Samara, e.g. Sycamore

Lomentum, e.g. Trefoil

Cremocarb, e.g. Hogweed

Carcerulus, e.g. Hollyhock

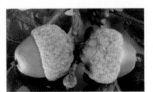
Nut, e.g. Acorn

Dry dehiscent

Capsule, e.g. Poppy

Siliqua, e.g. Wallflower

Silicula, e.g. Honesty

Legume, e.g. Lupin

Follicle, e.g. Monkshood

Seed dispersal

Eaten by animals e.g. Blackberry

Hooked, e.g. Burdock

Winged, e.g. Ash

Parachute, e.g. Dandelion

Censer, e.g. Antirrhinum

Schizocarp, e.g. *Geranium*

Figure 8.14 Fruit types and seed dispersal

▶ **Wind**: the seeds of poppy capsules (*Papaver*) are shaken from small pores in the fruit as the plant sways like a church censer. Other fruits have tiny feathery parachutes attached as in *Epilobium* (willow herb), *Clematis* and many members of the daisy family including *Senecio* (groundsel), *Taraxacum* (dandelion) and *Cirsium* (thistles). Many woody species such as *Tilia* (lime), *Fraxinus* (ash), *Acer* (sycamore and maples) produce winged fruit.

▶ **Animals**: mammals and birds can distribute fruits either externally or internally. Hooked fruits, e.g. *Galium aparine* (goosegrass) and *Arctium* (burdocks), become attached to animal's fur. The sticky succulent fruits of *Viscum album* (mistletoe) attach to birds' beaks and are rubbed off onto trees where they germinate. Squirrels may bury

fruits such as nuts, e.g. *Quercus robur* (oak), *Fagus sylvatica* (beech), *Castanea sativa* (sweet chestnut), in the ground far from where they were collected (scatter hoarding). Succulent fruits, e.g. *Solanum lycopersicon* (tomato), *Rubus fruticosus* (blackberry), *Prunus spinosa* (sloe), *Viburnum* and *Sambucus nigra* (elderberry), or those that are filled with protein, e.g. *Rumex* (dock), are eaten by birds and other animals, the seeds passing through their gut before being deposited elsewhere (**frugirory**).

▶ **Water**: many aquatic plants such as *Nymphaea* (waterlilies) or those growing close to rivers and seashores use water to disperse their fruits. The fruits of coconut palms can travel thousands of kilometres in ocean currents. The introduced weed

Reproduction in non-seed producing plants

In higher plants, seeds are the means by which they colonize new areas and reduce competition from the parent plant and other seedlings. However, simpler multicellular green plants such as mosses and ferns do not produce seeds, so other means of dispersal are needed. In these plants, during their life cycles, two stages of

quite distinct types of growth occur. In **ferns**, for example, the typical fern plant which we see growing is a vegetative phase of the life cycle. Spores are released from the underside of the fronds and are dispersed by the wind (Figure 8.15). With suitable damp conditions, they germinate to produce a second sexual stage. Each spore grows into a tiny leafy structure in which male and female organs develop and release sex cells which fuse. The cell resulting from this fertilization gives rise to a new fern plant and the small leafy structure withers away (Figure 8.16). Ferns can be produced in cultivation by spores if provided with damp sterile conditions to allow the tiny spores to germinate without competition (see p. 133).

8

Figure 8.15 Sori, brown spore-producing structures on the underside of fern fronds: left: *Asplenium scolopendrium* 'Cristata' (crested hart's tongue fern); right: *Dryopteris erythrosora*

Figure 8.16 Germinating fern spores and plantlets

Impatiens glandulifera (Himalayan balsam) has an explosive mechanism to disperse its seeds but they are also spread along waterways where they have become a serious threat to biodiversity in Britain and Ireland (see p. 8).

Some examples of fruit types and dispersal methods are illustrated in Figure 8.14.

Further reading

Hickey, M. and King, C. (1997) *Common Families of Flowering Plants*. Cambridge University Press.

Hodge, G. (2013) *RHS Botany for Gardeners*. Mitchell Beazley.

Holm, E. (1979) *The Biology of Flowers*. Penguin.

Stern, K.R., Bidlack, J.E. and Jansky, S.H. (2011) *Stern's Introductory Plant Biology*. McGraw Hill.

Please visit the companion website for further information:
www.routledge.com/cw/adams

CHAPTER **9**
Level 2

Plant growth

Figure 9.1 Pumpkins, representing the large amount of growth a plant can produce

This chapter includes the following topics:

- Photosynthesis
- Aerobic and anaerobic respiration
- Factors affecting photosynthesis and respiration
- Leaf structure and photosynthesis

Principles of Horticulture. 978-0-415-85908-0 © C.R. Adams, K.M. Bamford, J.E. Brook and M.P. Early.
Published by Taylor & Francis. All rights reserved.

What is meant by 'growth'? Growth is a difficult term to define because it really encompasses the totality of all the processes that take place during the life of an organism. However, it is useful to distinguish between the processes that result in an increase in size and weight, which we can call 'growth', and those processes that cause the changes in the plant during its life cycle, which can usefully be called 'development', as described in Chapter 5. All living organisms need food and energy to grow and plants obtain these through **photosynthesis**, **water** and **mineral uptake**. **Respiration** is the process by which this food and energy is converted into a form which can be used by the plant. Because one process makes food and the other breaks it down, for plants to grow, the right balance between photosynthesis and respiration is essential.

Photosynthesis

Photosynthesis is the process by which green plants manufacture 'food' in the form of high-energy carbohydrates such as sugars and starch, using light as an energy source.

'Food' is needed to build the plant's structure and to provide energy to fuel its activities such as the manufacture of: **proteins**, including enzymes which speed up chemical reactions in cells; **cellulose**, used to build cell walls at meristems; **oils and starches**, laid down in seeds to nourish the embryo when it germinates.

All the complex organic compounds, based on carbon, must be produced from the simple raw materials water and carbon dioxide. Green plants are able to do this through the process of photosynthesis (see producers p. 38). Many other organisms are unable to manufacture their own food, and must therefore feed on already manufactured organic matter such as plants or animals (see consumers p. 38). Since large animals predate smaller animals, which themselves feed on plants, all organisms depend directly or indirectly on photosynthesis as the basis of a **food web or chain** (see p. 39).

Photosynthesis (Figure 9.2) involves the conversion of water and carbon dioxide into glucose (a simple carbohydrate or sugar) and oxygen. Light energy from the sun is the fuel which drives the process and is captured by the green pigment **chlorophyll** in the **chloroplasts**, mainly in the leaf. Water is supplied by the roots and transported to the leaves while carbon dioxide is taken in by leaves from the air. The glucose produced is converted to another sugar, sucrose,

which is exported from the leaves to other plant organs and may also be converted to starch (a large carbohydrate) for storage until it is required. Oxygen is produced as a waste product and this is released to the air. All the oxygen that we breathe on earth has been produced by plants through photosynthesis over millions of years.

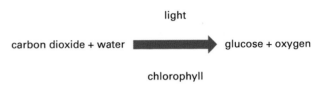

Figure 9.2 Equation for photosynthesis

> **Photosynthesis** is the process in the chloroplasts by which green plants trap light energy from the sun, convert it into chemical energy and use it to produce food in the form of carbohydrates such as sugars and starch. The raw materials are carbon dioxide and water. Oxygen is released as a waste product.

Factors affecting photosynthesis

The following environmental requirements for photosynthesis are explained in detail below:

- carbon dioxide
- light
- adequate temperature
- water (see also Chapter 10)
- mineral nutrients (see also Chapter 10).

Carbon dioxide

In order that a plant may build up organic compounds such as sugars, it must have a supply of readily available carbon. **Carbon dioxide** is present in the air in concentrations of around 400 ppm (parts per million) or 0.04%, and can diffuse into the leaf through the **stomata** (see p. 123). If no other factors are limiting, the rate of photosynthesis increases as levels of carbon dioxide in the surrounding air increase. The amount of carbon dioxide in the air immediately surrounding the plant can fall when planting is very dense, or when plants have been photosynthesizing rapidly, especially in an unventilated greenhouse. Ventilation can rectify this, replacing the carbon dioxide used up. Alternatively the atmosphere can be **enriched** in commercial glasshouses by supplying carbon dioxide at levels above that in the atmosphere. In fact the atmosphere within a glasshouse or polytunnel can be increased to levels well above ambient concentrations, typically three times greater, (e.g. up to 1,000 ppm (0.1 per cent) in lettuce), with a resulting increase in the rate of photosynthesis leading to improvements in yield and quality of many glasshouse crops.

Light

In any series of chemical reactions where one substance combines with another to form a larger compound, energy is needed to fuel the reactions. In plants this energy is provided by light from the sun. As with carbon dioxide, the amount of light energy present is important in determining the rate of photosynthesis; simply, the more light, or greater the **light intensity** supplied to the plant, the more photosynthesis can take place. Beyond a certain light intensity, however, the rate of photosynthesis levels off as the chloroplasts are fully engaged. This is called the **saturation point** and will vary from plant to plant, shade lovers such as *Ficus benjamina* having lower saturation points than those adapted to high light conditions. Light levels also affect stomatal opening: stomata close as light levels reduce, which restricts carbon dioxide uptake. Care must be taken to maintain clean glass or polythene and to avoid condensation that restricts light transmission. Light intensity can be increased by using artificial lighting (**supplementary lighting**) to boost light levels, particularly in the winter when light is the rate-limiting factor. The **duration** of lighting will naturally influence the length of time that photosynthesis can continue, longer in the spring and summer than during the winter months. Supplementary lighting can also be used to extend the duration of the daylight hours in winter.

As well as light intensity and duration, **light quality** is important in optimizing photosynthesis. Photosynthesis only utilizes certain wavelengths of light, those in the red and blue parts of the visible spectrum. Pigments such as chlorophyll absorb light of certain wavelengths, in this case the ones useful to photosynthesis, and reflect the rest such as yellow and green wavelengths, which is why chlorophyll appears green. If supplementary lighting is given in a greenhouse, the lamp chosen, as well as giving good light intensity, must produce the right wavelengths of light for photosynthesis (known as Photosynthetically Active Radiation or PAR).

Light wavelengths

Light, like other forms of energy, such as heat, X-rays and radio waves, travels in the form of waves, and the distance between one wave peak and the next is termed the wavelength. Light wavelengths are measured in nanometres (nm): 1 nm = one-thousandth of a micrometre. Visible light wavelengths vary from 800 nm (red light, in the long wavelength area) through the spectrum to 350 nm (blue light, in the short wavelength area). A combination of different wavelengths (colours) appears as white light. Photosynthetically Active Radiation or PAR contains wavelengths useful for photosynthesis, between 400 nm and 700 nm.

Other light-absorbing systems in the plant are responsible for developmental changes through the plant's life cycle. Blue light, with wavelengths around 400 nm, is important for vegetative growth, stimulating leafy growth and sturdy plants, and is involved in the directional growth responses to light (see phototropism p. 70). Red light, with wavelengths around 580 nm to 700 nm controls flowering. For successful plant growth, therefore, artificial lighting contains a mix of red and blue wavelengths which aim to mimic sunlight, optimizing overall increase in plant material (through photosynthesis) and correct development from early growth through to flowering and fruiting (through other light-absorbing systems).

Temperature

The complex chemical reactions that occur during the formation of carbohydrates such as glucose from water and carbon dioxide require the presence of chemicals called **enzymes** to accelerate the rate of reactions. Without these enzymes, little chemical activity would occur. Enzyme activity in living things increases with temperature from 0°C to 36°C, and ceases at around 40°C when the enzymes break down irreversibly. This pattern is mirrored by the effect of air temperature on the rate of photosynthesis which increases with increasing temperature up to an optimum (this varies with plant species from 25°C to 36°C) above which it slows again. At high temperatures, stomata may close to reduce water loss (see p. 123) thus preventing carbon dioxide uptake, while at very high temperatures leaves may be damaged and photosynthesis ceases altogether. As with light, in temperate countries, low winter

9

temperatures slow down photosynthesis and therefore slow the growth rate at this time of year.

To provide the optimum temperature conditions, glasshouses can be heated in winter using a range of methods such as thermostatically controlled electric heaters or paraffin burners together with insulation such as bubble wrap. To reduce temperatures which are too high, shading using blinds, netting or washes applied to the glass, ventilation or damping down (applying water to floor surfaces to evaporate) can be employed.

Water

Water is required in the photosynthesis reaction but this represents only a very small proportion of the total water taken up by the plant. Water supply through the xylem is essential to maintain leaf turgidity and retain fully open stomata for carbon dioxide movement into the leaf. In a situation where a leaf contains only 90% of its optimum water content, stomata will close to prevent further water loss and will reduce carbon dioxide entry to such an extent that there may be as much as a 50% reduction in photosynthesis. Changes in leaf angle in a wilting plant also reduce light interception. Wilting is most often associated with lack of water but can also be seen in waterlogged plants (see p. 116). A visibly wilting plant will hardly be photosynthesizing at all, therefore it is essential that plants should be supplied with the correct amount of water if the rate of photosynthesis to be optimized.

Mineral nutrients

Minerals are required by the leaf to produce the **chlorophyll** pigment that absorbs most of the light energy for photosynthesis. Production of chlorophyll must be continuous, since it loses its efficiency quickly. A plant deficient in iron, nitrogen or magnesium, especially, turns yellow (**chlorotic**) and loses much of its photosynthetic ability (Figure 9.3).

Variegated leaves (Figure 9.4), where parts of the leaf are pale or even white and are therefore lacking chlorophyll, have a lower rate of photosynthesis overall and therefore the plant will have a slower growth rate. Some variegated plants are prone to 'reversion' where the variegation is lost. The stronger growing green leaves can rapidly take over the plant and should be pruned out as soon as they appear if the variegated form is to be retained (see p. 266).

Supplying the correct balance and amount of nutrients through application of fertilizers is especially important in the vegetable plot and in lawns where plants are

Figure 9.3 Chlorosis due to magnesium deficiency in a lemon

Figure 9.4 Variegated leaves in *Pittosporum tenuifolium* 'Silver Queen'

harvested or the grass mown and removed as in these situations nutrients are not returned to the soil. Similarly plants in containers will eventually deplete the nutrients in the growing medium, so these will need to be replaced by feeding.

The law of limiting factors

The **law of limiting factors** states that the factor in least supply will limit the rate of photosynthesis. If, for example, carbon dioxide levels fall in an enclosed system such as a glasshouse, the rate of photosynthesis will decrease. In this situation, increasing light levels or temperature will not be advantageous if the carbon dioxide levels remain low. Similarly, if there is adequate carbon dioxide and a suitable temperature but light levels are reduced, as in the winter, there will be no advantage in increasing carbon dioxide levels or the temperature without giving additional light. In fact, if any one of these three factors (light, carbon dioxide or heat) is in short supply, then it will limit the rate of photosynthesis even

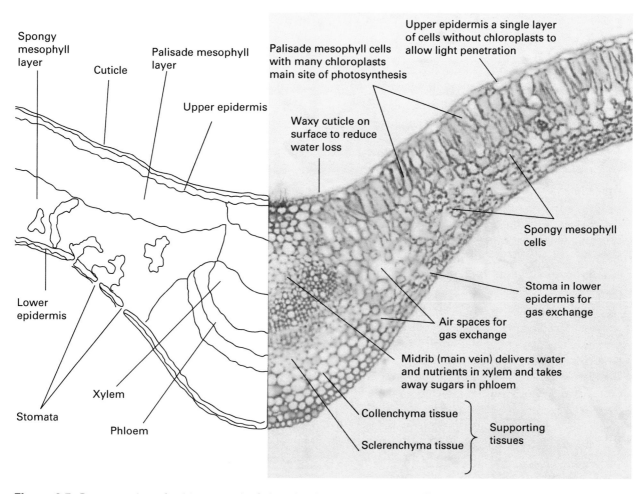

Figure 9.5 Cross-section of a *Ligustrum* leaf showing its structure as an efficient photosynthesizing organ

though the other factors may be plentiful. It would be wasteful, therefore, to increase the carbon dioxide concentration, light or temperature artificially, if the other factors were not proportionally increased.

> The **law of limiting factors** states that the factor in least supply will limit the rate of a process, e.g. photosynthesis.

Leaf structure and photosynthesis

Figure 9.5 shows the structure of the leaf and its relevance to the process of photosynthesis.

The leaf is the main organ for photosynthesis in the plant, and its cells are organized in a way that provides maximum efficiency. The upper **epidermis** is a thin transparent layer, without chloroplasts, permitting transmission of light into the lower leaf tissues. The cylindrical **palisade mesophyll** cells are packed together, pointing downwards, under the upper epidermis. The many **chloroplasts** within these cells absorb light to carry out the photosynthesis process

and can move to the top and bottom of the cells depending on light levels. The **spongy mesophyll**, below the palisade mesophyll, has a loose structure with many air spaces which allow for the two-way diffusion of gases. The carbon dioxide from the air is able to reach the palisade mesophyll and oxygen, the waste product from photosynthesis, leaves the leaf. The numerous **stomata** on the lower leaf surface (positioned here to reduce water loss) are the openings to the outside through which this gas movement occurs. Many small vascular bundles (**veins**) within the leaf structure contain the xylem vessels that provide the water and minerals for the photosynthesis reaction and phloem sieve tubes, for the removal of sugar to other plant parts.

The arrangement of leaves on the plant and the angle at which they are held maximizes light interception, as does the large surface area and thin structure of the leaf. A newly expanded leaf is most efficient in the absorption of light, but this ability reduces with age, so leaves may be constantly shed and replaced within a plant's life cycle, all at once in the case of

9

deciduous plants or at the end of each leaf's useful life for evergreen plants.

Respiration

In order that growth can occur, food manufactured by photosynthesis must be broken down in a controlled way to release the energy which was trapped from sunlight for the production of useful substances such as cellulose, the main constituent of plant cell walls, and proteins for enzymes. This energy is also used to fuel cell division and the many chemical reactions that occur in the cell. **Respiration** is the process by which these sugars and starches are broken down to yield energy, releasing carbon dioxide and water as waste products. It takes place in the **mitochondria** of the cells and is often referred to as 'cellular respiration'.

In order that the breakdown is complete, and the maximum energy is released, **oxygen** is required in the process of **aerobic respiration** (Figure 9.6).

The energy released by aerobic respiration is stored in a chemical form in a substance called **ATP**, which can be transported to wherever energy is needed in the cell, in effect acting like a 'battery'. Some energy is also released as heat, similar to the cellular respiration which produces heat in our bodies to keep us warm.

The energy requirement of cells within the plant varies: reproductive organs can respire at twice the rate of the leaves for example. Also, in apical meristems, the processes of cell division and cell differentiation require high inputs of energy to create new cells.

It would appear at first sight that respiration is the reverse of photosynthesis. This is correct in the sense that photosynthesis creates glucose as an energy-harvesting strategy, and respiration breaks down glucose as an energy-releasing mechanism. It is also correct in the sense that the simple equations representing the two processes are mirror images of each other. It should, however, be emphasized that the two processes have two notable differences. The first is that respiration in plants (as in animals) occurs in all living cells of all tissues, in leaves, stems, flowers, roots and fruits. Photosynthesis occurs predominantly in the palisade mesophyll tissue of leaves but not in other plant tissues such as the roots. Second, respiration takes place continuously whereas photosynthesis only operates when light is present, it cannot happen in the dark.

> **Respiration** is the process by which sugars are broken down to yield energy, the end products being carbon dioxide and water.

Factors affecting respiration

Two environmental factors which affect the rate of respiration are:

▶ oxygen
▶ temperature.

Oxygen

Oxygen is essential for **aerobic** respiration as it is needed to break down the carbohydrates to release

glucose + oxygen ⟶ carbon dioxide + water + energy

(ATP and heat)

Figure 9.6 Equation for aerobic respiration

the energy stored in them. It is analogous to the need for oxygen when a fire burns, where the energy stored in the fuel (trapped from the sun by plants millions of years ago) is released as heat.

In the absence of oxygen, inefficient **anaerobic** respiration takes place in the cytoplasm of cells (see Figure 9.7). Incomplete breakdown of the carbohydrates produces alcohol (ethanol) as a waste product, with much energy still trapped in the molecule. If a plant or plant organ such as a root is supplied with low oxygen concentrations, such as in a waterlogged or compacted soil, or an overwatered pot plant, the consequent alcohol production within the cells may prove toxic enough to cause root death. Furthermore, the small amount of energy released is insufficient for growth, repair or reproduction and only enables the plant to 'tick over' until aerobic respiration can be restored.

Figure 9.8 *Lysichiton americanus* (skunk cabbage) is a plant adapted to waterlogged soils and has many air filled spaces in its root and stem tissue (aerenchyma) which interconnect and enable oxygen to reach its roots

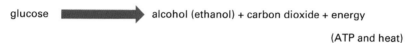

glucose ⟶ alcohol (ethanol) + carbon dioxide + energy

(ATP and heat)

Figure 9.7 Equation for anaerobic respiration

Sometimes, anaerobic conditions can be advantageous. For the plant it enables it to survive periodic inundations which would otherwise be fatal. For gardeners, the viability of stored seeds can be greatly increased if they are stored in a 'modified atmosphere' that is within sealed, airtight packets (see Figure 5.3). The oxygen is removed as the seeds respire and carbon dioxide levels rise which combine to reduce the respiration rate and inhibit germination, enabling longer storage.

In fruit and vegetable storage, inhibition of respiration is desirable to prevent produce going beyond the ripening stage and into senescence (see p. 74) with accompanying loss of quality. In addition, once harvested, photosynthesis ceases but respiration continues so carbohydrates will continue to be broken down and the dry weight of plants or produce will be reduced. Packaging may provide a 'modified' atmosphere similar to that of the seed packet with low oxygen and high carbon dioxide levels within. For some fruits such as apples, 'controlled atmosphere' storage in sophisticated large-scale airtight stores enables fine control of oxygen and carbon dioxide levels, which, along with temperature control, can extend the storage times well into the following year.

Temperature

As in photosynthesis, many of the reactions involved in respiration involve enzymes. The rate of respiration therefore shows a similar pattern of increasing rate with increasing temperature up to an optimum, beyond which the rate decreases. Plants adapted to high temperatures have a higher optimum than those from temperate or cooler regions.

In some horticultural situations, a high rate of respiration is desirable – for example, in propagation of cuttings or seed germination where new cell growth needs plenty of energy, heat may be given to speed up respiration (see Chapter 11). Alternatively, where low respiration rates are required, such as in seed storage (see 'Seed Storage and Respiration' on the companion website) or to delay ripening and senescence (see p. 74) in stored produce such as fruit and vegetables or cut flowers, temperatures are reduced, sometimes with control of the gaseous atmosphere as well. Growers, distributors, retailers and consumers have developed a 'cool chain' to keep produce at consistently low temperatures, between 0°C and 10°C along the supply chain, reducing waste and enabling longer shelf life through reducing respiration rates. As well as reducing respiration, cold storage can have other benefits. Cuttings stored at low temperature root more readily later, while strawberry runners kept in cold stores over winter maintain their quality and are also stimulated to flower the following year.

9

The balance between photosynthesis and respiration

The relationship between photosynthesis and respiration is crucial for both plants and growers. Photosynthesis converts carbon dioxide into sugars whereas respiration does the reverse. If the rate of photosynthesis is too low, then all the carbon 'fixed' may be lost again in respiration, leaving no surplus sugars for growth. The point at which photosynthesis exactly matches respiration, that is, there is no net gain in carbon, is called the **compensation point**. In temperate regions, low winter light levels mean that plants are operating below the compensation point, which is why growth rates are low and many plants 'shut down' for the winter, shedding their leaves and becoming dormant. Growers aim to provide light levels which keep photosynthesis rates above the compensation point, otherwise their crops will not grow sufficiently and yields will be low.

Further reading

Brown, L.V. (2002) *Applied Principles of Horticulture.* Butterworth-Heinemann.

Capon, B. (1990) *Botany for Gardeners.* Timber Press.

Hodge, G. (2013) *RHS Botany for Gardeners.* Mitchell Beazley.

Ingram, D.S., Vince-Price, D. and Gregory, P.J. (2008) *Science and the Garden.* Blackwell Science.

 Please visit the companion website for further information:
www.routledge.com/cw/adams

CHAPTER 10
Level 2

Transport in plants

Figure 10.1 Apple (Cox on M1 rootstock) excavated at 16 years to reveal distribution of roots. Note the vigorous main root system near the surface, with some penetrating deeply (source: Dr E.G. Coker)

This chapter includes the following topics:

- Diffusion and osmosis
- Uptake of water
- Water movement in the plant
- Transpiration
- Mineral nutrient uptake and movement in the plant
- Sugar movement in the plant

Principles of Horticulture. 978-0-415-85908-0 © C.R. Adams, K.M. Bamford, J.E. Brook and M.P. Early. Published by Taylor & Francis. All rights reserved.

Plants, like people, are dependent on the transport of substances around their structure. Sugars made in the leaves by photosynthesis must be transported to roots, stems, shoots and flowers where they can be used for immediate growth or stored until they are required for future growth. Water is essential for many processes in the plant and, since it is absorbed through the roots, it has to be moved to cells in other plant organs where it is essential for support, as a medium for chemical reactions and as a raw material for processes such as photosynthesis. The two 'superhighways' along which sugars and water flow, the phloem and xylem, respectively, are also responsible for moving other vital substances such as essential mineral nutrients and plant hormones around the plant body. In this chapter we examine how water is moved both through the plant and from cell to cell and how it is lost from the leaves. The transport of sugars and mineral nutrients is also described.

The structure of xylem and phloem is discussed in Chapter 6.

Water

As **water** is the major constituent of any living organism, then the maintenance of a plant with optimum water content is a very important part of plant growth and development. There is a tendency sometimes to overwater, but probably more plants die from lack of water than from any other cause (see p. 151).

Movement of substances in the plant

Diffusion and osmosis

Two ways in which substances move in the plant are described below:

▶ **Diffusion** is a process whereby molecules of a liquid or a gas move from an area of high concentration to an area of lower concentration of the diffusing substance. For example, sugar in a cup of tea will diffuse through the tea without being stirred – eventually! (Figure 10.2a) Examples of diffusion in the plant include the movement of gases such as water vapour (see transpiration p. 123), carbon dioxide and oxygen (see photosynthesis and respiration, Chapter 9) into and out of the leaf. Osmosis is a special kind of diffusion where water is the diffusing substance (see below).

▶ **Osmosis** is defined as the movement of **water** from an area of high water (low solute) concentration to an area of lower water

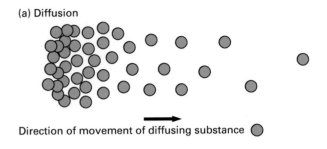

(a) Diffusion

Direction of movement of diffusing substance ⊙

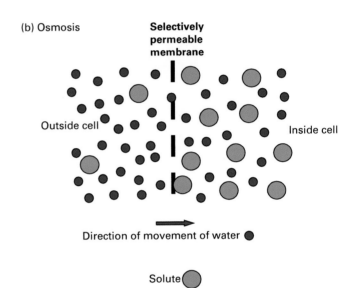

(b) Osmosis

Selectively permeable membrane

Outside cell

Inside cell

Direction of movement of water ●

Solute ⊙

Figure 10.2 Diagrammatic representations of: (a) diffusion; (b) osmosis

concentration (higher solute concentration), through a **selectively permeable membrane**, such as the cell membrane (Figure 10.2b). Osmosis is in effect the diffusion of water across a membrane and is the method by which water enters cells. A selectively permeable membrane allows passage of some dissolved substances but not others and the term 'solute' refers to the substances dissolved in the water.

> **Diffusion** is the movement of a substance from a high concentration to a lower concentration. **Osmosis** is the movement of water from a high water (low solute) concentration to a low water (high solute concentration) across a selectively permeable membrane.

When water moves into the cell by osmosis it swells like a balloon. This inner pressure, caused by the water pushing outwards, is called **turgor pressure** and is very important in providing support to young plants and non-woody herbaceous plants. The plant and

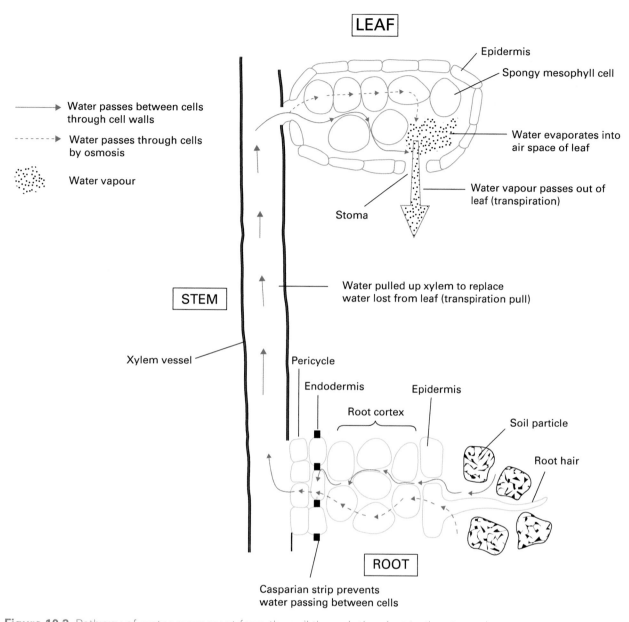

LEAF

Epidermis

Spongy mesophyll cell

Water passes between cells through cell walls

Water passes through cells by osmosis

Water vapour

Water evaporates into air space of leaf

Water vapour passes out of leaf (transpiration)

Stoma

Water pulled up xylem to replace water lost from leaf (transpiration pull)

STEM

Xylem vessel

Pericycle

Endodermis

Epidermis

Root cortex

Soil particle

Root hair

ROOT

Casparian strip prevents water passing between cells

Figure 10.3 Pathway of water movement from the soil through the plant to the atmosphere

its individual cells stay upright like a stack of inflated balloons. Turgor pressure is also the way in which new cells enlarge, contributing to growth by causing the cell to swell until the cell wall prevents further expansion. Without the cell wall the cell would explode!

The pathway of water movement through the plant falls into three distinct stages:

▶ water uptake from the soil by the **roots**
▶ movement up the **stem** in the xylem
▶ movement across the **leaves** and loss to the air by transpiration.

Water uptake from the soil

Soil water enters the root passing through the **cell walls** then across cell membranes and into cells by

osmosis. Whereas the cell wall is permeable to both soil water and its dissolved inorganic minerals, the cell membrane freely allows water through but is selective about passage of other dissolved molecules, somewhat like a sieve. Sucrose, for example, is too large to cross the membrane. A greater concentration of dissolved substances such as mineral and sugars (solutes) is usually maintained inside the cell compared with the soil water outside the cell. The water concentration outside the cell will therefore be greater than inside the cell so water will move in by osmosis (Figure 10.2). The greater the difference in concentrations of water, the faster water moves into the root cells.

If water is not available at the roots (see permanent wilting point p. 154) or if the plant is losing water from the leaves faster than it can be replaced, water

10

121

Figure 10.4 Reversible wilting: (a) the *Primula* is wilting because the water in the pot is frozen (physiological drought) – the roots are unable to replace the water lost by the leaves so turgor pressure is lost; (b) a few hours later the water in the pot has thawed, turgor pressure is restored and the plant has recovered

will cease to enter the cell and will eventually start to move out of the cells. **Turgor** pressure is lost and the plant will **wilt** (Figure 10.4). This will not be a problem if the water supply is only reduced temporarily, the plant will be able to recover once water loss is reduced or water supply is restored. However, if this continues, the cells may then become **plasmolysed**, a situation where the cell contents shrink away from the cell walls leading to irreversible damage and cell death. Plasmolysis can also occur if there is a build-up of salts in the soil or, for example, where too much fertilizer is added causing **root scorch**. Water moves out of the root cells because the solute concentration is greater outside the cell than inside. Such situations can be avoided by applying the correct dosage of fertilizer to soils (and leaves where patches of plasmolysed cells appear as leaf scorch if foliar feeds are too concentrated (see p. 173)).

Most of the water uptake takes place in the **root hair zone** where the root surface area is greatly enlarged (see p. 81). As well as passing into the root hair cells, water also flows through the cell walls surrounding them.

Functions of water

The plant consists of about 95% water, which is the main constituent of **protoplasm** or the living matter of cells. When the plant cell is full of water, or turgid, the pressure of water enclosed within a membrane or vacuole acts as a means of **support** for the cell and therefore the whole plant, so that when a plant loses more water than it is taking up, it may wilt. Aquatic plants are supported largely by external water and have very little specialized support tissue. In order to survive, any organism must carry out complex chemical reactions, such as photosynthesis and respiration, described in Chapter 9. Raw materials for these chemical reactions must be transported and brought into contact with each other by a suitable medium; water is an excellent **solvent**, that is, many substances are able to dissolve in it. One of the most important processes in the plant is **photosynthesis**, and a small amount of water is used up as a raw material in this process. Water may also be used for **seed**, **fruit** and **pollen dispersal** in aquatic plants and in more primitive plants such as mosses and liverworts, water is needed for **reproduction**.

Movement of water in the roots

It is the function of the root system to take up water and mineral nutrients from the growing medium, and the system is constructed accordingly, as described in Chapter 6. Detailed work by Dr E.G. Coker showed the extent of the root system in a mature apple tree, as revealed in the image in Figure 10.1.

Water enters the plant through the root primarily though the root hairs (Figure 10.3). Initially it crosses the epidermis either entering the root hair cells by osmosis or passing between the epidermal cells through the relatively porous cell walls. It then meets the **cortex** layer, which is often quite extensive, and moves across it to reach the transporting tissue that is in the centre of the root (see p. 80). Water movement is relatively unrestricted as it moves through the intercellular spaces and the lattice work of cell walls, although some will also pass by osmosis from cell to cell too (see p. 120).

The central region, the **stele**, is separated from the cortex by a single layer of cells, the **endodermis**, which has the function of controlling the passage of water and minerals into the stele. A waxy strip forming part of the cell wall of the endodermal cells (the Casparian strip) prevents water from moving between the cells and all the water now has to pass across the endodermal cell membranes and into the cells by osmosis. The cell membrane also acts as a control point for mineral uptake as only certain minerals are able to cross it. Water passes through the endodermis and pericycle to the **xylem** tissue (see p. 79), which transports the water and dissolved mineral nutrients up the stem to the leaves.

Root anatomy is described in Chapter 6.

Movement of water up the stem

Water is 'sucked' up the **xylem tissue** of the stem (Figure 10.3) by a process called **transpiration pull**, that is, as water is lost from the leaves it is replaced by water which is drawn up the stem carrying dissolved minerals with it (see below). The evaporation of water from the cells of the leaf means that in order for the leaf to remain turgid, which is important for efficient photosynthesis, the water lost must be replaced by water in the xylem. Pressure is created in the xylem and water moves up through the stem and leaf petiole by suction as long as the water forms a continuous column. If the water column in the xylem is broken, for example when a stem of a flower is cut, air moves into the xylem and may restrict the further movement of water when the cut flower is placed in a vase of water, much like an air bubble in a straw. However, by cutting the stem under water the column is maintained and water continues to enter and pass up the plant.

Stem anatomy is described in Chapter 6.

> **Xylem tissue** transports water and dissolved mineral nutrients from the roots, up the stem to the leaves and other plant organs.

Movement of water in the leaf

On reaching the leaf, water is distributed in the fine network of veins and passes out of the xylem (Figure 10.3). It flows between the leaf cells and also passes from cell to cell by osmosis as in the root. Eventually it evaporates from the cell surfaces into the air spaces of the leaf mesophyll. From here, the water vapour

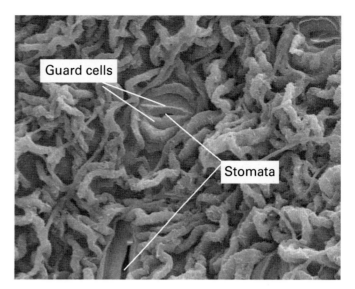

Figure 10.5 Scanning electron microscope image showing stomata on the surface of a *Betula pendula* (birch) leaf. Each small pore (stoma) is surrounded by a pair of guard cells (source: V. Vaslap and T. Jarveots)

diffuses out of the leaf into the surrounding air (see Figure 9.5) through the **stomatal pores** because there is a lower relative humidity in the surrounding air compared with inside the leaf. The loss of water vapour from the leaf is called **transpiration**.

> **Transpiration** is the evaporation of water vapour from the leaves and other plant surfaces.

Transpiration

Any plant takes up a lot of water through its roots – for example, a tree can transport about 1,000 litres (about 200 gallons) a day. Approximately 98% of the water taken up moves through the plant and is lost by transpiration; only about 2% is retained as part of the plant's structure, and a yet smaller amount is used up in photosynthesis. The seemingly extravagant loss through leaves is due to the unavoidably large pores in the leaf surface (**stomata**), essential for carbon dioxide uptake for photosynthesis. Stomata (singular: stoma) each consist of a central opening or pore surrounded by two sausage-shaped guard cells which control their opening. When the guard cells are fully turgid the stomata remain open whereas loss of turgor in the guard cells causes them to close to reduce water loss (Figure 10.5).

A remarkable aspect of transpiration is that water can be pulled ('sucked') such a long way to the tops of tall trees. Engineers have long known that columns of water break when they are more than about 10 m long, and yet tall trees such as *Sequoiadendrom*

10

Figure 10.6 (a) Hairy leaves of *Plectranthus argentatus* which appear silver; (b) water storage leaves of the jade plant (*Crassula ovata*)

giganteum, the giant redwoods, pull water up 100 metres from ground level. This apparent ability to flout the laws of nature is probably due to the small size of the xylem vessels, which greatly reduce the possibility of the water columns collapsing.

A number of environmental factors affect the rate of transpiration:

▶ **Humidity**. If the air surrounding the leaf becomes very humid, then the rate of diffusion of water vapour will be much reduced and the rate of transpiration will decrease. In contrast, **moving air** around the leaf will reduce the surrounding humidity so will increase the rate of transpiration. Windbreaks can help reduce windspeed and the risk of plants drying out.

▶ **Temperature** affects the rate at which water in the leaf evaporates and diffuses and thus determines the transpiration rate. As temperature rises, diffusion is speeded up, so the rate of transpiration increases. At high temperatures though, the stomata will close and transpiration will cease.

▶ **Light** affects the rate of transpiration due to the response of stomata to light levels. At night, for example, stomata close to conserve water loss and transpiration is reduced. In the daytime, stomata open to allow uptake of carbon dioxide for photosynthesis and transpiration increases. Shading may be used in glasshouses to reduce temperature and light levels in the height of summer, thus reducing the transpiration rate and preventing water loss. Similarly cuttings are kept out of direct sunlight to reduce transpiration until they develop roots and are able to take up water more efficiently.

Structural adaptations to the leaf occur in many species to enable them to reduce transpiration and withstand low water supplies. For example, conifer needles such as in *Pinus sylvestris* (see Figure 4.9) have a reduced surface area and stomata sunk into the leaf. Many evergreens such as holly (*Ilex aquifolium*) have a very thick, waxy cuticle, while other plants have many leaf hairs which trap humid air close to the leaf as in the felted leaves of sage (*Salvia officinalis*) and *Plectranthus argentatus* (Figure 10.6a). Others have leaves which store water, such as the succulent jade plant (*Crassula ovata*) shown in Figure 10.6b. In extreme cases (e.g. cacti), the leaf is reduced to a spine (see Figure 7.20f) and the stem takes over the function of photosynthesis and is also capable of water storage.

Mineral nutrient uptake and movement in the plant

Essential mineral nutrients are inorganic substances necessary for the plant to grow and develop (see Chapter 14). They are dissolved in the soil water and are taken up when this is absorbed by the root. At the endodermis, nutrients have to cross the cell membrane (see above) and since the concentration of nutrients inside cells is almost always greater than the concentration in the soil water, uptake is **against a concentration gradient**. Nutrients therefore cannot enter the cell by simple diffusion and have to be taken in by a process called **active transport** which requires energy. This uptake is also **selective**, the plant only absorbs the mineral nutrients it requires and rejects others.

Essential mineral nutrients are inorganic substances necessary for the plant to grow and develop.

> **Active transport** is the movement of a substance into a cell across the cell membrane against a concentration gradient. It requires energy and substances are taken up selectively.

Mineral nutrients are taken up predominantly by the extensive network of fine roots that grow in the top layers of the soil (Figure 10.1). Damage to the roots near the soil surface by cultivation should be avoided because it can significantly reduce the plant's ability to extract nutrients and water. Care should be taken to ensure that trees and shrubs are planted so their roots are not buried too deeply, and many advocate that the horizontally growing roots should be set virtually at the surface to give the best conditions for establishment. Over-enthusiastic hoeing of weeds can damage crop roots near the soil surface and also cause increased loss of water from the soil by bringing more moisture to the surface.

Having crossed the roots, mineral nutrients are transported up the xylem to the leaves and are also redistributed in the phloem to other plant organs such as flowers and fruits.

Movement of sugars in the plant

Phloem tissue (see Chapter 6) is responsible for transporting sucrose from the leaves as a food supply for the production of energy through respiration in plant cells. Unlike xylem, where the direction of flow is always from the root to the leaves, sugars can flow in the phloem sieve tube cells both up and down the plant moving to the plant organs where it is needed, such as growing points, shoots, roots, flowers, fruits or storage organs. Moving the sugars in the phloem, like the uptake of minerals in the roots, is an **active process** requiring energy, as the concentration of sugars in the phloem is much greater than in the leaf cells. The **companion cells** which accompany each **sieve tube cell** are thought to control this process. The flow can be interrupted by the presence of disease organisms such as club root (see p. 256).

> **Phloem tissue** transports sugars made in photosynthesis from the leaves to other plant organs where it is used in respiration to release energy or stored as starch for later use.

Further reading

Capon, B. (2005) *Botany for Gardeners*. Timber Press.

Hodge, G. (2013) *RHS Botany for Gardeners*. Mitchell Beazley.

Ingram, D.S., Vince-Price, D. and Gregory, P.J. (2008) *Science and the Garden*. Blackwell Science.

Lack, A.J. and Evans, D.E. (2005) *Instant Notes in Plant Biology*. Taylor & Francis.

Stern, K.R., Bidlack, J.E. and Jansky, S.H. (2011) *Stern's Introductory Plant Biology*. McGraw Hill

10

CHAPTER 11
Level 2

Plant propagation

Figure 11.1 Nursery

This chapter includes the following topics:

- propagation from seeds
- seed germination
- harvesting and storing different types of seed
- growing seeds in containers
- growing seeds in open ground
- propagation of ferns
- vegetative propagation
- methods of vegetative propagation

Principles of Horticulture. 978-0-415-85908-0 © C.R. Adams, K.M. Bamford, J.E. Brook and M.P. Early.
Published by Taylor & Francis.

Figure 11.2 Containers in a garden centre

Gardeners are familiar with the choices they have available to add new plants to their gardens. They commonly collect the plant they want in a container from a garden centre (Figure 11.2). This will have been created in a nursery from seed or by vegetative means such as divisions, cuttings, bulbs or grafting. Many undertake to grow their plants from these sources themselves especially when it comes to annuals, such as vegetables and bedding plants.

A plant's life cycle is detailed in Chapter 5. The length of time plants live for may be a few weeks (ephemerals) or hundreds of years (many woody perennials such as trees), but before it dies the plant ensures continued life by either sexual or asexual reproduction. Many plants employ both methods to produce offspring. Sexual reproduction leads to the formation of seeds in higher plants (see p. 101). The ability of plants to reproduce asexually is made use of in horticulture and is known as vegetative propagation (see pp. 91–96).

Propagation from seeds

The internal and external structure of seeds is detailed in Chapter 8.

> **Sexual reproduction** is the formation of new individuals through fusion of male and female gametes (sex cells). It results in variable offspring.

Sexual reproduction provides a means of ensuring **variation** in a species. For the grower this makes it an important source of new cultivars and commercial hybrids. Other horticultural benefits of propagating plants from seed include:

▶ it is the only method of propagation for some species such as annuals generally e.g. *Phacelia*

tanacetifolia (see green manuring p. 165) and sterile plants
▶ for many species, it is possible to get large numbers from each plant
▶ it can be easily stored
▶ it can be a means of avoiding virus transmission (see p. 264)
▶ guaranteed seed sources are available
▶ provenance can be carefully sourced.

> **Provenance** is the place of origin.

It is generally considered a cheap method where seeds are easily available, such as native trees and shrubs. The main horticultural uses of seeds include annuals notably vegetables such as carrots (*Daucus carota*), French beans (*Phaseolus vulgaris*), tomatoes (*Solanum lycopersicon*); decorative flowers such as *Nigella damascena*; bedding (*Lobelia erinus*); and the production of lawns (*Lolium perenne, Poa pratensis, Festuca rubra*).

Some limitations of propagating plants from seed should be noted:

▶ some plants may not produce viable seed or viability, especially from some sources, is poor
▶ lack of uniformity (i.e. there is variation), plants may not breed true
▶ dormancy problems
▶ difficult germination, in the open there can be high loss from the natural 'field factors' (see p. 132)
▶ time to maturity
▶ some seeds do not store easily.

Seed germination

Only **viable seed** will germinate (see p. 66); the seed has to have a living embryo and have the potential to germinate when conditions are right.

> **Seed germination** is the emergence of the young root or radicle through the testa, usually at the micropyle.

> A **viable** seed has the potential for germination when the required external conditions are supplied.

Viable seeds germinate provided with the right conditions regarding:

▶ water
▶ air (oxygen)
▶ temperature suitable for the specific plant (see Table 5.1).

Also, for some, an exposure to light, or, for others, an absence of light (see p. 67). However, there are many plants that do not allow their viable seed to germinate until the seed has had its 'dormancy' broken. Typically dormancy ensures that the seed does not germinate until favourable germination conditions occur in the following spring. **Overcoming dormancy** can be problematic for amateurs and professionals alike; the commonest types of seed dormancy with the usual methods of overcoming them are given in Chapter 5.

> **Dormancy** is the condition when viable seed fails to germinate even when all germination requirements are met.

Once the viable seed is able to germinate, it takes up water leading to processes resulting in the **emergence of the seedling** (see p. 66):

▶ increasing respiration rate
▶ breakdown of food store to release energy
▶ rapid cell division
▶ splitting of the seed coat (testa)
▶ emergence of the radicle.

This is followed by the development of the plumule which reaches the surface by one of two main ways: epigeal or hypogeal germination (see Figure 5.4).

Harvesting and storing seed

Most organisms including plants cannot survive drying out, but many seeds are an exception, which makes them readily storable (e.g. *Phaseolus vulgaris* and *Nigella damascena*). Once dry, these **orthodox seeds** age slowly and the ageing process is slowed even more if they are kept cold as well as dry; sometimes they remain viable for hundreds of years. However, some seed is intolerant of drying and cannot be frozen, so storage has to be in cool, moist conditions and then only for short periods of time. This **recalcitrant seed** (also known as 'unorthodox') includes *Quercus robur* (oak), *Aesculus hippocastanum* (horse chestnut), *Castanea sativa* (sweet chestnut), willow, elm, avocado, mango, rubber, cocoa and many climax species.

Normally immature dry seed will not germinate, but there are some plants where it is collected when well developed but still immature and green such as *Anemone nemorosa*, *Calendula* and *Ranunculus*. This sort of seed germinates best if sown as soon as it ripens.

For most species such as carrot (*Daucus carota*), French bean, kidney bean, haricot bean, (*Phaseolus vulgaris*), *Nigella damascena*, *Lobelia erinus* and *Lolium perenne* the seed can be stored dry and cool (but not in a plastic bag) and then sown at an appropriate time. Such seed can be harvested, cleaned and stored following these guidelines ensuring that **labelling** is in place from start to finish. See the companion website: www.routledge.com/cw/adams.

Dried and cleaned seed can be stored in a range of packaging such as paper bags or plastic boxes. These should be labelled and kept in a dry and cool environment; silica gels can be used to help ensure dryness in the packaging.

The ability of seeds to germinate depends greatly on their storage factors including:

▶ moisture content of seed when put in store
▶ temperature in store
▶ length of time held in store.

Purchasing seeds, especially vegetable and flower seeds, has the advantage of convenience and the protection of the regulations (Plant Varieties and Seeds Act 1964). A check of the date should always be made to ensure that the seeds are from the last seed harvest. The seeds are usually supplied in foil packets. Once opened the seeds deteriorate rapidly so should be sown immediately but, so long as they are kept dry and cold in a resealed packet, most seeds will remain viable for a year and some, often the larger seeds, for many years (see p. 66).

There are difficulties when it comes to seeds from trees or shrubs because there are fewer regulations to protect the buyer. In the preparation of seeds for sale, the drying process used often increases the dormancy effect (harder coats), adversely affects the energy reserves and damages the embryo, so reducing seed viability (see p. 129).

Growing seeds in containers

The ideal conditions for raising plants from seed can be achieved in a protected environment such as a glasshouse or cheaper alternatives such as polythene tunnels or cold frames.

Containers

Most seeds grown in protected culture are sown into **containers** (Figure 11.3):

Figure 11.3 Range of containers for growing plants: (a) traditional clay pots; (b) standard seed tray and half tray; (c) standard plastic pots in range of sizes, compared with (d) 'long toms' and (e) half pots; (f) biodegradable pots; (g) compressed blocks; (h) square or (i) round pots in trays; (j) various 'strips' in trays; (k) typical commercial polystyrene bedding plant tray

▶ seed trays
▶ half trays
▶ standard pots (as deep as wide)
▶ pans ('half pots')
▶ 'long toms' – longer pots, i.e. deeper than the standard size.

These must have adequate drainage to allow excess water out or for the water in the capillary matting or sand to be drawn up into the compost. Square shapes utilize space better, but it is harder to fill properly in the corners. Although more expensive, rigid plastic trays are easier to manage. Rims on containers give yet more rigidity and make them easier to stack and clean. Gardeners can make use of plastic food containers so long as they are given sufficient drainage holes. All containers should be clean before use (see hygiene p. 194). There are also disposable pots made of compressed organic matter, paper or synthetic 'whalehide' through which roots will emerge which makes them

useful for the planting out stage. For production horticulture, where cost and presentation of the plants becomes the main consideration, there is a wider range of materials including polystyrene for once-only use.

Too large a container is a waste of compost and space, whereas one that is too small can lead to the seedlings having to be spaced out before they are ready; if left, they become overcrowded and susceptible to **damping-off diseases** (see p. 254).

Throughout the whole process of sowing seeds, care should be taken to ensure **hygienic** conditions starting with:

▶ clean containers
▶ clean growing environment
▶ 'sterile' growing media (see p. 181).

Sowing density should be appropriate for the size of seed, but a thick layer of seedlings should always be avoided as this makes damping off diseases more likely.

Seed composts

These are commonly equal parts peat or peat alternative and sand or perlite or vermiculite mixes. They include lime except for 'lime-hating' specimens (see p. 182) and a source of phosphate. Seedlings are transferred to **potting composts**, where the young plants are established, which have a higher proportion of peat or alternatives with lime and a full range and higher concentration of nutrients. Increasingly alternatives to peat are being utilized (see p. 182) and many advocate the use of sterilized loam because it makes the nutrient management easier (see compost formulations, p. 181).

Sowing seeds

The sowing of medium seed such as *Lactuca sativa* (lettuce) and *Solanum lycopersicon* (tomatoes) is shown on the companion website. After the container has been filled with compost it is lightly firmed to just below the rim of the container using an appropriately sized presser board. Seeds are then sown on the surface at the rate recommended. Many advocate that when using trays, half the seed is sown then, to achieve an even distribution, the other half is sown after turning the container through 90 degrees. The seeds are then covered with sieved compost or fine grade vermiculite to their own thickness. Fine seeds such as *Begonia semperflorens* are sown in equal parts of fine dry sand to help achieve an even distribution. They are lightly pressed into the surface and then left uncovered. Large seed such as *Cucurbita pepo* (courgettes) or pelleted seed tends to be 'space or station sown', that is, placed at recommended distances in a uniform manner.

The seed containers should be labelled with name of plant and the sowing date. The compost is then watered from a can fitted with a fine rose. Water is more gently delivered with the holes of the rose pointing upwards. Standing pots in water is likely to create a problem because the compost becomes fully saturated and air is driven out when there is a need for oxygen. The popularity of perlite or vermiculite to cover seeds is largely because they offer moist conditions whilst allowing oxygen to reach the developing seeds and seedlings.

The moist conditions around the seed must be maintained, which is most easily done by covering with a sheet of glass, clear plastic or kitchen film. The container should be kept in a warm place (approximately 20°C). If necessary, a sheet of paper can be used to shade the seeds from the direct

Figure 11.4 Propagator. These can range from the very simple ones providing just a humid atmosphere to those providing some ventilation and warmth

sunlight and to minimize temperature fluctuations. There are advantages in placing the seed containers in a closed propagator (Figure 11.4).

Covers on the container should be removed as soon as the seedlings appear and they must now be well lit to avoid **etiolation** (see p. 68), but not exposed to strong sunlight. Watering must be maintained but without waterlogging the compost.

In production horticulture, much of the work is done by machines. Pots are rarely filled by hand and increasingly the whole process is automated including the seed sowing.

Pricking out

When the seedlings are large enough to be handled, they should be transplanted into a potting compost prepared as for the seed tray. Each seedling is eased with a dibber, lifted by the seed leaves, dropped into a hole made in the new compost, gently firmed and watered in. The seedlings are normally planted in rows with space, typically 24 to 40 per seed tray, for them to grow on to the next stage (Figure 11.5)

Many will be planted out when ready but others will continue in containers so they are '**potted off**' (moved into another container) to give them more space. They are then '**potted on**' when they outgrow this container or planted out. As they are moved to the next size container, it is a mistake to go to a much larger container, the compost, especially the nutrient level, is chosen to suit the stage reached (see JIP p. 181).

Hardening off (acclimatization)

This is required to ensure that the seedlings raised in a protected environment can be put out into the open

Figure 11.5 Bedding plants in nursery

Figure 11.6 A frameyard: a collection of cold frames each covered with 'lights' which are sheets of glass (or clear plastic) in a frame

ground without a check in growth caused by the colder conditions, wind chill and a more variable water supply. As the pricked-off seedlings become established, they are moved to a cooler situation, typically a cold frame (Figure 11.6), which starts off the process of hardening off by providing a closed environment without heat. After a few weeks the cold frame is opened up a little by day and closed at night. If tender plants are threatened by cold or frosts, they can be given extra protection in the form of easily handled insulation such as bubble wrap or coir matting put over the frame. Watering has to be continued and usually the plants will use up the fertilizer in the compost so need applications of liquid fertilizer. The hardening process then continues with the frame lid ('light') gradually being opened up more to allow air circulation day and night. Ideally the plants have been fully exposed to the outdoor conditions by the time it is ready to plant them out.

The young plants are very susceptible to fungal diseases while in the cold frame because of the high density of planting and the difficulties with keeping

the humidity level right. The need to maintain air circulation is essential as the opportunity arises. Excessive feeding with high nitrogen fertilizers should also be avoided because it can create soft growth which makes them vulnerable to disease and excessively soft and vigorous plants can be checked (their growth interrupted) when planting out.

Growing seeds in open ground

The success of sowing outdoors depends greatly on preparing the seedbed; the **tilth** needs to be matched to the type of seed, soil texture and the expected weather conditions (see soil structure p. 147).

> **Tilth** is the crumb structure of the seedbed.

Weeds need to be dealt with by creating a stale (or false) seedbed, hoeing or using weedkillers (see p. 194). Nutrients, especially phosphate fertilizer, are worked in and the ground levelled to receive the seed. The process of sowing seeds in the open can be by broadcasting or sowing in drills (hand or machine) and is shown on the companion website.

The **sowing rate** will depend on the species and the likely losses that can be estimated from:

▶ **seed viability** (see p. 128)
▶ **germination percentage** – a measure of potential emergence under ideal conditions; the proportion of seeds that produce healthy seedlings under laboratory conditions
▶ **field factors** – seed emergence under field conditions, i.e. allows for quality of the seedbed and weather conditions.

Sowing dates in the open vary according to the plants concerned (see germination temperatures, p. 68) and the location where they are to be grown. The general rule is not to sow before the soil warms up to the appropriate temperature.

Aftercare

There are advantages in providing protection for the developing plants including windbreaks or fleece (see Figure 11.7). Where residual herbicides are not used there needs to be ongoing control of emerging weeds while they are in competition with the seedlings and young plants (see p. 208). If the seedbed was well watered then there should be no further need to add more water; indeed there are advantages in not doing this in terms of water conservation (see p. 156) to encourage deeper rooting and the prevention of soil capping (see p. 150).

Figure 11.7 Protection for plants sown outdoors: (a) windbreaks, natural hedging or artificial structures; (b) fleece or similar covering

Propagation of ferns

Ferns produce spores rather than seeds (see p. 52). These spores develop in spore cases (sori) to be found on the back of some of their fronds (see Figure 8.15) from which the spores can be collected when ripe (usually dark brown). The propagation of ferns from spores is described on the companion website.

Vegetative propagation

Many plant species use their ability for vegetative propagation in their normal pattern of development. The production of these vegetative **propagules**, as with the production of seed, is to increase numbers and provide a means by which the plant survives

adverse conditions. The energy storage for this purpose often makes them attractive as food for us such as potatoes, onions and carrots.

> **Asexual reproduction** is the formation of new individuals without fusion of gametes resulting in genetically identical offspring.

> **Vegetative propagation** involves asexual reproduction and results in a clone.

> A **propagule** is any part of a plant that can be used to create a new plant.

Natural vegetative propagation includes bulbs, corms, rhizomes, stolons/runners, suckers, stem and root tubers (see pp. 91–96 and Figures 7.21, 7.18, 7.16, 7.15h, 7.14g and 7.15g). **Artificial vegetative propagation** is also used in horticulture to produce numbers of plants from a single parent plant including divisions, layers and cuttings. Success is greatly increased by taking material from **juvenile** plant material (see pp. 70–72). In commercial horticulture, **stock plants** are used to provide large amounts of propagating material which are managed to ensure they remain juvenile.

> **Juvenile growth** is non-reproductive (vegetative) growth, whereas **adult growth** is reproductive (flowering).

All the plants from a single parent, a **clone**, have the same genetic characteristics. This is important for gardeners, but other reasons for using natural or artificial vegetative propagation include:

▶ avoids seed dormancy problems
▶ quicker to reach maturity compared with growing from seed
▶ there are some cultivars that can only be reproduced by vegetative means.

However, it should be noted that there are several limitations of propagating plants by vegetative means including:

▶ lack of variation
▶ transmission of disease (e.g. viruses see p. 264)
▶ limited availability of material
▶ different skills to seed sowing
▶ requirement for different propagation environments, e.g. use of mist units, cold frames
▶ stable clones need to used, i.e. ones that are unlikely to revert.

Methods of vegetative propagation

All vegetative propagation is a form of division and has been exploited by mankind for a very long time. In many cases little more than breaking up the plant or taking the natural propagules is involved e.g. bulbs, corms, rhizomes and runners. However, there are several techniques which are less straightforward including taking cuttings of many kinds.

Divisions

Most gardeners will be familiar with dividing **herbaceous perennials** such as hardy geraniums, asters, *Achillea*. This usually arises because the shoots become overcrowded and the thick clumps that develop often become woody or have bare centres (Figure 11.8).

Iris has **rhizomes** (see p. 92) which lend themselves to dividing. Normally about 10 cm of non-woody rhizome is cut off with leaves that are reduced to a third to minimize water loss and wind rocking when re-planted. Borders are rejuvenated by carefully lifting the clumps ('crown') of herbaceous plants preferably with a generous ball of soil. Most of the soil is teased off carefully to maximize the amount of root for the new plant. Pieces from the clump are removed each with a shoot, bud and roots. The unsightly woody or bare middle section is discarded. Whilst some specimens can be pulled apart by hand, some are so tough as to require knives. Splitting some specimens can be very difficult and, if necessary, they can be prised apart with back to back forks for good leverage. Others such as *Phormium* might need an even more rugged approach. However, care should to be taken not to cause damage to the propagating material. The fleshy rooted specimens such as *Hostas* need to be treated very carefully. In general, division can be undertaken in the autumn as plants die down, but is usually better done in the spring as the new shoots appear. The younger sections with strong shoots can be replanted in prepared ground (normally with many surplus pieces to give away or sell). This should be done before the roots have dried and the plants then watered in. However, there is usually a best time and method for each type of plant and this is summarized on the companion website.

Some species such as *Bergenia* flower in the early spring so it is best to lift the plants in mid-winter and remove the rhizomes. These should be washed and the dormant buds found. Sections with a bud can be taken off and rooted by burying them horizontally to half their depth in potting compost in trays. As for propagation generally, it is advantageous to provide

Figure 11.8 Iris with bare centre

'bottom heat' by standing the trays on soil warming cables. The plants are not ready to go out until the fibrous roots have emerged from the bottom of the tray and have been '**hardened off**'. This acclimatization to the outdoor temperature is usually achieved by two or three weeks in a cold frame during which time ventilation is gradually increased until the covers are left off even through the whole night.

Layering

Layering is method of propagation that uses the ability of some plants to produce roots from stems still attached to the mother plant (see companion website).

Simple layering is a cheap means of producing large new plants quickly, and in the nursery trade, it remains a good way to produce high-value specimens such as *Hamamelis*. It is also useful for species that are otherwise difficult or slow by other methods such as *Cotinus* spp. and *Magnolia* spp. It is usually carried out in the spring (March to early June). Vigorous shoots are **wounded** (by cutting half through or a by giving a sharp twist) to interrupt the sap flow and induce new root cells, leading to the development of adventitious roots. These are pegged down around the mother plant where they can root into loosened soil with wire pegs. Each stem should then be covered at the wounded area with fine crumbs of soil (5 to 10 cm deep) to prevent drying out. The remaining stem should be tied to a cane to help produce an upright specimen. All but the slowest (e.g. *Magnolia*) can be lifted with a fork and cut free of the mother plant when dormant in the autumn or left to the following spring. Typically these can be bedded out in a sheltered area. In nurseries, they would be containerized and grown on in shade tunnels.

Figure 11.9 Air layering on tree

Serpentine layering is essentially the same as simple layering but several new plants are created from each shoot, that is, wounded, pegged down and covered in more than one place. Care needs to be taken to ensure that there is at least one bud exposed and one bud covered with soil for each section. This method works well for vine-type species such as *Vitis* spp. (grape) and *Clematis* spp.

French layering is another variation on the theme which yields yet more new plants, but success tends to be limited to vigorous specimens such as *Cornus* spp. This method depends on pegging down stems on the soil surface which, by the following spring, have new stems growing upwards along their length. Mounds of soil crumbs can then be placed at the base of each new stem to encourage rooting at the node (junction) and treat as for serpentine layering.

Air layering can be used when the stem of the subject such as *Ficus elastica* (Rubber plant) which, despite its name, is not flexible enough to bend down to the soil. This is usually done in the spring. The main stem or a branch (at least of pencil thickness) can be **wounded**. This area is wrapped with moist vermiculite which is kept moist and held in place with polythene secured with twist ties or similar; more details are given on the companion website. This method is often used for many hardy plants because they are particularly difficult to root by other means (Figure 11.9).

Cuttings

Cuttings are parts of plants that have been carefully removed from the parent plant and used to produce a new plant. Different methods may be needed for different species. Factors to consider are:

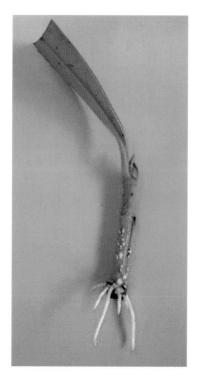

Figure 11.10 Rooted cutting

▶ where to make the cuts (e.g. below a node or internodal)
▶ which composts to use (e.g. neutral or ericaceous see p. 181)
▶ the environment for the different stages (e.g. temperature, humidity)
▶ the use of rooting hormones or not.

Cuttings are taken from juvenile growth and with vascular cambium (see p. 79) that gives rise to adventitious ('wound' roots). Only healthy parent plants should be used and the hygienic use of knives, compost and containers is strongly recommended (see p. 194).

Rooting is usually more reliable from younger plants and when taken when the parent plant is turgid. In general, rooting is more successful when cuttings are provided with 'bottom heat' but when their aerial parts are kept cool to reduce foliage growth and water loss. These special conditions ('cool top, warm bottom') are provided by propagation benches in a greenhouse or heated propagators (see Figure 11.4).

Stem cuttings can be taken from stems that have attained different stages of maturity: softwood, semi-ripe or hardwood.

Softwood cuttings are taken in spring from stems before there is any woodiness. They are used for the propagation of a wide range of species such as *Fuchsia*, *Pelargonium* and *Chrysanthemum* (Figure 11.10). This material is very vulnerable to drying out so collect from well-watered parent

Figure 11.11 Mist propagation unit

Figure 11.12 Hardwood cuttings

plants, keep in polythene bags and use as quickly as possible.

The area of leaf on these cuttings should be kept to a minimum to reduce water loss. Misting (spraying the plants with fine droplets of water to increase humidity and reduce temperature) can further reduce this risk by slowing down the transpiration rate. Automatic misting (Figure 11.11) employs a switch attached to a sensitive device that assesses the evaporation rate from the leaves.

For plants such as *Chrysathemum* spp. it is often better for amateurs and small growers with limited propagation space to obtain rooted cuttings from specialist propagators because then all that is required is an area of reduced light where high humidity (misting) can be maintained for 2–3 days until root growth begins. If they cannot be 'stuck' immediately, the cuttings can be stored in a refrigerator (at 1–5°C) for several days.

Leaf cuttings include leaf lamina and leaf petiole cuttings. There are several ways by which leaf lamina cuttings may be taken. One method suitable for *Begonia rex* is to take whole leaves or just small squares of the leaf and pin down so the veins make good contact with a free-draining compost in a tray, keeping them humid and at 20°C; plantlets develop along the length of the veins in 6–8 weeks. A similar method is used for *Streptocarpus* cvs but the whole

leaf is split along the midrib, which is removed, and the cut edges are inserted into trays of compost.

Leaf petiole cuttings are taken from *Peperomia* spp., using mature leaves, and *Saintpaulia* cvs, with full-grown young leaves, with 5 cm of leaf stalk. This is inserted into the compost; rooted plantlets can be detached from the petioles after about eight weeks.

Semi-ripe cuttings are taken from stems that are just becoming woody (partially mature wood of the current season's growth); typically, in mid-summer to early autumn. They are used for many broadleaved evergreen shrubs and some conifers such as *Chamaecyparis* spp. These cuttings should be 5–10 cm long with the base firm but the tip soft active growth. Rooting in a sand/compost mixture may be achieved in cold frames or, more quickly, in a heated structure at about 18°C. Some species such as *Elaeagnus* spp. (oleaster) will root only if heat is provided.

Heeled cuttings are preferred for many shrub and tree species (e.g. *Ceanothus*, holly and conifers), by which the semi-ripe cutting is taken in such a way that about a centimetre sliver of the previous year's wood, the **heel**, is still attached. The heel cambium facilitates root formation and, hence, easier establishment. However, it is a method which is not attractive for commercial production because it removes basal buds from the mother plant and in so doing prevents any future regrowth.

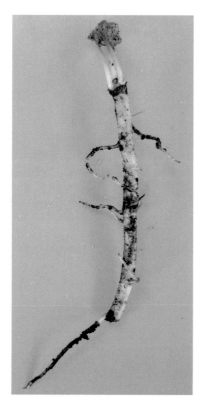

Figure 11.13 Rooted root cutting of *Acanthus spinosus*

Leaf-bud cuttings comprise a short piece of semi-ripe stem, an axillary bud (see p. 85) and the leaf (or leaves) at this node. Several cuttings can be made from each stem which is cut with sharp secateurs between nodes, such as for *Clematis* spp. and *Lonicera* spp. (honeysuckles), or, for difficult species such as *Camellia* spp. (susceptible to rotting), just above one node and below the lower one.

Hardwood cuttings are from pieces of dormant woody stem containing a number of buds, which grow out into shoots when dormancy is broken in spring. Some plants such as *Salix* spp. and *Populus* spp. root so easily that little more is needed than pushing the well-ripened wood of the current season into the soil. The more usual procedure, such as for *Buddleja* spp., *Cornus* spp. (dogwoods), *Deutzia* spp., is for the cuttings to be taken in late autumn. The base of the cutting is cut cleanly at 45° to expose the cambium tissue from which the adventitious roots will grow; the top is cut at right angles to the stem to give a length of 15–25 cm. These are usually placed with half their length immersed in deep container with a cuttings compost (Figure 11.12) or in a Dutch Roll (see companion website). In *Hydrangea* spp. and *Ribes* spp. (currants), the stems show evidence of pre-formed adventitious roots (root-initials) that help the process of root establishment. A 12-month period is often necessary before the cuttings can be lifted.

Root cuttings are used for species such as *Papaver orientalis* cvs (poppy), *Primula denticulate* cvs, *Phlox paniculata*, *Acanthus spinosus* and *Anchusa azurea* (alkanet). For most species, young roots about a centimetre in thickness are removed close to the crown of the plant in late autumn/winter. These are cut into 3–5 cm lengths and inserted vertically into a cutting compost. It is important that these cuttings are not put inadvertently upside-down as this prevents establishment; the convention is that the top is cut off square whereas the bottom cut is made on the slant. However, the thinner-rooted species such as *Phlox* are placed horizontally in a tray then covered with a thin layer of compost.

These root cuttings are lightly watered and put in a cold frame. When there are signs of growth, the root development can be checked carefully and potted on individually when well rooted (Figure 11.13).

Further information on vegetative propagation can be found on the companion website: www.routledge.com/cw/adams.

Further reading

Hartman, H.T. et al. (1990) *Plant Propagation, Principles and Practice.* Prentice-Hall.

Lamb, K. et al. (1995) *Grower Manual 1 – Nursery Stock Manual.* Grower Books. Swanley.

Macdonald, B. (2006) *Practical Woody Plant Propagation for Nursery Growers.* Timber Press.

McMillan Browse, P. (1999) *Plant Propagation.* 3rd edn. Mitchell Beasley.

Toogood, A. (ed.) (2006) *RHS Propagating Plants.* Dorling Kindesley.

Toogood, A. (2002) *Growing from Seed.* Dorling Kindersley.

Toogood, A. (2003) *Plants for Cuttings.* Dorling Kindersley.

Please visit the companion website for further information:
www.routledge.com/cw/adams

CHAPTER 12
Level 2

Physical properties of soil

Figure 12.1 Rocks

This chapter includes the following topics:

- Plant requirements
- Origin of soil
- Soil formation
- Topsoil and subsoil
- Composition of soils
- Soil texture
- Soil structure
- Cultivations
- Soil water
- Drainage
- Applying water
- Water conservation

Plant requirements

As gardeners, most of us are familiar with plants, our specially selected ones and the weeds, doing well in one place but not in another. Indeed, in some situations, our preferred plants perform very badly. Often the difference in their performance can be attributed to the soil. It can be because of the nutrient problems and these are dealt with in Chapter 14. Quite often it is simply the physical properties of the soil that provide the challenge. Although out of sight, roots play a vital role in supplying **water** and **nutrients** to the plant. They must access a large enough volume of soil to supply the plant's needs and reach a depth that helps to maintain a water supply as the surface layers dry out. Within a full season, a single plant growing well in open ground develops some 500–1,000 km of root, with most plants penetrating at least half a metre below the surface. This vast root system is usually more than is required to supply the plant in times of plenty, but the extent of the network is indicative of what is needed in unfavourable conditions. It also serves to remind us of what we undertake to provide when we restrict root growth either accidentally in soils or deliberately when we grow plants in containers (see Chapter 11).

Plants do not grow until the growing medium is warm enough: usually above 5°C for most temperate plants and 10°C for those from tropical areas; optimum temperatures are in Table 5.1. Growing media also need to be free of harmful substances, which can include some nutrients applied in excess quantities.

The growing tip of the root wriggles through the growing medium following the line of least resistance. Roots are able to enter cracks that are, or can be readily opened up to, about 0.2 mm in diameter, which is about the thickness of a pencil line. Once into these narrow channels, the root is able to overcome great resistance to increase its diameter. Compacted soils severely restrict root exploration, which in turn limits plant growth. When this happens, action should be taken to remove the obstruction to root growth or to supply adequate air, water and nutrients through the restricted root volume.

The root ball normally provides the **anchorage** needed to secure the plant in the soil. Plants, notably trees with a full leaf canopy, become vulnerable to the effect of the wind if their roots are in loose material, in soil made fluid by high water content or are shallow, such as roots over rock strata close to the surface. Until their roots have penetrated extensively into the surrounding soil, transplants are very susceptible to wind rocking; the plant may be left less upright unless secured. Furthermore, water uptake remains limited after transplanting until the delicate root hairs damaged in the process are replaced and 'plugged in' to the soil.

In order to grow and take up water and nutrients, the root must have an energy supply. Efficient energy production is only possible if **oxygen** is brought to the site of uptake (see respiration, p. 116). Consequently, the soil around the root must contain air as well as water. To ensure the supply of oxygen is constantly replenished, and for the carbon dioxide to be taken away, there needs to be good gaseous exchange between the atmosphere around the root and the soil surface. This is usually achieved by managing the soil particles in the rooting zone, so the spaces between them contain air as well as water. A lack of oxygen or a build-up of carbon dioxide will reduce the roots' activity. Furthermore, in these conditions anaerobic bacteria will proliferate and produce toxins such as alcohols. In warm summer conditions, roots can be killed after just a few days in waterlogged soils. There are plants that can grow successfully in waterlogged soil, or in water, because they have adaptions that make that possible, such as aerenchyma (see p. 79). Plants can also grow very successfully in water if the water around the roots is kept oxygenated (see hydroponics p. 183). A naturally **fertile** soil or a well-managed one provides the plant with:

▶ water
▶ air (oxygen)
▶ nutrients
▶ anchorage.

Origin of soil

Most soils in Britain and Ireland are mineral soils, that is, derived from **rocks** (Figure 12.1). Many of the remainder are peats which are made solely from organic matter, that is, dead plants. The mineral soils are the product of three types of weathering: physical, chemical and biological activity.

Weathering of rocks

Soils form in the layers of rock fragments over the Earth's surface. The parent rocks that provide the mineral material for soil formation are weathered by physical, chemical and biological forces.

Parent rock is the rock from which a soil is made.

Weathering is the breakdown of rocks.

Figure 12.2 Weathering and erosion by water. Note that the faster moving water on the outside of the bend removes soil and rocks (weathering) and carries these until the water slows down (erosion). The larger stones that were carried downstream when the stream was in spate are left on the inside of the bend where the water is moving more slowly. Note the meandering stream has cut into the alluvial soil deposited across the valley floor in earlier times

> **Erosion** is the movement of rock fragments and soil.

Physical weathering

Water. Moving water, whether in streams, rivers or the sea, is able to carry particles in it and the faster it moves, the more it can carry. This leads to the rocks it flows over being abraded. As the speed of the water increases it is able to carry disproportionally more both in total quantity and in size; large boulders can be bounced along in water that is in spate. It is at these times when most of the scouring of rock occurs, producing large quantities of debris which is further ground up in the moving water (Figure 12.2).

Wind. In a similar way, the wind carries abrasive particles that 'sandblast' exposed rock; typically this happens in hot dry climates where sand particles are readily picked up.

Heat. Also in hotter regions, rock is broken down as they are heated up in the sun. The rock surface expands while being attached to cooler rocks setting up strains within the material. This leads to surfaces breaking away; the 'onion skin effect'.

Frost. In temperate areas, it is the action of frost that does much of the weathering. Water that gets into any cracks or when taken up in porous rocks such as chalks and limestones expands when it freezes. The pressures set up to accommodate the expanded water cracks the rock and, like the same effect on frozen pipes, the thaw reveals the damage done.

Glaciers. Ice in a glacier tends to stick to the adjacent rock, but as the glacier moves downhill, the enormous forces involved pluck rock away and gradually the glacier becomes a huge mix of rock debris and ice. As this moves downhill, the large rocks embedded in the ice scour away the rock with which it comes into contact; the 'scrubbing brush' action.

Figure 12.3 Lichen on rock. The respiration of this and other organisms living on or near rocks increases the carbon dioxide levels. This generates more carbonic acid that reacts with some of the minerals in the rock and leads to its disintegration

Chemical weathering

Many of the chemicals that make up the rocks react with the elements around them. Some are soluble in water and others react with oxygen to form new compounds. Rainfall is one of the main causes of rock weathering because the water combines with carbon dioxide in the air to form **carbonic acid**. This dissolves away chalk and limestone and reacts with many other minerals making up rock, resulting in their disintegration.

Biological weathering

This is the action of all organisms. Plants and animals, large and small, lead to rocks being broken up from the patches of lichens growing on rocks (Figure 12.3) to the mining and quarrying by mankind that creates rock debris in which soil forms.

It is the respiration process of aerobic organisms that provides the carbon dioxide (see p. 116) which, when combined with water, forms the carbonic acid. This is particularly concentrated in the plant rootzone. It is particularly significant in the continued breakdown of rock, rock fragments and soil particles that tend to be protected against physical weathering because they are down below the soil surface. Roots also contribute physically by growing into cracks and opening them up; most gardeners come across this sort of damage that can be done to paths and driveways.

The fragments from the weathered rock are the basis of soils that form in the loose material over the Earth's surface. The type of soil formed and how useful it is for the growing plants depends greatly on the proportions of the different sizes of these particles, which are familiar to us as clay, silt, sand,

Table 12.1 Soil particle sizes

Soil Particle	Diameter (mm)
Stones	greater than 2
Coarse sand	0.6–2.0
Medium sand	0.2–0.6
Fine sand	0.06–0.2
Silt	0.002–0.06
Clay	less than 0.002

stones, boulders and so on. In the study of soils, these have been classified according to their diameter (Table 12.1). The weathering process also shapes the particles, which are normally angular with sharp edges when first formed but become more rounded as they are weathered further, such as river washed pebbles, or wind blown sand.

Distribution of rock fragments

The loose material produced by the weathering of rocks is likely to be eroded, that is, moved to other places. The main agents of erosion are gravity, water, wind and glaciers.

Gravity

Loose rock fragments on slopes are moved downhill. On steep slopes they fall under the effect of gravity to form heaps of rock, '**scree**', where they come to rest (Figure 12.4). On gentler slopes material is gradually moved downhill under the influence of rainsplash

Figure 12.4 Scree. The rock weathered on steep hillsides falls under the influence of gravity. The eroded rock continues to break down to smaller and smaller pieces by continued physical and chemical weathering

Figure 12.5 Soil creep on hillsides. On the gentler slopes, gravity causes the loose rock debris and soil particles disturbed by rainsplash to move gradually downhill. This process slows as the particles are covered with grasses but continues creating the characteristic striations on the hillside

(Figure 12.5) leading to deep layers of debris at the base of slopes.

Water

As the water with its load of rock fragments slows down, it is unable to carry as much. Initially, it is the larger particles that drop to the bottom first. Typically, it is in the streams that boulders, pebbles and coarse sands can be seen, whereas the fine sands are not dropped until the water is in the slower-moving rivers. Rivers in flood spread out over a valley bottom or across a plain beyond their banks so when the water level goes down, layers of fine sand and silt are left behind along with much organic material. This gives rise to the fertile 'alluvial' soil (see Figure 12.2). Much of the silt is still in the very slow-moving river until approaching the discharge point and is dropped around estuaries. The clay fraction is made up of such small particles that it stays suspended in the water to be carried out into a lake or out to sea where it gradually settles in still waters.

Wind

The wind also carries particles that are small enough, typically sands in hot, dry conditions. Again the larger particles are dropped first as the wind slows down or when the particles touch water. This leads to large areas of sandy material in which form the 'loess' soils.

Glaciers

Large areas of Britain and Ireland were covered by glaciers in, geologically, recent times. When the glaciers melt, they drop their load, which comprises everything from boulders to finely ground material. The soils that develop in this 'till', also known as 'boulder-clay', are very variable and much is not easy to cultivate.

Soil formation

Initially there is only a shallow layer of rock fragments that remains subject to erosion until something starts to grow there. Usually it is the lower plants such as mosses, algae and lichens (see p. 52) that are able to colonize such inhospitable areas and in doing so they modify the effect of water and wind erosion. The vegetation that develops tends to protect the loose material by covering it and by reducing wind speeds. The reduced wind speed also leads to particles being dropped from the air. Along with this additional material, the soil deepens as the parent rock continues to be broken up below. Furthermore, organic matter is added as plants die, which decomposes to release plant nutrients. A hole dug in such a 'sedentary' soil reveals the characteristic horizons that gradually emerge: typically an organic-rich litter layer at the surface with different horizons down to the unweathered parent rock at the bottom (Figure 12.6). Initially this may be only a few centimetres deep but over time, as the original plants are succeeded by higher plants with roots, eventually shrubs and trees, the soil profile can become more than a metre deep. Their roots bind the soil and generally reduce further erosion. However, the soils that form on the steeper slopes tend to remain very shallow.

> **A soil horizon** is a specific layer in the soil seen by digging a soil pit.

'Transported' soils form in essentially the same way, but in material that comes from another area. This means they are not made up of the rock found underneath the site but rather from particles that have been eroded, that is, transported, often from many miles away (Figure 12.6). **Alluvial soils** form in the material brought by rivers and **loess** in the wind-blown deposits. Average agricultural soils develop in much of the variable '**boulder clay**' left behind by the glaciers; very little of which makes good horticultural soil unless substantially improved or if used to grow crops that need very little cultivation, such as orchards.

Organic matter is added at the surface – for example, leaves, dead annuals and perennial 'tops' giving rise to the 'organic layer' (also known as the 'litter layer'). This is food for many small animals such as earthworms that start the decomposition process and incorporate

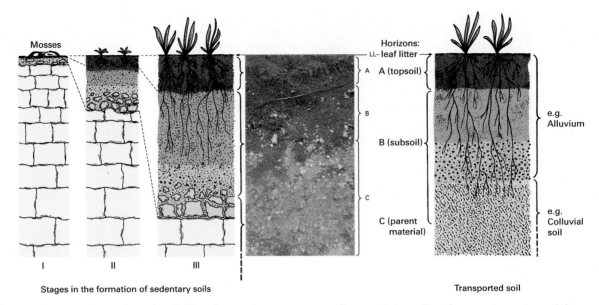

Figure 12.6 The formation of soil. The change from a young soil comprising a few fragments rock particles to deep sedentary soil is shown alongside a transported soil. Subsoil, topsoil and leaf litter horizons can be identified in each soil. Simple plants such as lichens and mosses establish on rocks or fragments to be succeeded by higher plants as soil depth and organic matter levels increase

much of it into the layers below (see primary decomposers p. 38). It is also a source of nutrients for the next generation of plants. The natural development is for there to be a steady decline in the organic matter levels away from the surface until none is found below the root zone. This can be seen because the organic matter gives rise to a black 'jelly' (see humus p. 158) that coats all the particles. However, once a soil has been cultivated, the disturbance of the top layers gives rise to a distinct boundary between light-coloured **subsoil** and the darker **topsoil**. While this spreads the beneficial effect of organic matter to cultivation depth, it should be noted the disadvantage is that it reduces the concentration of it at the surface.

Topsoil and subsoil

There are several important differences between topsoil and subsoil that affect growing plants. The differences are largely because of the topsoil being nearer the surface (see Table 12.2).

Topsoil is the uppermost layer of soil normally moved during cultivation. It is typically 10–40 cm deep and darkened by the decomposed organic matter it contains.

Subsoil is the layer below the cultivated layer and lighter in colour because of its low organic matter level.

The topsoil is richer in organic matter because most of the roots are near the surface, particularly in the top 15–30 cm. The plant remains that fall on the surface are taken down into the top layers by earthworms and many types of insect. The presence of undecomposed organic matter and the burrowing of earthworms tend to keep the topsoil 'open', that is, a high proportion of air spaces. The topsoil is exposed to more extreme effects of weather such as freezing and thawing, wetting and drying whereas the subsoil tends to be buffered from such. As it is near the surface, topsoil is within the range of cultivating equipment but this also makes it vulnerable to compaction by feet or wheeled traffic. All the dead organic matter that accumulates there is a source of food that attracts a vast population of living organisms (see Chapter 13). The topsoil is richer in nutrients as a result of the decomposition of the organic matter or because fertilizer is added at or near the surface of the soil. The colouring of **subsoil** depends greatly on the chemicals coating the particles, commonly iron oxide. In well-drained soils, the iron oxide present is in an oxidised form, giving the soil a rusty brown colour, but in waterlogged (anaerobic) conditions the iron oxide is reduced to a form that has a grey or bluish tone. Subsoils tend to accumulate the finer soil particles washed down from above, making them 'heavier', that is, with a higher clay content. This accumulation and the reduced biological activity (especially roots and earthworms) lead to poorer water and air movement.

Table 12.2 The main differences between topsoil and subsoils in typical garden soil in Britain and Ireland

Characteristic	Topsoil	Subsoil	Further details
Colour ***	Dark brown/black; lighter if high chalk content	Light browns or grey	see p. 144
Texture		Higher in the finer particles especially clay	see p. 146
Organic matter content (%):	2–5	<1	see p. 158
Living	Enormous numbers especially near the surface Many roots	Comparatively low numbers Few roots	see p. 158
Dead organisms	Large quantity	Very little	see p. 158
Humus	Present	None	see p. 158
Pore space	Naturally more 'open' Can be maintained near 50% + Can be improved by cultivating	More compressed Too deep for cultivation**	see p. 144 see p. 144
Aeration	High proportion of large pores so good aeration	Limited large pores so poorer aeration*	see p. 144
Water content	Depends mainly on soil texture but improved by the OM content	Depends on soil texture	Table 12.3 see p. 152
Nutrient content	In nature, the site of nutrients Enriched by addition of fertilizers	Low nutrient content * Below main feeding roots	see p. 168 see p. 172
Suitability for plants	Can be ideal, primary source of nutrients and water	Poor*, but important water reserve	
Effects of weather	Exposed to extremes of freezing and thawing; wetting and drying	Protected from the extremes of weathering by topsoil	

Note:

* Avoid mixing with topsoil.

** Cultivation depth normally considered to be a spade depth ('a spit') or plough depth.

*** Colours depend on the coating on the particles; most commonly iron oxide which exists in two forms: when plenty of oxygen present, it is in the ferric oxide form (rust); but if waterlogged, has limited oxygen supplies during soil development so the ferrous oxide form prevails (grey, bluish). The degree of whiteness because of the chalk in the soil depends on how much is present. Humus on the particles darkens the soil.

Composition of soils

Although we tend to think of soil particles as 'too small to see', there are huge size differences between coarse sand, fine sand, silt and clay. This has a significant effect on root penetration, water-holding properties and aeration (Figure 12.7). If a basketball is used to represent the size of fine sand, then coarse sand would be size of a room in a house, whereas marbles would represent silt particles and sugar grains the largest of the clay particles.

Sand

Sand grains are soil particles between 0.06 and 2.0mm in diameter.

The shape of the particles varies from gritty to more weathered, rounded grains ('soft'). Colour varies

according to the iron oxide coating from very pale yellow to rich, rusty reddish brown. Silver sand has no such coating. Sand grains are inert; they neither release nor hold on to plant nutrients. They are not sticky.

Soils dominated by coarse sand are usually free draining but have poor water retention, whereas those composed mainly of fine sand can hold much larger quantities of water against gravity. All the water held on all sand particles is readily removed by roots.

Clay

Clay particles are those less than 0.002 mm in diameter.

Clay particles tend to be platelets that pack together closely. Their combination of very small size and chemical characteristics makes clay soil sticky when

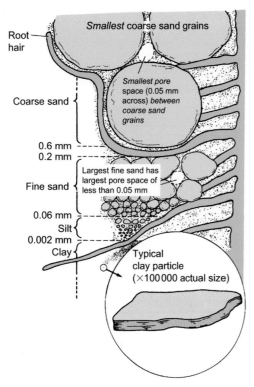

Figure 12.7 Relative sizes of sand, silt and clay (based on SSEW classification) with root hairs drawn alongside for comparison. Note that even the smallest pore spaces between unaggregated spherical coarse sand grains allow water to be drawn out by gravity and so allow some air in at field capacity, whereas most pores between unaggregated fine sand grains remain filled

wet and hard when dry. Unless there are cracks between these blocks of packed clay particles, water movement is very restricted. However, many types of clay shrink on drying so introduce cracks. Alternatively, cracks are introduced by cultivation or the action of soil organisms (e.g. earthworms) to allow gravitational ('excess') water to leave the rooting zone.

These clay particles are also porous, so hold water inside as well as on their surface. Consequently, clay-rich soils are able to hold on to a lot of water (see Table 12.3) but some of it is too tightly bound to be released to plant roots (about 15%).

Clay continues to weather and release plant nutrients especially potash. It also has the ability to hold on to some nutrients in such a way that they remain available to plants but protected against being leached (washed down the profile) below the reach of roots and eventually into water courses. This property gives clay the ability to maintain nutrient levels in the soil solution even when being taken up by plants. In contrast, sandy soils are less well buffered against the loss of nutrients by the leaching from the root zone.

Silt

> **Silt** particles are those between 0.002 and 0.06 mm in diameter.

Most particles in this size range are inert and non-porous like sands, so they behave like very fine sand. They have good water-holding capacity and plants can take up a high proportion of this water (see Table 12.3). However, some of the particles (usually around 15%) have the properties of clay, so soils dominated by silt do yield some nutrients. Similarly, they can hold on to some nutrients like clay does, but to a lesser extent.

Stones and gravel

> **Stones** are particles larger than 2 mm in diameter.

Particles bigger than sand are commonly known as grit, gravel, pebbles, cobbles and boulders, according to size and shape. The effect of stones on cultivated areas depends on the type of stone, their size and the proportion in the soil. In general, they make soils difficult to cultivate. Digging is harder and the spade is more easily blunted. They have detrimental effects on mechanized work: tines and tyres are worn more quickly especially if the stones are hard and sharp such as broken flint. Stones interfere with drilling of seeds and the harvesting of roots. Close cutting of turf is more hazardous where there are protruding stones. A high proportion of stone reduces the fertility of the soil as it dilutes the soil components that supply nutrients and hold the water. The amount of soil that roots can explore is reduced according to the volume of stone present.

Soil texture

Soil texture describes the mineral composition of a soil. In most cultivated soils the mineral content forms the framework and exerts a major influence on its characteristics.

> **Soil texture** can usefully be defined as the relative proportions of the sand, silt and clay particles in the soil.

A soil dominated by

▶ sand particles is called '**a sand**' and feels 'gritty' or abrasive

- silt particles is called 'a silt soil' and feels 'silky' (or 'soapy') when wetted
- clay particles is called 'a clay' and feels 'sticky' when wetted.

A loam is an idealized soil for growing because its proportions of sand, silt and clay particles are such that none of their individual properties (grittiness, soapiness and stickiness) is evident. Soils like this are relatively easy to manage. In practice, when the soil is wetted and moulded between the fingers, it is usually possible to detect at least a little

- grittiness, making it a sandy loam
- soapiness, indicating a silt loam
- stickiness, indicating a clay loam.

Soil texture can be considered to be a fixed characteristic and provides a useful guide to a soil's potential.

Soil water-holding properties

In general, fine-textured soils such as clays, clay loam, silts, silty loam and fine sands have good water-holding properties (see p. 152) but poor water movement unless improved by cultivation and/or the addition of organic matter. In contrast, coarse-textured soils (coarse sands and coarse sandy loams) have good water movement but low water-holding capacity unless improved by addition of organic matter.

Soil temperatures are closely related to soil texture because water has a much higher specific heat value than soil minerals, i.e. it takes more energy to heat up wet soil than the same volume of dry soil. Consequently, well-drained coarse sands warm up more quickly in the spring compared with other soils. Darker soils (e.g. those enriched with organic matter) also warm up quicker than light ones. Conversely, plants growing on darker and drier soils are more vulnerable to frost damage.

Nutrient levels

Soils with high clay content continue to release nutrients as they weather and they have good nutrient retention. In contrast, nutrients are not released from sand because most of it is inert and those that are present in sandy soils are readily lost by leaching. This constant need for nutrients is often met by adding manures (see p. 169), which need frequent replenishment because they decompose rapidly in the well-aerated sandy soils. Hence sandy soils are often referred to as 'hungry soils'.

Ease of cultivation

The differences in the effort required to dig various soil is familiar to most gardeners. The power requirement to cultivate a clay soil is much greater than that for a sandy soil. The expression 'light' and 'heavy' reflects the working properties rather than the actual weight of the soil, that is, for sandy soil requiring only one horse to pull a plough in sandy soil (light) but four horses needed on clay (heavy). The high water content of clays does make them heavier than sands when both are wet.

The texture of a soil also influences the soil structure and cultivations.

Soil structure

In order to provide a suitable root environment for cultivated plants, the soil must be constructed in such a way as to ensure:

- gaseous exchange (oxygen in, carbon dioxide out)
- adequate reserves of water available to plants.

> Soil structure is the arrangement of particles in the soil.

There should be:

- a high water infiltration rate
- free downward movement of water ('drainage')
- an interconnected network of spaces allowing roots to find water and nutrients without hindrance.

There should be no large cavities that:

- prevent good contact between soil and seeds or roots
- dry out seeds or roots.

Soil crumbs

The plant roots and soil organisms live in the spaces (pores) between the solid components of the growing medium (sand, silt, clay, organic matter). In the same way that a house is mainly judged by the living accommodation created by the solid material (bricks, wood, plaster, mortar), so a soil is evaluated by examining the spaces created between the particles. By aggregating the particles correctly, a mixture of pore sizes, allowing free water movement, gaseous exchange and thorough root exploration can be created. This is achieved with a 'crumb soil', which is ideal for most horticultural purposes. The best arrangement of small and large pores for establishing plants is illustrated in Figure 12.8 alongside a 'dusty' tilth with too few large pores and a 'cloddy' tilth that has too many large pores. The interior of a soil crumb is made up of many small pores holding water against the pull of gravity, whereas the bigger gaps between

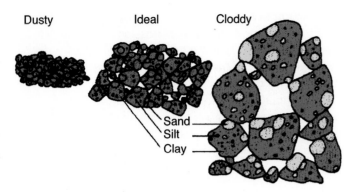

Dusty   Ideal   Cloddy

Sand
Silt
Clay

Figure 12.8 Soil crumbs – the 'ideal' arrangement of small and large pores for establishing plants is illustrated alongside a 'dusty' tilth with too few large pores and a 'cloddy' tilth that has too many large pores

the crumbs can be large enough to allow water to be pulled out by gravity. This means that after being fully wetted and allowed to drain, there will be mainly water in the crumbs and mainly air between the crumbs; ideal for plant growth and for the beneficial organisms living in the soil.

> **Tilth** is the crumb structure of the seedbed.

The fineness of a seedbed should be related to the size of seeds, so it usually consists of crumbs between 0.5 and 5 mm in diameter. Cloddy surfaces lead to poor germination of all but the largest seeds and make weed control difficult. If made too fine, then there will be too few soil pores that hold both air and water. Fine crumbs tend to form a 'soil cap' when wetted (see p. 150).

Soil conditioners such as manure and compost (see organic matter, p. 158) help the soil to form a good crumb structure. When fresh they can 'open up' the soil (i.e. improve aeration and drainage) and the humus created from it improves crumb formation in very sandy soil and in heavy clays. Lime is added to remedy soil acidity, which ensures that the beneficial soil organisms are encouraged, but it also contributes 'calcium' that encourages clay to form crumbs. When clay needs to be improved without raising pH, gypsum (calcium sulphate) can be added to the soil.

Timeliness

The right time for soil cultivation depends on the weather, but more specifically the **soil consistency** (sometimes referred to as the 'workability' of the soil). It influences the timing and effect of cultivations on the soil, the **cultivation window**. Many soils are too sticky when wet; soil clings to the equipment and walking on the ground at this stage damages the soil surface and

the footprints left behind are evidence of compaction. As a soil dries out, it becomes ideal for cultivating: **load-bearing and friable**. If they dry out further, the lighter soils can fragment too easily when raked and produce dusty tilth, whereas heavier soils can become too hard to break down without a great deal of effort.

> **Friable** is the consistency of the soil when it is easily cultivated, i.e. readily forms crumbs.

The delay to work on the land can be reduced by the use of boards to walk on: the pressure on the ground is spread and compaction is minimized. Equipment with very wide tyres (or tracks) does the same thing for mechanized work on a larger scale.

Good crumb structure can be destroyed by cultivating at the wrong time, but also by working the soil too much, that is, 'over cultivation'. Soil crumbs are vulnerable to collapse when wetted especially if they are low in organic matter. Puddles on seedbeds lead to crumbs collapsing and the particles released fill the gaps between remaining crumbs; on drying these can form 'soil caps', which is a crusty layer in the soil that reduces gaseous exchange and can hold back seedlings.

The crumbs at the surface collapse if they stand in puddles and are easily broken by large droplets of water from rainfall or irrigation, particularly when there is no protection from plants or mulching. Care should be taken when applying water (irrigation).

Cultivations

In temperate areas, the conventional preparation of land for planting starts with thorough disturbance of the top 15–30 cm of soil (primary cultivation), usually by digging on small plots and ploughing on larger areas. Mainly rakes or harrows are used to prepare the final surface (secondary cultivation).

Digging

Digging inverts the soil and this arduous activity is undertaken to loosen the soil and bury trash. The land is broken up into clods and an increased area is exposed to weathering. As the soil is inverted, weeds, plant residues and bulky manures are incorporated. The depth of routine digging should be related to the depth of topsoil because bringing up the subsoil reduces fertility in the vital top layers, seriously affecting germination of seeds and the establishment of plants. Single digging, no more than the depth of the spade, is usually undertaken on an annual basis

Figure 12.9 Aeration of turf: (a) by hand with a simple corer; (b) removing cores with machinery

for most crops. This can be hard work, especially on heavy soils, so the choice of equipment is important and the blade should be kept sharp. Smaller and lighter versions of the normal-sized spades (even children's size) are readily available. Double digging is undertaken when a deeper rooting depth is required for long root vegetables or to eliminate a known deep-lying soil structure problem. Again, care is needed to make sure subsoil is not brought into the topsoil horizon. Traditionally this involved the inversion and loosening of two spade depths with the lower subsoil spit being returned to the subsoil zone.

Forking

Forking can be adopted where loosening or breaking up the soil is the major requirement, but when the burying of weeds and trash is not a priority. It does have the advantage of reduced soil inversion and consequently is less arduous. A fork is often used instead of a spade in a modified 'double digging' in which the top spit is inverted by the spade, but the second spit is loosened by forking.

Forks can also be used to improve the structure of soil that cannot be dug over, such as established flower beds, taking care not to damage shallow feeding roots. The fork can be pushed into the turf to improve aeration and the infiltration of water. An improvement on this is the hand-held hollow tine, which is pushed into the ground to extract a 10 cm core of soil (Figure 12.9); pedestrian-powered versions are available for dealing with larger areas.

Raking

Raking is used on roughly prepared ground for removing stones and debris, producing a suitable tilth, incorporating fertilizers and leveling.

There are many types of rake available and suitable ones for cultivating should be selected and used appropriately – for example, if levelling is the prime requirement, then a wide one should be considered. The traditional rake is used in two main ways:

▶ as a means of reducing soil aggregates to the size of the crumbs required for the seedbed. For this the rake should be kept low to the ground with soil pushed back as much as forward. The impact of the tines breaks the clods, which is made much easier if the soil is in a friable condition (see p. 150). Weathering over the winter helps the working of heavier soils.

▶ as a means of removing stones and unwanted vegetation when the rake is acting as a sieve. For this the rake should be pulled through the seedbed held at 45° and in one direction leaving the soil crumbs behind.

If load-bearing is poor, then planks should be used to avoid compaction on the loose soil.

Rotary cultivating

Rotary cultivators are used to create soil crumbs on uncultivated or roughly prepared ground instead of digging, forking and raking. Small pedestrian types are available for use in gardens and on allotments; there are much larger ones for use in nurseries, market gardens and in commercial glasshouses.

The type of tilth produced depends not only on the soil conditions but also on the adjustment of forward speed, rotor speed, blade design and layout, shield angle and depth of working.

Creating the seedbed

Usually sandy soils are easily broken down to the right size with cultivation equipment but there is a risk of 'over cultivating' resulting in 'dusty' tilth. Heavier soils are more difficult to cultivate. Traditionally, clay soils are dug over (or ploughed) in the autumn and exposed to wetting and drying and especially freezing and thawing to produce a 'frost mould' (crumbs). In this way, clods are easier to 'knock down' to a crumb structure and level in the spring. **Timeliness** is essential when cultivating because not only does it make cultivation easier when the soil is **friable** but going on to the soil at the wrong time can damage the good soil structures.

Once the seedbed is created, care is needed to maintain it. Rain hitting the surface causes crumbs to break up further. The particles released fill in the gaps between the broken crumbs, reducing water infiltration rates. As infiltration is reduced, puddles form and the crumbs collapse even more quickly. When the surface dries, a soil **cap** is formed (Figure 12.10) which hinders gaseous exchange and can trap germinating seeds.

In general, fine tilths should be avoided outdoors until well into spring when conditions are becoming more favourable and seedling emergence through any developing cap is rapid. The surface can be protected by using mulches (see p. 160) and a leaf canopy reduces the problem.

'No dig' methods

The traditional preparation of a seedbed, involving as it does the inversion of the soil and creating the appropriate tilth, is very demanding on energy, labour and time. Furthermore, the cultivations tend to interfere with the natural structure-forming agents, such as earthworms (see p. 159), and when undertaken at the wrong time, they create pans or leave a bare, loose soil vulnerable to erosion (see soil structure, p. 147). It also reduces the organic matter levels at the top of the profile where it is most useful. 'No dig' methods have increasingly been adopted in growing in order to eliminate these weaknesses especially as weeds can be controlled with herbicides. In gardens, on allotments and horticulture generally

Figure 12.10 Soil 'capping' is caused by the collapse of the soil crumbs at the surface as a result of being saturated with water and/or hit by raindrops

the need for so much cultivation is avoided by the use of 'bed systems' or growing in containers (see Chapter 15).

Bed systems

The compaction problems can be overcome by cultivating in beds, which confines traffic to well-defined paths between the growing areas. Effective weed control methods, including modern herbicides, have enabled the inversion of soil to be eliminated in more situations than before.

On a garden scale, beds are constructed so that all parts of the growing area can be reached from a path; this eliminates the need to step on the growing area. They can be laid out in many ways, but should be no more than 1.2 m across. The paths should be minimized while allowing access for all activities through the growing season. Beds are usually raised above the normal ground level to improve drainage and ease of working. Much of this extra height results from the addition of large quantities of suitable bulky organic matter. Good topsoil is often taken from this sacrificed path area to be used to raise the beds up higher.

On larger areas, the width of the beds is adjusted to the distance between the wheels of the vehicles used so that the growing area is unaffected by the traffic passing during the life of the plants. The equivalent of this is done in farming and is easily seen from the roadside in fields where cereals are being grown.

Soil water

The healthy growth of a plant requires a constant supply of water, which is taken up through the roots from the growing medium (see p. 121). Problems occur if there is too little water for the plants so watering (irrigation) becomes a consideration. However, if there is too much water which stops us working on the land or harms the plants, the possibility of improving drainage should be considered.

Water infiltration

Most rain falling on the soil surface soaks in, but if it exceeds the rate it can infiltrate, then water accumulates on the surface. This **standing water** (also known as 'ponding' or simply as puddles) leads to soil capping because soil crumbs tend to collapse when wet. Soil surfaces can be protected with mulches (see p. 160) and become less vulnerable as they become covered by plants. As the soil becomes **saturated** (full of water, waterlogged), the standing water (puddles) becomes more extensive and long lasting so damage to soil structure at the surface increases.

> A **saturated** soil is one which has all the soil pores filled with water.

On slopes there is **surface run-off**. Depending on the steepness of the slope and the intensity of the rain, soil can be carried away (erosion) along with seeds, fertilizers and mulches (Figure 12.11). This is wasteful for the gardener, but also it is bad for water courses that receive the fertilizer. Growing on steep slopes is a

Figure 12.12 Terracing is used to reduce 'run-off' and to create flat areas on slopes that are otherwise too steep for cultivation

major problem worldwide and is commonly overcome by the creation of terraces. There is less pressure to do this in Britain and Ireland, but there are situations where it is appropriate (Figure 12.12).

As water soaks into the dry soil, all the pore spaces are filled with water, that is, they become saturated (also known as 'waterlogged'). The roots in this saturated zone can only obtain water from what is dissolved in the water.

> The **saturation point** of a soil is when water has filled all the soil pores (i.e. no air in the pores).

Field capacity

When rainfall ceases, the water in the larger soil pores continues to move downwards under the influence of gravity. As **gravitational water** (sometimes referred to as 'excess water') is removed, air returns in its place bringing with it a fresh supply of oxygen. On sandy soils this may take a matter of hours after the rain has stopped, but far longer on clay where this process may continue for many days. The soil is then said to be at **field capacity** (FC).

> **Field capacity** is the amount of water the soil can hold against the force of gravity.

Figure 12.11 Soil erosion is a result of water 'run-off'; the moving water carries particles downhill. The faster the water moves, the more soil carried and larger 'rills' are created eventually forming gullies

The amount of water held at field capacity is known as the **water-holding capacity (WHC)** or moisture-holding capacity (MHC) of the soil (Table 12.3). Most

Table 12.3 Soil water-holding capacity of different soils

Soil texture	Water held in 300 mm soil depth (mm)		
	At field capacity (FC) i.e. water-holding capacity (WHC)	At permanent wilting point (PWP)	Available water (AW)
Coarse sand	26	1	25
Fine sand	65	5	60
Coarse sandy loam	42	2	40
Fine sandy loam	65	5	60
Silty loam	65	5	60
Clay loam	65	10	55
Clay	65	15	50
Peat	120	30	90

soils in the lowland areas of Britain and Ireland hold about an average month's rainfall (60 mm) in a topsoil of 30 cm. Coarse sandy and gravelly soils hold far less and peaty soils hold much more.

> **Gravitational water** is the water that can be removed from the soil by the force of gravity.

Drainage

The continued addition of water eventually brings all the root zone to field capacity. In many soils the gravitational ('excess') water can drain naturally into the lower depths or because some form of artificial drainage has been installed. This ensures that **saturation** in the root zone is temporary. However, when more water is added to soils overlying impervious material such as non-porous rocks and some subsoils (e.g. wet clay and clay loams), it leads to poorly drained soils. In these air only gets into the top layers of the soil during the drier parts of the year. A **water table** marks the level below which the soil is saturated with water. This usually varies over the year with the water table falling to a minimum in the summer. A hole dug in the ground will fill up with water to the water table.

> **Drainage** is the removal of gravitational (excess) water.

Saturated (waterlogged) soils can be recognized by:

▶ standing water and surface run-off
▶ grey or mottled soil colours (Figure 12.13)
▶ smell of hydrogen sulphide – 'bad eggs'
▶ indicator plants (such as rushes, mosses)
▶ weed problems
▶ restricted rooting
▶ reduced working days for cultivation
▶ pest and disease problems, e.g. clubroot (see p. 256).

Draining soils

Wet soil problems should be tackled according to the cause or causes. Many of the symptoms of poor drainage occur when there are **compacted layers** in the cultivation zone. This is essentially a soil structure problem that should be dealt with by appropriate cultivation to remove the obstruction. If the problem is as a result of water being held back by the conditions below cultivation depth, then drainage may be installed to advantage.

Improving poor drainage by removal of gravitational ('excess') water trapped in the root zone should be undertaken before any other remedial work on a soil. It is normally achieved on large areas by laying drainage pipes, such as 'clays' (Figure 12.14) or perforated plastic pipes, in parallel lines under the ground about a metre deep and taking the drainage water to a ditch.

In domestic situations, the most common problem is the large amount of water running off buildings and hard areas such as patios and driveways on to borders and lawns. A **French** ('rubble' or 'interceptor') **drain** can be placed across where the run-off occurs. A trench is dug where it can intercept run-off and lined with woven fibre. A line of clay (tile) drains, perforated plastic pipe or even 'rubble' is laid with a slope and the trench is back filled with gravel. It can be planted up if the trench is part filled with gravel and 'blinded' with sand or woven fibre before putting on a layer of soil. The pipe then leads the water away for disposal in a ditch or a soakaway. Note that 'bye-laws' normally prevent you discharging water into one of the many pipes servicing your house. Finding a discharge point below the level of the drainage pipe is a problem in a garden surrounded by other gardens. Most people have to resort to a soakaway which works very well if the bottom of the hole is permeable; much less so if it just a large water reservoir.

Figure 12.14 Drainage pipes. A range of 'clays' (earthenware pipes) is shown. Most commonly they are 30 cm long and either 75 mm or 100 mm diameter; these are butted up close to each other to form straight lines to intercept and carry away 'excess' water

Figure 12.15 Swales

Figure 12.13 Poorly drained soils. Soils that have developed in aerobic conditions have bright orange, yellow or brown colours (top), whereas those developed in waterlogged conditions have grey, green or blue tones (bottom). Fluctuating waterlogged and aerobic conditions as the water table rises and fall leads to mottled soil (middle sample)

On a larger scale, French drains can be used to intercept water coming off higher ground, water running down valley sides or from springs. They lead water around or under areas to be protected. **Swales** are shallow ditches, often with vegetation in them, that slow run-off speeds and allow the water to soak away over their length. They are commonly used in urban situations as a very shallow, safe ditch (Figure 12.15).

Loss of water by evaporation

Water drains from the soil profile, but also leaves from the surface directly or indirectly through plants (transpiration). The rate of drying depends on the drying capacity of the air; increased water loss will be depend on the same factors that make up a 'good drying day' for clothes on the washing line or good haymaking weather, that is, sun (or air temperature) and wind.

If it is a good drying day, a wet bare soil surface gives up water very quickly. This water is slowly replaced from below and evaporation from the surface continues. The replacement water has to come from deeper in the soil and gradually the surface layers begin to dry out and give up water much more slowly. There comes a point when virtually no water can reach the surface. This happens for most soils when a layer of about 10 cm becomes dried out. This is sometimes referred to as a 'dry mulch'; so long as moist soil from below is not brought to the surface, further water loss does not occur except though roots. Once there are plants growing in the soil, water can be taken from below the surface. Although

Table 12.4 Potential transpiration rates. The calculated water loss (mm) from a crop grown in moist soil with a full leaf canopy based on weather data collected in different areas in Britain and Ireland

Area	April	May	June	July	Aug	Sept
Ayr	50	80	90	80	65	40
Cheshire	50	75	80	90	75	45
Channel Isles	50	85	90	100	85	45
Glamorgan	50	80	85	85	75	45
Hertfordshire	50	80	90	95	80	45
Northumberland	45	65	80	75	60	35
N Ireland (coast)	45	70	80	80	70	40

water loss from an area covered with leaves is not as rapid as from a wet bare soil surface, it is now being lost from the whole root zone. The losses from ground covered by foliage for different lowland areas are shown in Table 12.4. Compare your local rainfall figures with the losses of water from full crop cover that occur in your area and you can see what sort of soil moisture deficits you might have in your area over the summer months.

As the drying of the soil in the root zone continues, it becomes more and more difficult for plants to extract and the plant's cells begin to lose turgor (see p. 120). The leaves begin to look 'stressed' and they wilt. Growth is affected because flaccid (wilted) leaves are less effective at intercepting sunlight. It also leads to the stomata closing to reduce the water loss (see p. 123), which, in turn, affects photosynthesis (see p. 112). The inability of the plant to keep up with water loss may occur even in ideal soil conditions with the roots taking water up efficiently. A **temporary wilt of a plant** commonly occurs when there are very strong drying conditions, especially hot and windy conditions. Typically in Britain and Ireland this occurs on summer afternoons; then the plants recover as the temperature drops and the rate of transpiration falls. However, as more water is lost from the root zone, temporary wilt becomes more frequent. If water is not added to the soil, there comes a point when no more water can be taken up by the plant so there is no recovery of turgor overnight. This is when the **permanent wilting point of the soil (PWP)** has been reached. At this point most sandy soils have virtually no water left in them whereas clay loams may have 15% left inside the clay particles and in the very smallest pores (see Table 12.3).

> The **permanent wilting point (PWP)** is the water content of the soil when a wilted plant does not recover overnight.

Available water

The water that plants are able to take from a soil is that held between field capacity and at the permanent wilting point (see Table 12.3). The amount of water is made available for plants by increasing:

▶ root exploration by eliminating poor structure
▶ rooting depth
▶ organic matter levels.

Roots remove the water at field capacity very easily. Even so, plants can wilt temporarily and any restriction of rooting (e.g. caused by soil pan, phosphate deficiency, root disease) makes wilting more likely. Water uptake is also reduced by high soluble salt concentrations such as excess use of fertilizers (see osmosis, p. 120) and by the effect of some pests (see vine weevil p. 242) and diseases (e.g. vascular wilt diseases).

> **Available Water Content** (AWC) is the water held in the soil between field capacity and the permanent wilting point.

It is after about half the available water content has been removed that temporary wilting becomes significantly more frequent. Watering the soil before it reaches this point helps to maintain growth rates.

Adding water

For most deep-rooted plants (trees and shrubs) there is usually no need for water after they are established. Ideally, once seeds and plants have been 'watered in' there should be no need for more unless significant amounts of water are lost from the soil, that is, about half the available water in the root zone. Plants with roots down to 50 cm in loamy soils will have access to an available water content of about 100 mm (see Table 12.3); half this is about 50 mm of water. This is a **soil moisture deficit** of 50 mm, that is, the amount of water to add to return the soil to field capacity.

The **soil moisture deficit (SMD)** is the amount of water needed to return the soil to field capacity.

For most lowland parts of Britain and Ireland, this is roughly a month without rain in April, 20 days in May, 16 days in June and July, 20 days in August but over 60 days in September and October. If this water loss has occurred at a time when water shortage affects the performance of the plants involved, the 'response periods', the soil should be returned to field capacity in one go (so long as rain is not forecast), that is, add 50 mm water. If a sprinkler is being used, the quantity of water added can be checked as you deliver it by having a straight-sided container receiving the same amount of water as the plants: switch off as the water level comes up to 50 mm.

Response periods are when the plant/crop benefits from the addition of water.

The general rule about adding water is avoid 'little and often', which encourages shallow rooting and maximizes water loss. Instead adopt the principles outlined above. Hold off watering and then apply large quantities (the soil moisture deficit but no more) to return the soil to field capacity. An alternative approach is to drip water close to plants that benefit from continuous water supplies, but wet only a small proportion of the soil surface so losses by evaporation are minimized. The distribution of water as it soaks in depends mainly on the soil texture with very little spread in sandy soils, so delivery points need to be closer together (Figure 12.16).

Irrigation plans

Soil moisture deficit (SMD) becomes important when trying to follow an irrigation plan that identifies how much water to add and when. The plan should also identify the response periods of the plant when there are benefits, that is, when it is worthwhile adding water (after the plants or seeds have been watered in).

A typical plan for runner beans growing on a soil with a medium Available Water Content would be to irrigate from early flowering onwards (June to August) with 50 mm water when a SMD of 50 mm had built up. Similarly, first early potatoes grown on light, sandy soils (low AWC) benefit from irrigation after tubers reach 10 mm diameter (May–June) with 25 mm water when

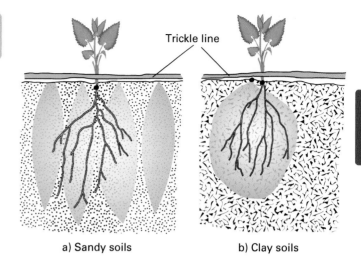

a) Sandy soils   b) Clay soils

Figure 12.16 Trickle or seep irrigation – the pattern of water spread below the surface is shown in (a) sandy soil and (b) clay soil. Note the dry soil at the surface between the delivery points

SMD reaches 25 mm. Note that 25 mm is the equivalent to 25 mm rain which is about a third of typical May rainfall in many parts of the SE England. Adding 25 mm means the equivalent of covering the whole area to be treated to a depth of 25 mm ('an inch of rain'), which can be checked by placing a straight-sided container within this area. The rate of irrigation should be adjusted so as not to exceed the infiltration rate of the soil. Clearly irrigation should be held off if rainfall is imminent; any extra water can be detrimental as it leads to standing water, surface run-off and the leaching of nutrients as well as being wasteful.

Methods of applying water

Applying water should be carefully related to plant requirements, climate and soil texture. The following methods are commonly used:

▶ **Watering cans** are a sound choice in most gardens. The use of 'dipping ponds' can save time waiting for cans to be filled making it almost as quick as using hoses when water is being added to small areas or a limited number of plants. Care should be taken to avoid damaging the crumb structure of the soil or disturbing the growing medium in containers. Consideration should be given to the use of fine roses on cans. For watering seeds and cuttings, a fine rose turned upwards is recommended in order to minimize any disturbance by droplets.

▶ **Hoses fitted with trigger lances** with means

to adjust flow rate and fineness of spray are an alternative to using watering cans. Care is needed because poor adjustment of flow can lead to damage to soil, growing media and plants especially if a powerful jet of water is selected.

▶ **Sprinklers** become an obvious choice, when allowed, to deliver water to a large area rather than to individual plants or containers. Bare soils are vulnerable to droplet damage. Large and fast-moving droplets damage soil crumbs in seedbeds as well as disturbing seeds. Standing water develops when the delivery rate exceeds infiltration rates; standing water leads to soil cap formation (see p. 150). On slopes, run-off can cause erosion of soil and even losses of seeds and fertilizer. There is high water loss by evaporation associated with water in the air and on soil surfaces. It also creates a humid atmosphere.

▶ **Trickle lines** or **seep hoses** (Figure 12.16) deliver water very slowly to the soil, leaving plant foliage and most of the soil surface dry, which ensures a drier atmosphere and reduced water loss. However, care is needed because there is very little sideways spread of water into coarse sand, loose soil or a growing medium that has completely dried out.

▶ **Drip irrigation** is a variation on the trickle method, but the water is applied through pegged-down thin, flexible 'spaghetti' tubes to exactly where it is needed – for example, the base of each plant (see Figure 15.6b). This minimizes water loss because no water is put in the air and only a small proportion of the growing media surface is wetted.

Water conservation

The need to manage water efficiently is a major concern in the use of scarce resources. Responsible action is increasingly supported by legislation and encouraged by the higher price of using water. The major factors that determine the level of water use are related to the choice of plant species and the reasons for growing. There are many ways by which water use can be reduced if certain principles are kept in mind and acted upon appropriately:

▶ **Plant selection**. Choose drought-tolerant rather than water-intensive plantings.

▶ **Use recycled water**. Whenever possible, the use of mains water should be avoided. The capture of rainwater is an important consideration in the choice of water source.

▶ **Minimize evaporation of water**. This is best achieved by not spraying water into the air and by minimizing the time when the soil surface is moist. When water does have to be applied overhead, this should be undertaken in cool periods of the day.

▶ **Increase the water reservoir of the soil**. The application of water can be reduced by increasing the available water for the plants; by increasing rooting depth, adding organic matter and/or adding water-holding gel supplements to composts.

▶ **Improve soil structure**. Help roots explore the maximum amount of soil. Compacted layers in the soil (soil pans) should be eliminated and good soil structure maintained to increase the effective rooting depth.

▶ **Plant root system**. Plants should be encouraged to establish as quickly as possible but, after the initial watering-in, infrequent applications will encourage the plant to put down deeper roots by searching for water.

▶ **Minimize water lost through drainage**. This is not only to avoid losses of water from the root zone but also the associated leaching of soluble nutrients. Adding water to an outdoor soil (irrigation) should be to field capacity and no more. Avoid adding the full soil moisture deficit when rain is imminent.

Further reading

Ashman, M.R. and Puri, G. (2002) *Essential Soil Science*. Blackwell Science.

Bell, B. and Cousins, S. (1997) *Machinery for Horticulture*. 2nd edn. Old Pond Publishing.

Culpin, C. (1992) *Farm Machinery*. 2nd edn. Blackwell Science.

Davies D.G., Eagle D.T. and Finley J.B. (1993) *Soil Management*. 5th edn. Farming Press Books and Videos.

Ellis, S. and Mellor, A. (1995) *Soils and Environment*. Routledge.

Ingram, D.S. et al. (eds) (2008) *Science and the Garden*. 2nd edn. Blackwell Science.

McIntyre, K. and Jakobsen, B. (1998) *Drainage for Sportsturf and Horticulture*. Horticultural Engineering Consultancy.

Munns, D.N. and Singer, M.J. (2005) *Soils: An Introduction*. 6th edn. Prentice Hall.

Prentice Baily, R. (1990) *Irrigated Crops and Their Management*. Farming Press.

Please visit the companion website for further information:
www.routledge.com/cw/adams

Organic matter in the root environment

Figure 13.1 Garden compost

Summary

This chapter includes the following topics:

- Types of organic matter in the soil
- The role of organic matter in the soil
- Decomposition of organic matter
- Mulching
- Composting
- Organic fertilizers and compost teas
- Bulky organic matter
- Green manures

Principles of Horticulture. 978-0-415-85908-0 © C.R. Adams, K.M. Bamford, J.E. Brook and M.P. Early.
Published by Taylor & Francis. All rights reserved.

Types of organic matter in soil

Most gardeners in Britain and Ireland work with mineral soils that contain between 2 and 5% organic matter. The main types of organic matter in the soil are:

▶ living organisms
▶ dead organic matter
▶ humus.

Soil organic matter is derived from living organisms such as plants, earthworms, insects, fungi and bacteria. On death these plant and animal remains decompose and are recycled along with any other organic matter that is added. The remains of cultivated plants are incorporated directly by digging them in or indirectly having been used as a mulch or been composted. More bulky organic matter can be added in the form of material imported from elsewhere such as farmyard manure (FYM), garden compost, mushroom compost, composted municipal waste, leaf mould, chipped bark, composted straw and green manure.

The role of organic matter in the soil

The organic matter in its many forms has important effects on the soil colour (see p. 144), soil structure (see p. 147), water availability (see p. 154) and the cultivation window (see timeliness p. 148).

Most of the **living organisms** contribute to the decomposition of organic matter and the formation of humus, but many also affect the soil structure by moving soil and creating a network of interconnected tunnels thereby improving aeration and water movement.

Benefits of living organisms include:

▶ plant and animal debris are converted to minerals and humus
▶ *Rhizobia* and *Azotobacter* spp. fix gaseous nitrogen
▶ soil structure is improved by the activity of plant roots, earthworms and other burrowing organisms
▶ detoxification of harmful organic materials such as pesticides is undertaken by many bacteria species.

Dead organic matter is a source of food for the living organisms, but has two other important roles in the soil. In its undecomposed state it 'opens up' the soil, that is, creates bigger gaps between the soil aggregates, increasing air and water movement. As it decomposes, it can yield plant nutrients. In general, 'green' (succulent, leafy) organic matter decomposes very rapidly, if conditions are right. Consequently, it tends to have just a short-term physical effect, but yields nutrients, especially nitrogen compounds. The 'brown' (fibrous/woody) plant material tends to

decompose very slowly, so its physical effect persists, but nutrient contributions are low. The distinction between the 'green' and 'brown' organic matter is crude, but a useful one when selecting materials for composting (see p. 161 and Table 13.1).

Benefits of dead organic matter in the soil:

▶ microbial activity is increased because it is food for soil organisms
▶ soil is physically opened up and aeration improved by the dead but recognizable organic matter
▶ water-holding capacity of the soil is improved by fine (unrecognizable) organic matter
▶ it is a dilute source of slow-release nutrients as the organic matter decomposes and releases minerals.

Humus is one of the end products of decomposition in aerobic soils. Large quantities of organic matter yield tiny amounts of this black jelly that coats soil particles, darkening the soil where it is present. It is also sticky when wet so it helps in the creation of crumb structure in soils where there is little clay to bind particles together, but it also forms clay-humus complexes which helps heavy (i.e. high clay) soils to crumble more readily. Humus also has 'buffering capacity' so, along with clay particles, plays an important part in holding on to the nutrients that would otherwise be leached from the soil profile while continuing to release them to plants.

Benefits of humus in the soil:

▶ in sandy and silty soils it helps to form stable crumbs
▶ in clays the surface charges on humus are capable of combining with the clay particles, thereby making them less sticky and more friable
▶ 'buffering capacity' is increased, which reduces the leaching of some nutrients from the profile
▶ soils warm up more quickly in the spring because darker soils absorb more of the sun's radiation
▶ improves water-holding capacity.

> **Buffering capacity** is the ability of the soil to retain nutrients including lime against loss by leaching.

Decomposition of organic matter

As in any other plant and animal community, the organisms that live in the soil form part of the **food web** (see p. 38). The organic matter derived from dead plants and animals of all kinds is digested by a succession of species: large animals by crows, large trees by bracket fungi, small insects by ants, roots and fallen leaves by earthworms, mammal and bird faeces by dung beetles, and so on. Subsequently, progressively smaller organic particles are consumed by millipedes, springtails, mites, nematodes (eelworms), fungi and bacteria to leave, eventually, just water, carbon dioxide, minerals (including plant nutrients) and **humus**. This recycling process makes available water, carbon dioxide and nutrients to a new generation of plants. The production of plant nutrients such as ammonia, nitrates, phosphates, potash and sulphates from dead organic matter is sometimes referred to as **mineralization**. Mineralization yields nutrients that are readily taken up by plants from the soil solution (see Chapter 14).

The type and **activity level of the living organisms** depends greatly on the food supply, organic matter levels, but also water content, temperature, pH and air (oxygen) levels. In general, decomposition is favoured by aerobic, warm, moist and neutral conditions – for example, earthworms are most active and abundant in neutral, aerobic organic matter rich soils when temperatures are about 10°C (i.e. spring and autumn). In anaerobic (e.g. waterlogged), cold and/or very acid areas there are fewer and completely different organisms; decomposition is slower and the organic matter is not fully broken down so it accumulates, leading to the development of peaty soils or peat bogs.

Earthworms affect soil structure directly as they create a network of burrows which significantly improves aeration and water movement. This activity is particularly important in uncultivated areas including 'no-dig' cultivation (see p. 150) and organic growing (see p. 10). Many species feed on soil and digest the organic matter in it before returning the waste into the tunnel they have created. The crumbly nature of this soil which is now an intimate mix of partially digested organic matter and mineral matter is evident from the worm casts left by one of the few species that casts on the surface. Other species, including those that are involved in composting, only eat organic matter. Earthworms digest organic matter and play an important part in incorporating leaf litter from the surface; often dragging whole leaves underground (see Figure 13.2).

Plant roots also create tunnels that improve aeration and water movement; they move soil as they penetrate the soil and grow in size. Roots leave large quantities of the organic matter in the soil profile. When they die, the channels they have created are stabilized by the coating of their decomposed remains. Roots take up water and as the heavier soils dry out,

Figure 13.2 Earthworm cast on the surface, which is a mix of organic matter and finely divided soil. Most casts are left underground. Those on the surface can be a problem on fine turf. Earthworms play a major role in burying organic matter such as leaves

they shrink and crack thereby helping to develop and improve their structure (see p. 147). A good demonstration of the effect of abundant roots on soils is seen under long-term grassland.

Bacteria (see p. 262) are present in soils in vast numbers. About 1000 million or more occur in each gram of fertile soil. Consequently, despite their microscopic size, the top 150 cm of fertile topsoil carries about one tonne of bacteria per hectare. There are many different species of bacteria to be found in the soil and most play a part in the decomposition of organic matter. This bacterial activity also affects soil structure in so far as it leads to the release of nutrients and hence more vigorous root growth as well as by the creation of humus that improves soil crumb development and stability.

Saprophytes are organisms that live on dead plant material.

The majority of **fungi** (see p. 62) live saprophytically on soil organic matter. There are many that are tolerant of acid soils and they are responsible for much of the decomposition of organic matter in these conditions. Some fungi species are amongst the few organisms that can digest and breakdown wood. The white rot fungi are very unusual as they are able to digest lignin in wood, leaving the white cellulose, such as bracket fungi, Bootlace or Honey fungus (*Armillaria* spp., see p. 261). Soft rot fungi attack and break down the cellulose in wood and along with the brown rot fungi (formerly 'dry rot') can damage wooden structures.

Mulching

Mulches are materials added to the soil surface in order to provide one or more of the following:

▶ decorative finish (e.g. chipped bark on borders)
▶ weed suppression
▶ moisture retention
▶ protect the soil surface and reduce erosion
▶ maintaining/increasing soil organic matter
▶ stimulating beneficial soil organisms
▶ modifying soil temperatures (insulates)
▶ protecting edible crops from soil contact/splash (e.g. straw under strawberries).

> **Mulches** are materials applied to the surface of the soil to suppress weeds, modify soil temperatures, reduce water loss, protect the soil surface and/or reduce erosion.

Many organic materials are used as mulches including farmyard manure, garden compost, mushroom compost, composted municipal waste, leaf mould, chipped bark, composted straw and green manure. There are several non-organic materials used for mulches including minerals (pebbles, slate, stone chippings), tumbled glass and sheets such as woven polypropylene/fibre, polythene, paper and old carpet.

Except for the sheet mulches (woven fibre, polythene, paper, carpet), the materials need to be laid thickly to be effective. Most can be applied at any time when the soil is moist in accordance with their function, but those that are insulatory should be applied after the soil has warmed up. Care should be taken to remove perennial weeds before applying and to keep mulch away from the base of woody stems.

Composting

Compost is a dark, soil-like material made of decomposed organic matter (see Figure 13.1). Many gardeners depend on composting as a means of recycling their garden and kitchen waste to maintain organic matter levels in their soils. Many councils now collect 'green waste' and supply composting equipment to encourage householders to recycle garden waste as well as their paper, glass and metals. This is not as environmentally friendly as home composting as there is significant transport involved. Horticulturists are increasingly concerned with the recycling of waste and attention is being given to composting methods to deal with the large quantities of material generated on their units. Composting is fundamental to successful organic growing (see p. 10).

> **Composting** is the decomposition of organic matter (including plant residue) in a heap before it is applied to soils.

Conditions for successful composting must be favourable for the decomposing organisms (see p. 157):

▶ **Air (oxygen)** – the beneficial organisms are aerobic (see p. 114) so they require well aerated conditions throughout.
▶ **Water** – the decomposers require a moist material to eat and to live in, most are inactive when too dry. If too much water is added, the air (oxygen) is driven out (in the waterlogged conditions the anaerobic organisms take over leading to poor compost). Once the heap is moist enough, a roof/cover is advantageous by keeping out rain (excess water) and retaining heat.
▶ **Organic matter mix** – in order to achieve a mix that provides the right balance of nutrients, water and aeration, it is convenient to distinguish between 'green' (leafy/succulent/tender) and 'brown' (fibrous/woody/tough) materials and combine them in approximately equal measure (Table 13.1).
▶ **Accelerators or activators** can help when there is not enough 'green' material to supply the full range of nutrients especially nitrogen. They are essentially nitrogen fertilizers and not normally needed in garden composting. Also available but even less likely to be needed are materials that make good a shortage of 'brown' material, such as sawdust (see Table 13.1).
▶ **Shredding** – the rate of decomposition is speeded up by reducing the size of the material put on the compost heap. Shredding increases the surface area so making more of the material accessible to the organisms. The degree of shredding should be

Table 13.1 Compost ingredients

Proposed ingredient	Category
Cardboard	Brown
Farmyard manures	Intermediate
Fibrous prunings	Brown
Haulm (old plants)	Brown
Hedge clippings	Intermediate
Herbaceous plants (old)	Brown
Grass mowings	Green
Grass – long	Brown
Kitchen (plant) waste	Green
Leaves – young	Green
Leaves – autumn	Brown
Nettles – young	Green
Nettles – old	Brown
Paper	Brown
Sawdust	Brown
Seaweed	Green
Straw	Brown
Woody prunings	Brown

Figure 13.3 Compost bins. A typical set large enough for efficient composting; slatted to allow in air and removable so as to allow easy access to add new material and for regular turning of the contents

such that aeration is maintained and waterlogging avoided.

▶ **pH** (see p. 174) – the compost mix should not be too acid. Thin layers of lime (see p. 000) can be added as the heap is built.

▶ **Temperature** – the rate at which the organisms decompose the organic matter also depends on the temperature of their environment. The composting process gives off heat (exothermic reaction); under ideal circumstances the temperature can rise to over 70°C within seven days. However, most garden composting does not go to a high enough temperature to kill harmful organisms and weed seeds.

▶ **Heap size** – heat is generated within the volume of the heap by the exothermic process but it is lost from the surface so the degree to which the compost heats up depends on the surface area to volume ratio. Small heaps (less than a cubic metre) have a disproportionally large surface area which dissipates the heat generated within the small volume so the heap does not heat up ('cold composting'). Insulation reduces heat loss from the surfaces but this also reduces the flow of air (oxygen) into the heap.

Organic waste brought together in large enough quantities under ideal conditions heats up quickly, but the process then tends to slow down because the availability of oxygen becomes reduced. Rather like stirring up the dying embers of a fire (i.e. getting oxygen around unburnt fuel), decomposition

rates can be restored by 'turning' the compost heap on a regular basis to improve aeration. This is continued until all that is left is a crumbly material with no recognizable plant material and an 'earthy' smell.

Garden (home) composting

Most gardeners are not usually able to obtain enough components at any one time to create the ideal compost heap. It is difficult to be successful with batches less than one cubic metre at a time; the cooling at the surfaces is greater than the heat generation at the centre. Compost bins can be purchased but slatted wooden sided bins can be made (see Figure 13.3). An open base over soil allows organisms and air in. It is advantageous to have a second bin alongside so that the compost heap can be turned from one to the other frequently (slats between the two should be removable to make 'turning' easier). A suitable cover (e.g. old carpet) is needed to keep warmth in and rain off.

This is a 'cold' method that can produce good compost but tends to take many months or even a year or two to complete. When making compost this way, care should be taken to avoid perennial weeds or infected plant material which is unlikely to be rendered harmless.

Compost tumblers

These are containers that can be rotated on an axis to provide an easy method of turning small batches to create compost in a relatively short time (see Figure 13.4). Batches can heat up sufficiently to kill off

Figure 13.4 Compost tumbler. Turning frequently made easy by the use of a handle to rotate the load.

weeds and diseases and the enclosed nature deters vermin. The compost ingredients should be gathered together and the tumbler filled in a short space of time. Nothing is added until the batch is completed.

Worm composting

This lends itself to handling small quantities which can be added as they arise, such as kitchen waste, especially over the winter period when there is little plant material to accumulate. Brandling or tiger worms (*Dendrobena* spp. and *Eisenia foetida*) are compost worms that feed on organic matter. These can be purchased, but they are readily found in rotting vegetation such as compost heaps (see Figure 13.1) and quickly multiply. Each kilogram of worms eats 1–2 kg of waste food per day.

The compost container can be a plastic dustbin usually equipped with a tap to drain off liquids, which can be diluted and used as liquid feed for plants. Smaller containers, wide rather than narrow, can be made out of wood, ideally with some insulation to maintain temperatures. The worms eat the vegetation as it starts to rot, which means that once in balance there is no smell.

A 10 cm layer of sand is used at the base and it is covered with a polythene sheet. Bedding material such as well-rotted compost or farmyard manure is needed for the worms to live in until the process begins. Chopped waste is spread to a depth of 5 cm. Around 100 worms are added and covered with wet newspaper to keep out the light and maintain moisture levels. A lid is needed to keep out the rain. Ideally, temperatures should be maintained between 20 and 25°C and the pH kept between 6 and 8; lime can be sprinkled on if the compost becomes too acid.

The compost is removed when ready; the decomposing top layer is separated off and used to start the next run. The compost is spread out to dry in the sun and the worms are recovered by placing a wet newspaper on the compost, under which they will congregate.

On a larger scale, wormeries are used to compost farmyard manure, with continuous systems available that separate the composted material from the worms, which can be recycled with surpluses being available as animal feed.

Composted municipal waste

When very large quantities of green waste from households and parks departments are available, the ingredients can be heaped up on a concrete base. This makes it easy to use large powered equipment to turn the ingredients to maintain good aeration. The cooler outer layers can be mixed in to ensure all parts are heated up and decomposed rapidly. Specialist composting vessels with automatic turning equipment and biofilter beds ensure that the exhaust vapour is 'scrubbed' (cleaned). The main advantage of the composted municipal waste is that high temperatures can be achieved and so the material is less likely to contain viable weed seeds, perennial weed material or plant diseases. 'Windrow turners' are used in big plants to turn and loosen the green waste. These tend to dry out the material and so some are in housed areas to confine the dust created.

Hot beds

Using the composting process to generate heat has been used since at least Roman times. The heat generated in the hot bed process was much exploited by Victorian gardeners to produce exotic fruits such as melons and pineapples and to ensure a continuity of supply of fruit and vegetables for the kitchen even through the coldest months of the year. The Victorians also had access to large quantities of fresh horse manure that still remains the best material for a hot bed system.

The emphasis of this decomposition is the generation of heat so there has to be sufficient quantity of fresh manure to complete the process in a short period of time, days rather than weeks. Compare this with the cold process where the build-up of a typical garden compost heap is over months. The process is started by building a loose heap, that is, plenty of oxygen circulating. Within two to three days the pile heats up and generates steam. At this point the pile is inverted and 'fluffed up'. This is repeated three days later sprinkling on water if necessary to keep the mix moist.

After nine to ten days the decomposing organic matter is made into the hot bed at the site where plants are to be grown. The material in the heap is forked into a layer about 10 cm deep and tamped down. This is repeated with further layers to a depth of 20 cm or more. This can all be done within a brick construction on which 'lights' can be placed like a 'cold frame'. Alternatively, a free-standing frame can be put on top of the block created. The size of the frame is usually made to be a multiple of the 'lights' being used (see Figure 11.6). Traditionally heat was retained by earthing up the sides, but modern materials such as polystyrene can be used to line the frame. The 'lights' are put on but kept open for ventilation.

In less than a day the material starts to heat up significantly. This can be tested with a thermometer or, more crudely, with a cane pushed into the centre of the block and tested by hand. Temperature should be monitored over the next three days when the maximum is usually achieved. As soon as the temperature begins to drop, a 15 cm layer of moist soil is put on top and is ready to use. Once seeds or plants are in place, the temperature is controlled by opening and closing the 'lights' in the same way as growing in cold frames. The heat provided from the decomposition process will last until the days warm up in the spring and growing in the frame can continue until the autumn when the well-rotted manure can be used as a mulch or dug in to the soil.

Organic fertilizers and compost teas

Most organic fertilizers are of animal origin – for example, blood meal, hoof and horn, bone meal. However, concentrated sources of nutrients can be produced from plant material. They are particularly useful as a liquid feed (see p. 171) and can be made from a wide range of plants including compost, comfrey, nettles, borage, clover and bracken. More biologically active 'teas' can be made from these by steeping in oxygenated water. Further details can be found on the companion website: www.routledge.com/cw/adams.

Using comfrey is particularly attractive to organic growers as it is deep rooted and able to extract nutrients below that of most plants. These deep roots also make it difficult to get rid of so choosing it to put in a garden needs to be thought through. The leaves are harvested to produce a nutrient-rich plant food. There are many different methods of making the liquid feed from comfrey and other plants, all of which should be diluted before use. Both comfrey and nettle leaves can be used to in the following way:

- ▶ Use gloves to protect against the irritants produced by the leaves of both these plants.
- ▶ Chop up leaves.
- ▶ Pack tightly into a container with a lid (this is to contain the smell which develops as leaves decompose).
- ▶ Keep warm (e.g. in the sun; a black container absorbs heat well).
- ▶ Mix daily (i.e. aerate), foaming indicates leaves are decomposing.
- ▶ When foaming stops (after about a fortnight) draw off liquid (stock).
- ▶ Dilute the stock liquid 1:10 for use as a potassium rich liquid feed (see p. 167).

Bulky organic matter

Bulky organic matter is an important means of maintaining organic matter and humus levels in the soil and includes:

- ▶ farmyard manure
- ▶ garden compost
- ▶ composted municipal waste
- ▶ spent mushroom compost
- ▶ leaf mould
- ▶ chipped bark
- ▶ green manures.

These materials also 'open up' the soil, that is, improve aeration (see soil structure p. 000). The main problem is obtaining cheap enough sources because their bulk makes transport and handling a major part of the cost. They can be evaluated on the basis of their effect on the physical properties of soil and their (small) nutrient content.

Farmyard manure

This is the traditional material used to maintain and improve soil fertility. It consists of straw, or other bedding, mixed with animal faeces and urine that provides the nitrogen that offsets the problem of using straw alone. The exact value of this material in nutrient terms depends on the proportions of the ingredients, the degree of decomposition and the method of storage. Samples vary considerably. Much of the manure is rotted down in the first growing season but almost half survives for another year, and half of that goes on to a third season, and so on. A full range of nutrients is released into the soil and the addition of the major nutrients should be allowed for when calculating fertilizer requirements. The continued release of large quantities of nitrogen can be a problem, especially on unplanted ground in the autumn, when the nitrates formed are leached deep into the soil over the winter and can be lost from the root zone.

Farmyard manure is most valued for its ability to provide the organic matter and humus for maintaining or improving soil structure. It must be worked into soils where conditions are favourable for its continued decomposition. If fresh organic matter is worked into wet and compacted soils or deep into clay, the need for oxygen outstrips supply and anaerobic conditions prevail to the detriment of any plants present. These soils develop grey colouring and a foul ('bad eggs') smell.

Garden (home) compost

The main advantage of garden compost is that it is the most convenient way of recycling garden and kitchen waste without the need for transport. If well made, an attractive crumbly dark brown material becomes available for use as a soil conditioner, ideally as a mulch (see p. 160). This helps to protect the surface from soil capping (see p. 150), maintains organic matter levels in the soil and supplies nutrients in a slow-release form (see p. 173). It can be the basis of 'no-dig' and bed systems of growing (see p. 150).

Alternatively, it can be worked into the soil, providing many of the same advantages but requiring further work to incorporate. This is a better option if the compost has not been heated sufficiently to kill weed seeds. It has too many problems to be used alone as a compost for seed production, but well-made material can be used mixed with other materials to make potting composts.

Instead of composting in a bin, the garden waste can be shredded and distributed directly on to the soil as a mulch where it will decompose given the right conditions (as in the natural world). This achieves the same result, often more quickly, and without the need for managing the process in bins.

Composted municipal waste

This is essentially the same as garden compost, but because it can be composted in bulk it has normally been heated sufficiently to inactivate harmful weeds and plant diseases. The product is usually too high in nutrients to be used alone in growing composts. The high wood content gives the final product good stability (see p. 179), but can lead to nitrogen lock-up. It has a high pH (see p. 172), but it can be mixed with composted barks, coir and so on to create good container compost for a wide range of plants. However, unless the composting is done in high-cost composting vessels, the most common issues are contamination with plastics and glass.

Spent mushroom compost

This compost is a by-product of the mushroom industry, that is, it is the 'spent' (i.e. used) material that becomes available direct from the growers or some garden centres. It used to be made from well-rotted horse manure but now it is almost all composted straw capped with chalk, which gives its characteristically high pH, that is, basic reaction (see p. 174). This makes it useful for raising soil pH as an alternative to liming (see p. 176). It works well with the growing of calcicoles (see p. 174) especially when growing brassicas (cabbages, cauliflowers, Brussels sprouts).

It is an excellent source of organic matter with added nutrients left over from mushroom growing with which to mulch or to incorporate in the soil but its use is limited. It must not be used in plants that prefer acid conditions (see calcifuges, p. 174). If used too freely, nutrient deficiencies in plants can be induced (see iron-induced chlorosis, p. 171) leading to poor performance. Leftover fertilizer levels in the sample can make it unsuitable for young plants particularly when growing in containers.

Leaf mould

Leaf mould is made of the rotted leaves of deciduous trees and makes a highly prized compost. The leaves are often composted separately from other organic matter and much valued in ornamental horticulture for a variety of uses such as an attractive mulch or, when well-rotted down, as a compost ingredient. They are commonly composted in mesh cages, but many achieve success by putting them in polythene bags well punched with holes. The leaves alone have a high brown to green ratio so decomposition is slow (see organic matter mix p.160). Usually it is not until the second year that the dark brown crumbly material is produced, although the process can be speeded up by shredding the leaves first.

It is low in nutrients because nitrogen and phosphate are withdrawn from the leaves before they fall and potassium is readily leached from the ageing leaf. Unless they are from trees growing in very acidic conditions, the leaves are rich in calcium and the leaf mould made from them should not be used with calcifuge plants (see p. 174).

Pine needles

These are covered with a protective layer that slows down decomposition. They are low in calcium and the resins present are converted to acids. This extremely acid litter is almost resistant to decomposition. It is valued in propagation, for growing calcifuge plants such as rhododendrons and heathers, and as a material for constructing decorative pathways.

Chipped bark

This is mainly used as a mulch either alone on the soil surface or one of the non-organic sheet mulches such as woven fibre to make an attractive finish. It is valued in many situations but there are several characteristics that need to be noted. Bark is nitrogen deficient so it does not decompose readily, which means it can last a long time on the surface. However, once in the soil, it tends to rob plants of nitrogen (see p. 168). It is light when dry, so tends to be blown around and floats on water, so is difficult to manage on slopes. Birds throw it around when looking for food underneath, making adjoining paths untidy.

Green manures

This is the practice of growing plants to:

▶ cover bare ground
▶ compete out weeds
▶ reduce soil erosion
▶ capture soluble nutrients that would otherwise be leached
▶ add organic matter
▶ increase micro-organism activity in the soil
▶ develop and maintain soil fertility and structure.

It is mainly used in vegetable plots and in organic growing systems (see p. 10). The plants used are typically agricultural crops that cover the ground quickly and yield a large amount of leaf that can cover the ground (see Figure 13.5). The seeds are normally

Figure 13.5 Plants used for green manuring: (a) mustard; (b) *Phacelia*; (c) clover; (d) lupins

165

broadcast sown in the autumn when there are no overwintering plants. The green manure is then dug in or cut, left to wilt then dug into soil. While it can be left as a mulch, this can make it difficult to put the next crop in.

Table 13.2 Plants used for green manuring

Legumes	Non-legumes
Lupinus angustifolius (Bitter blue lupin)	*Fagopyrum esculentum* (Buckwheat)
Medicago lupulina) (Trefoil)	*Phacelia tanacetifolia* (Phacelia)
Trifolium hybridum (Alsike clover)	*Secale cereale* (Grazing rye)
Trifolium incarnatum (Crimson clover)	*Sinapis alba* (Mustard)
Trifolium pratense (Essex red clover)	
Trigonella foenum-graecum (Fenugreek)	
Vicia faba (Winter field bean)	
Vicia sativa (Winter tares)	

Further reading

Brinton, W.F. (1990) *Green Manuring: Principles and Practice of Natural Soil Improvement*. Woods End Agricultural Institute.

Brown, L.V. (2002) *Applied Principles of Horticulture*. Butterworth-Heinemann.

Caplan, B. (1992) *The Complete Manual of Organic Gardening*. Hodder.

Dowding, C. (2012) *Vegetable Course*. Frances Lincoln.

HDRA (2005) *Encyclopaedia of Organic Gardening*. Henry Doubleday Research Association (Garden Organic). Dorling Kindersley.

Killham, K. (1994) *Soil Ecology*. Cambridge University Press.

Lowenfels, J. and Lewis, W. (2006) *Teaming with Microbes: A Gardener's Guide to the Soil Food Web*. Timber Press.

Pears, P. and Sticklands, S. (2007) *The RHS Organic Gardening*. Bounty Books.

Readman, J. (2004) *Managing Soil Without Using Chemicals*. Dorling Kindersley.

Please visit the companion website for further information:
www.routledge.com/cw/adams

CHAPTER 14
Level 2

Plant nutrition (nutrients, pH and fertilizers)

Figure 14.1 Wild flower arable meadow

This chapter includes the following topics:

- The nutrient requirements of plants
- Nitrogen (N)
- Phosphorus (P)
- Potassium (K)
- Magnesium (Mg)
- Calcium (Ca)
- Sulphur (S)

- Iron (Fe)
- Providing plant nutrients
- Fertilizers
- The importance of soil pH
- Causes of soil acidity
- Soil testing
- Adjusting soil pH

Principles of Horticulture. 978-0-415-85908-0 © C.R. Adams, K.M. Bamford, J.E. Brook and M.P. Early.
Published by Taylor & Francis. All rights reserved.

The nutrient requirements of plants

Gardeners are familiar with ensuring that they need to provide a **fertile soil** to grow most of their plants, especially if they are intending to produce vigorous productive vegetables and fruit. They have also seen the consequences of having an impoverished soil; a shortage of one or more nutrients has led to disappointing plant growth. However, it can be puzzling to find that in some circumstances gardeners are advised to reduce the fertility of their soil to get better flowering. Those who have considered establishing a wild flower meadow might well have wondered why it usually involves stripping the good topsoil away (see Figure 14.1). In the feeding of plants, so much depends on the soil available, the plants involved and the intentions of the gardener. Objectives can be as varied as high production from vegetables and fruit, prolific flowering or attractive wild flower meadows leading to very different approaches to nutrition. Rather like the use of medicines or herbicides, twice the recommended dose does not produce better results; nor should it be assumed that poor plant performance will be solved by adding more nutrients.

Most of what is required to build the plant is taken in as water or from the air as carbon dioxide (see photosynthesis p. 112). However, a small but significant amount of a plant's requirements is made up of minerals commonly referred to as 'plant nutrients'. These are normally taken up in soluble form through the roots from the soil solution, but can also be taken in through other parts of the plant, such as foliar feeding.

> **Essential minerals** are inorganic substances necessary for the plant to grow and develop and are often referred to as nutrients.

Most gardeners who have used fertilizers will have noted that the main nutrient contents listed on the containers are nitrogen, phosphorus (also known as phosphate) and potassium (also known as potash). These are the major nutrients because they are needed in relatively large quantities. Also needed in quite large quantities are magnesium, calcium (usually supplied in lime) and sulphur. Iron is an example of a nutrient that is essential but needed in much smaller quantities: a minor nutrient (micro- or trace element) along with manganese, boron, zinc, copper and molybdenum (see Table 14.1).

Essential minerals have very specific functions in the plant cell processes. When they are deficient (in

Table 14.1 Major and minor nutrients in healthy plants

Element	Chemical symbol	grams per kilogram of plant dry matter
Major nutrients		
Nitrogen	N	20–50
Phosphorus	P	2–5
Potassium	K	20–50
Magnesium	Mg	2–10
Calcium	Ca	5–2
Minor nutrients		
Iron	Fe	0.1–0.3

short supply), the plant shows certain characteristic symptoms usually related to their role in the plant.

Nitrogen (N)

The element nitrogen is needed by plants to form chlorophyll and is associated with leafy (vegetative) growth. Consequently, large dressings of nitrogen are given to plants grown for their leaves, such as cabbages, lawn grasses. A mobile nutrient.

Deficiency causes slow, spindly growth in all plants. There is usually a general yellowing (chlorosis) of the leaves due to lack of chlorophyll (see p. 112), often preceded by a bluing of the foliage appearing first on the older, lower leaves. When nitrogen shortages occur in the plant, the chlorophyll in the oldest leaves is broken down to release nitrogen for use in the new young efficient leaves (Figure 14.2); mobile in the plant.

Excess nitrogen produces soft, lush leafy growth, making the plant vulnerable to pest attack and more likely to be damaged by cold. Very large quantities of nitrogen are undesirable because they can harm the plant by producing high salt concentrations at the roots (see osmosis, p. 120) and can easily be lost by being leached.

Nitrogen fertilizers commonly used by gardeners and their nutrient content are given in Table 14.2.

Phosphorus (P)

This element is important in the production of the major chemical required for energy transfer in the plant (adenosine triphosphate, see ATP p. 116). Consequently, large amounts are concentrated in seeds and the meristems of roots and shoots. Phosphorus supplies at the seedling stage are critical: the growing root has a high requirement and the plant's ability to establish itself depends on the roots being able to tap into supplies in the soil before the reserves in the seed are used up. The presence

Figure 14.2 Nitrogen deficiency. Note that it is the lower, older leaves that have been affected before the younger ones at the top of the plant

of more than it needs can be detrimental as this interferes with the development of mycorrhiza (see p. 42). Phosphorus in the plant is constantly being recycled from the older parts to the new growing points. This means that older plants have a low

Figure 14.3 Phosphate deficiency

phosphorous requirement compared with quick-growing plants that are harvested young.

Deficiency symptoms are related to poor root development, which leads to reduced growth of stem and root. There can also be bluish or purplish stem and leaf colourings with or without speckling associated this deficiency (Figure 14.3).

Phosphates fertilizers (containing phosphorus) commonly used by gardeners and their nutrient content are given in Table 14.2.

Potassium (K)

This is associated with successful flowering and fruiting. The element is present in relatively large amounts in plant cells in solution where it acts as an osmotic regulator – for example, in stomata (see p. 123). It also has a role in providing hardiness, resistance to chilling injury, drought and disease in plants. **Mobile** in plant.

Deficiency results in brown, scorched areas on leaf tips and margins of dicotyledonous plants (Figure 14.4). Monocotyledons show similar brown markings at the growing tip of the leaves which spread down the leaf as the deficiency continues. Low potassium levels are usually associated with poor performance of flowers and fruit made worse if large quantities of nitrogen are added.

Table 14.2 Nutrient content of common fertilizers

	N %	P$_2$O$_5$ (P) %	K$_2$O (K) %	Mg %	Ca %	S %
Ammonium nitrate	33–35					
Ammonium sulphate	20–21					24
Bone meal *	3	20 (9)				
Dried blood *	12–14					
Hoof and horn*	12–14					
Keiserite				16		25
Meat and bone meal*	5–10	18 (8)				
Potassium chloride			59 (49)			
Potassium sulphate			50 (42)			17
Superphosphate		18–20 (8–9)			20	12–14
Triple superphosphate		47 (20)			14	
Urea	46					

* an organic fertilizer

Figure 14.4 Potash deficiency

Figure 14.5 Magnesium deficiency

Potash fertilizers (containing potassium) commonly used by gardeners and their nutrient content are given in Table 14.2.

Magnesium (Mg)

This has many roles in the plant, including being required in large quantities to make chlorophyll (see p. 112). **Mobile** in plant.

Deficiency symptoms appear as a characteristic yellowing between the leaf veins (interveinal chlorosis) appearing on the lower, older leaves (Figure 14.5) because the inefficient old leaves release magnesium from their chorophyll to build the chlorophyll in the new young leaves. Consequently, other than spoiling the look of a plant, the deficiency has little or no effect on the performance. Continued shortage leads to more of the leaves becoming chlorotic and plant growth starts to be affected. Gradually the affected areas become brown or reddened.

Magnesium fertilizers commonly used by gardeners and their nutrient content are given in Table 14.2.

Calcium (Ca)

Calcium is a major constituent of plant cell walls as calcium pectate, which holds the cells together after cell division (see p. 81). It also influences the activity of meristems, especially in root tips.

Deficiency symptoms tend to appear in the younger tissues first because calcium is immobile in the plant. It causes weakened cell walls, resulting in inward curling, pale young leaves and sometimes death of the growing point. Specific disorders include 'topple' in tulips, when the flower head cannot be supported by the top of the stem, 'blossom end rot' in tomato fruit and 'bitter pit' in apple fruit (see p. 269).

In general, growing plants within their recommended pH range ensures adequate calcium is available for most plants (see lime p. 174).

Sulphur (S)

This is required by plants in large quantities but is rarely in short supply. It is another nutrient needed for the synthesis of proteins including chlorophyll (see p. 000). Consequently, a deficiency produces a yellowing (chlorosis) of leaves appearing in the younger leaves first.

Figure 14.6 Lime-induced iron deficiency on *Elaeagnus x ebbingei*. Note that it is the younger leaves that have chlorosis (yellowing) while the older leaves are still green

Iron (Fe)

Iron is involved with chlorophyll production. Although it does not form part of the chlorophyll molecule, it is a component of some enzymes required to synthesize it.

Deficiency of iron results in yellow ('chlorosis') or even white leaves appearing in the younger leaves first. The deficiency is commonly caused by the presence of large quantities of lime (see p. 176). This '**lime-induced**' chlorosis (Figure 14.6) occurs on over-limed soils and chalky soils; calcicoles are adapted for the chalk and limestone areas, but other plants grown in such conditions do not fare well and have a typically yellow appearance. Good drainage and soil structure should be maintained as a water-logged root zone can contribute to the problem of an iron deficiency.

Iron fertilizer or supplement in the form of ferrous sulphate is commonly added in small quantities to amenity turf to 'green up' its appearance without the need to add more nitrogen. Chelated iron is a supplement that can be used as a precaution or to treat affected plants being grown on an unsuitable growing medium.

Other trace nutrients

Copper, boron, zinc, manganese and molybdenum are essential nutrients that are required in minute quantities and are rarely deficient in cultivated soils in Britain and Ireland. They are added as supplements to some growing media made up of materials with little or no trace nutrients present.

Providing plant nutrients

In nature the recycling of nutrients ensures the continued growth of plant communities (see p. 34) unless there are net losses through leaching. This can also be the situation in some gardens, especially where plantings are similar to natural ones, such as predominantly trees and shrubs. In practice, even in decorative gardens there is a need to import some nutrients. There are significant differences in maintaining a lawn where mowings are allowed to stay on the lawn (recycling) compared with one where the arisings are 'boxed off' (i.e. collected and not put back). Most gardeners make good the nutrient losses with their garden compost (see p. 164), brought in as composted municipal waste, manures (see p. 172) and/or fertilizers.

Organic gardeners and growers put the emphasis on the recycling of organic matter including composting (see p. 160) and minimizing the use of artificial (synthetic) fertilizers. They can only make good nutrient loss by crop removal through the importation of manures (under strict rules about the sources if organic status is to be maintained) or the use, in strictly prescribed circumstances, of a limited range of ('permitted') fertilizers.

> **Fertilizers** are concentrated sources of plant nutrients that are added to growing media.

> **Manure** is a source bulky organic matter comprising animal faeces and bedding.

Organic sources of nutrients

Organic in its original sense meant chemicals derived from organisms, that is, living things. The idea that there are chemicals that could only be made this way was shown to be false when urea was synthesized from inorganic ingredients in 1828. 'Organic chemistry' is now more commonly referred to as 'Carbon Chemistry', that is, the chemistry of all compounds containing carbon atoms. See companion website: www.routledge.com/cw/adams.

14

Organic sources of nutrients include the different types of bulky organic matter that release nutrients many of which are better known as '**manures**'. **Organic fertilizers** are normally considered to be the materials with concentrated sources of nutrients derived from organisms, such as hoof and horn, bone meal, blood meal (see Table 14.2). Urea presents an anomaly in so far as when originating from urine it is clearly 'organic', but the same chemical when synthesized is not in an 'organic gardening' sense.

Fertilizers

Organic fertilizers include those derived from organic sources such as bone meal; hoof and horn; dried blood; fish, blood and bone; comfrey feed; or nettle tea.

Inorganic fertilizers are those commonly referred to as 'synthetic', 'artificial' or 'bag' fertilizer such as ammonium sulphate, triple superphosphate, potassium chloride and National Growmore. Some are mined/quarried materials such as rock phosphate.

The quantity of nutrient supplied by a fertilizer is expressed in terms of a percentage of the contents.

▶ Nitrogen fertilizers are given terms of percentage of the element nitrogen in the fertilizer, i.e. %N.
▶ Phosphate fertilizers provide the element phosphorus and given as %P and more usually in Britain and Ireland described in terms of the 'equivalent amount of phosphoric oxide', i.e. %P_2O_5.
▶ Potash fertilizers provide the element potassium and given as % K; the 'old' name for potassium is 'Kalium'; and more usually in Britain and Ireland described in terms of 'equivalent amount of potassium oxide', i.e. %K_2O.
▶ Magnesium fertilizers are described in terms of %Mg.

The percentage figures show the quantities of nutrient in each 100 kg of fertilizer – for example, ammonium sulphate is 20% N so there is 20 kg of the element nitrogen in every 100 kg of the fertilizer so by proportion there is 5 kg in a 25 kg bag or 1 kg in a 5 kg bag.

Fertilizers are a concentrated source of nutrients as illustrated by the following materials that supply the element nitrogen (N) – for example:

1kg N is supplied in a 5kg bag of ammonium sulphate.

1kg N is supplied in 2-5 tonnes of fresh farmyard manure.

i.e., the nutrients in bulky organic matter are more diluted (or non-existent).

Besides the major nutrient content, fertilizer regulations require that details of trace elements, pesticide content and phosphorus solubility should appear on the packaging. For **organic gardening** it is necessary to look at specific requirements including the limitations on fertilizer use.

There are many different types of fertilizers in terms of their

▶ content (straight/compound)
▶ formulation (granules/powders, quick/slow/controlled release)
▶ the ways they are used (base/top dressing, liquid feed, foliar feed).

Straight fertilizers are those that supply only one of the major nutrients: nitrogen, phosphorus, potassium or magnesium such as ammonium sulphate (supplying N), triple superphosphate (P), potassium chloride (K).

Compound fertilizers are those that supply two or more of the nutrients nitrogen, phosphorus and potassium, such as Growmore. The accepted convention for describing the fertilizer content of compounds is to label the content in the order N P K so Growmore is described as 7:7:7, that is, 7% N: 7% P_2O_5: 7% K_2O. Most of the proprietary fertilizers for gardeners are compounds including those specifically for tomatoes, roses, orchids and cacti (see Figure 14.7).

Quick release or soluble fertilizers are those that dissolve immediately in water before or after application to soil – for example, ammonium sulphate and urea both of which yield nitrogen that can be taken up by plants within days of application. The

Figure 14.7 Range of fertilizers. Some are sold as straights or general compounds, but many for gardeners are compounds formulated for specific plants

downside of quick release materials is that if too much is applied, the plants can be harmed or even killed; often referred to as 'root scorch' or 'burnt plants' because water is drawn out of the root by the high salt concentration (see osmosis p. 120 and disorders p. 271). Furthermore, care needs to be taken to ensure ground water is not polluted by applying more than can be taken up by plants or watercourses contaminated by run-off following top dressings.

Organic gardeners avoid the use of quick release fertilizers because they identify the elevated level of solubles in the soil water as being responsible for reducing the effect of beneficial soil organisms.

Slow release fertilizers do not dissolve immediately in water but provide nutrients in soluble form over a long period. Many are organic fertilizers such as bone meal and hoof and horn that are decomposed by micro-organisms, so the rate at which nutrients are released depends on temperature. This fits well with plant requirements which increase as they grow larger over the late spring and early summer. However, a downside is that there is little or no nutrient to take up when plants start to grow after the winter when temperatures have been too low for micro-organism activity.

There are several fertilizers that dissolve very slowly in the soil water. These include Urea- Formaldehyde (UF) often used on turf because it has the advantage of releasing nitrogen even in cold conditions. Rock phosphate releases phosphate in a form for plant uptake very slowly (over many years).

Controlled release fertilizers are slow release fertilizers that are formulated to release nutrients in a controlled way over a specified long period. One group comprises quick release fertilizers held within a permeable resin coating that lets water in and allows nutrients to diffuse out. The other group is soluble fertilizer coated with sulphur which is broken down by micro-organisms so nutrients are released. In both cases, the thickness of the coatings can be varied to enable fertilizers to be designed so they release nutrients in line with particular plant needs and for the required period of time – for example, a three-month formulation for bedding plants or a nine-month one for container roses.

Base dressings are the fertilizers that are applied to the soil and worked into the seedbed, or incorporated in composts, before sowing/planting.

Top dressings are granular fertilizers applied to the surface of soils because nutrients are needed after plants have been established for ongoing nutrition often in the form of a compound fertilizer – for example, autumn treatment for lawns, spring dressing for established borders.

In order to reach the root zone, they are usually quick release fertilizers – for example, applications of nitrogen, often with ferrous sulphate, to 'green up' lawns. This means that care needs to be taken to avoid harming plant leaves ('scorch'). An alternative approach is to use liquid feed.

Liquid feeds are fertilizers dissolved and watered on to soils or composts to provide ongoing plant nutrient requirements over growing season typically for pot plants, hanging baskets, houseplants, bedding and in greenhouse production. When used in conjunction with irrigation systems ('fertigation'), nozzle blockages can occur unless pure materials are used. As with top dressing, care needs to be taken to avoid damaging leaves ('scorch').

> **Fertigation** is supplying nutrients in the irrigation water.

Foliar feed is a liquid feed diluted sufficiently so that it can be applied to leaves without causing 'scorch'. It can be used for routine feeding and has the advantage of immediate uptake to treat deficiency symptoms. Application is usually undertaken first thing in the morning by spraying the leaves until there is 'run-off'.

Fertilizers are provided mainly in the following forms for use in the above situations:

▶ **Granules** which have the advantage of being easy to deliver by hand or by fertilizer spreaders when applying base or top dressings.
▶ **Powders** (or crystals) are provided for those making up liquid or foliar feeds.

The importance of soil pH

Plants adapted to grow on more acid soils (low pH) are seen on typical upland moorland, dry heaths and wet peat bogs in Britain and Ireland. They feature **calcifuges** such as bilberry, ling and acid-tolerant heathers on the moors, or gorse (furze), sheep sorrel, broom, harebell and tormentil on the dry lowland heaths. The peat bogs (see p. 7) are wet usually acid areas often dominated by sphagnum moss with cotton grass, cranberry, bog myrtle and sundews. **Calcifuge** plants commonly seen in the garden include many *Rhododendron* spp., *Camillia*, *Pieris*, Blueberries and some heathers.

173

Calcifuge plants are those adapted to grow on acid soils below pH 5.5.

Calcicoles such as Clematis and *Prunus padus* have adapted to the calcium-rich soils (high pH). Cultivated plants that thrive on these soils include *Agapanthus*, Clematis, *Geranium*, *Echinacea*, *Jasminum*, *Lonicera*, *Parthenocissus*, *Rudbeckia* and *Verbascum*.

Calcicoles are plants that are adapted to grow on calcareous soils (calcium-rich, chalky).

Acids, bases and the pH scale

▶ **Acids** are a group of chemicals that have a sour taste and are corrosive.
▶ **Bases** are those chemicals that neutralize acids. Bases include lime (calcium carbonate).
▶ **Alkali** ('lye') is a are soluble base. They have a soapy feel and tend to be irritants; strong bases/alkalies are corrosive.

When bases (alkalies) and acids are mixed they neutralize each other and form 'salts' – for example, common salt which is sodium chloride made from sodium hydroxide and hydrochloric acid. Many fertilizers are formed this way, such as ammonium sulphate and potassium chloride.

The **pH scale** is a means of expressing the degree of acidity or alkalinity (see Figure 14.8).

Very low pH values are associated with the strong acids such as sulphuric, nitric and hydrochloric acids; strong bases (alkalies) such as caustic soda have very high values. As they are diluted with more and more water, their pH moves closer and closer to pH 7 which is neutral – for example, pure water. Weak acids and bases (alkalies) have values nearer to pH 7 (neutral) even when they are concentrated.

The soils of Britain and Ireland are usually between pH 4 and 8; the vast majority being between 5.5 and 7.5. Although there does not appear to be much difference across this range, the significance for organisms living in the soil is considerable. The pH scale is logarithmic (like measuring sound in decibels or earthquake energy release on the Richter Scale): each 'unit' is ten times larger or smaller than the one next to it, that is, pH 5 is ten times more acid than pH 6 and pH 4 is one hundred times more than pH 6.

The ideal growing condition for most plants is a soil of pH 6.5 because at this point all the essential plant nutrients are available for uptake by the roots of most plants. Although the majority of plants grow well in soils between pH 6 and 7, there are considerable differences in the tolerance of plants to soil pH conditions (see Table 14.3). Potatoes are considerably more tolerant of soil acidity than most plants and are still productive down to pH 4.9. In contrast, the yield of celery falls significantly in soils below pH 6.3.

The pH scale

The **pH scale** expresses the amount of acidity or alkalinity in terms of hydrogen ion concentration.

▶ pH 7 = neutral, e.g. pure water
▶ below pH 7 = acids, e.g. rainwater, carbonic acid, fizzy drinks, lemon juice (contains citric acid), vinegar (dilute ethanoic acid), sulphuric acid
▶ above pH 7 = bases, e.g. limewater, caustic soda

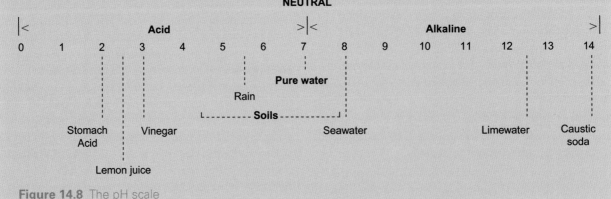

Figure 14.8 The pH scale

Table 14.3 Soil pH and plant tolerance

a) *pH below which plant growth may be restricted on mineral soils*			
Celery	6.3	Rose	5.6
Daffodil	6.1	Raspberry	5.5
Bean	6.0	Cabbage	5.4
Lettuce	6.1	Strawberry	5.1
Carnation	6.0	Tomato	5.1
Chrysanthemum	5.7	Apple	5.0
Carrots	5.7	Potato	4.9

b) *Growth adversely affected in soils greater than pH 5.5*

▶ *Rhododendron* spp, *Camellia* spp., *Pieris* spp.
▶ Blueberries (*Vaccinium corybosum*)
▶ Some heathers such as *Daboecia spp.* (some species are more tolerant of lime e.g. *Carnea* spp.)

c) *Hydrangea*

▶ Tend to be pink in soils above pH 5.9
▶ Are blue when grown in soils below pH 5.5
▶ In between tend to have a transitional mix of neither one or the other (see Figure 14.9)

Figure 14.9 *Hydrangeas*. The blue colour of the flowers depends on the availability of aluminium to the plant; this element is readily available in growing media with a pH less than 5.5 and gradually less so at higher pH levels

Basic conditions (alkaline) are usually created by the presence of large quantities of lime ('calcium') which interferes with the uptake and utilization of several of the plant nutrients. Calcicoles are adapted to these conditions and thrive in the limey/chalky soils. In contrast, calcifuge plants are adapted to growing in soils with a pH below 5.5; above that the lime (calcium) present interferes with the uptake and utilization of many of the nutrients they require.

Soil pH also has important effects on other organisms besides plants. Beneficial soil organisms are affected by soil acidity and liming. A few soil-borne disease-causing organisms tend to occur more frequently on acid, lime-deficient soils (see clubroot, p. 256), whereas others are more prevalent in well-limed soils, such as common scab of potatoes.

The addition of lime ('calcium') usually improves soil structure because raising pH can create conditions more favourable for beneficial organisms. This leads to the decomposition of organic matter, yielding humus that helps crumb formation in both very sandy and heavy clay soils (see humus p. 158). Furthermore, calcium-rich clays tend to crack and crumble more readily.

Causes of soil acidity

In Britain and Ireland, where over a year the rainfall exceeds evaporation, the soils tend to become too acid (creating 'sour soils'). This because the rain is carbonic acid (see p. 142) which dissolves basic material, lime, which is then leached from the soil. Lime is very readily leached from free-draining sandy soils in high rainfall areas, so they tend to go acid very rapidly. Calcareous soils (i.e. those containing pieces of chalk or limestone) do not become acid until all these base reserves are used up. In addition to the

175

effect of rainfall (carbonic acid), there are several other factors that increase the rate at which soils become more acid:

▶ **Acid rain** (polluted rain and snow) is directly harmful to vegetation and also contributes to the fall in soil pH.
▶ **Organic acids** derived from the microbial breakdown of organic matter, e.g. humic acids, also lead to an increase in soil acidity.
▶ **Fertilizers**. Some fertilizers such as ammonium sulphate increase the rate at which soils become acid.
▶ **Crop removal**. Some plant nutrients such as calcium, magnesium and potash are bases so when they are taken up by plants but not recycled the soil acidity can increase.

Soil testing

Soil testing kits are available for the gardener which can give a useful guide to the nutrient status and soil pH with guidance as to how to use the results for liming, fertilizer and/or manure application. Growers of valuable crops would more normally use laboratory testing.

The testing of outdoor soils is normally undertaken in the autumn (but not within a few months of applying nutrients or lime/sulphur) in readiness for the following year. Nutrient testing is usually limited to determining phosphate and potash levels; 'nitrogen' testing is of limited value at this time because so much of the soluble nitrogen (nitrate) is leached from the soil during the winter.

Soil pH can be measured accurately in the laboratory with pH meters. Gardeners usually use either very simple pH meters (sticks) or colour-indicator methods to do their own pH testing, but these methods are not usually better than a half unit either side of the correct value. A method of testing soil pH is given on the companion website. Above all, the usefulness of any soil analysis depends on the degree to which the sample taken is representative of the area from which it is taken (and to be treated).

The **variability of the soil** makes it difficult to obtain a result on which to base any calculations for nutrient or lime application. That is why it is recommended that several cores of soil are taken (ideally 20 cores down to 15 cm) over the area to be treated, such as 'the lawn', but avoiding any abnormal areas, such as a new area of the lawn that was formerly being used to grow vegetables (this should be tested and treated separately). On a large scale, areas with different textures of soil (sandy loam, clay loam and so on)

should be sampled separately. Finding a satisfactory target area in a garden or allotment can be more difficult because of the variable treatment of all the small areas leading up to the test: different history of crops/plants, fertilizer and organic matter additions.

Nevertheless, there is merit in testing the soil when:

▶ taking on a new allotment or garden
▶ establishing a new lawn
▶ establishing plantations (soft fruit, cane fruit, orchards)
▶ planting valuable specimens
▶ planning to grow plants that require an unusual soil pH range to survive, e.g. calcifuges or calcicoles
▶ there appears to be a nutritional problem, e.g. deficiency symptoms or general poor growth.

Adjusting soil pH

There is merit in selecting plants that will grow in the soil without adjusting it especially if it is at either of the extremes of the pH range. However, soil pH does change over time (see above) and plants may suffer as a result. More particularly, growing fruit and vegetables requires having soil in the appropriate pH range in order for the plants to be productive.

Soil pH can be raised ('sweetened') by the addition of **lime** normally as ground chalk or ground limestone (both are calcium carbonate, garden lime). An alternative commonly bought from garden centres is hydrated lime (calcium hydroxide, slaked lime, builder's lime). All should be applied as a fine powder because coarse material takes too long to affect the soil. Care should be taken to protect the eyes as the powders are easily blown around and gloves should be worn to handle hydrated lime as it can cause skin irritation and chemical burns. Wood ash, the result of burning organic matter, is rich in potash, which has a similar effect as calcium and magnesium so it can also be used to help raise the pH.

> **Lime requirement** is the quantity of calcium carbonate required to raise the soil pH to pH 6.5.

A rough guide to how much lime to add can be found in Table 14.4. This table gives **the quantities of calcium carbonate required to raise the pH of different soils to pH 6.5** (the recommended level to return mineral soils when its pH is too low for a future planting of 'normal plants'). 'Over-liming' a soil must be avoided because this can reduce the availability of plant nutrients; easily done on very sandy soils.

Table 14.4 Raising soil pH guidelines: quantities of calcium carbonate* (g/m² of ground chalk or limestone*) required to raise soil to pH 6.5

Starting pH	Soil texture		
	Light soils	'Loams'	Heavy soils
	(very sandy, low clay)		(very high clay content)
5.0	1,000	1,200	1,400
5.5	700	800	1,000
6.0	400	500	600

* If using hydrated lime, these application rates should be reduced by a quarter.

Note that even fine lime needs to be added many months in advance to achieve the pH required for the plants to be grown. The amount of lime to add depends on:

▶ final soil pH to be achieved (normally raised to pH 6.5)
▶ soil pH from which it is to be lifted (as found in the soil test)
▶ texture of the soil (it is the clay content that resists the effect of lime)
▶ strength of the lime used ('neutralizing value')
▶ fineness of the lime.

Soil pH can be lowered by the addition of sulphur which is converted to sulphuric acid by soil micro-organisms. The amount of sulphur required depends on essentially the same factors as for lime requirement: the pH change needed and the soil texture, that is, more is required for clay soils to have the same effect as on sandy soils.

Acid fertilizers such as ammonium sulphate reduce soil pH over a period of years in outdoor soils, such as on lawns. They can also be used in liquid feeding to offset the tendency of hard water to raise pH levels in composts.

Further reading

Brown, L.V. (2008) *Applied Principles of Horticulture*. 3rd edn. Butterworth-Heinemann.
Cresser, M.S. (1993) *Soil Chemistry and Its Applications*. Cambridge University Press.
Ingram, D.S. et al (eds) (2008) *Science and the Garden*. 2nd edn. Blackwell Science.
Pratt, M. (2005) *Practical Science for Gardeners*. Timber Press.

14

Please visit the companion website for further information:
www.routledge.com/cw/adams

Growing in containers

Figure 15.1 Tomatoes growing in a hydroponics system

This chapter includes the following topics:

- Growing in a restricted root volume
- Composts
- Loam-based composts
- Plant containers
- Loamless composts
- Hydroponics
- Nutrient film technique (NFT)
- Aggregate culture
- Green walls

Principles of Horticulture. 978-0-415-85908-0 © C.R. Adams, K.M. Bamford, J.E. Brook and M.P. Early.
Published by Taylor & Francis.

Figure 15.2 Hanging baskets to attract customers

Growing in a restricted root volume

Most gardeners have at least some of their plants in containers usually pots, tubs, troughs or hanging baskets. For some, their plant growing is confined to window boxes. Many containers are used for indoor plants, including quite large ones in conservatories. For others their interest lies in the production of tomatoes, cucumbers and peppers in their greenhouses using 'grow bags'. Many non-horticultural businesses make good use of the colourful displays in containers that are there to create a pleasant working environment and also to attract customers (Figure 15.2). Across the horticultural industry there are the equivalents of these methods of growing, and many more, but on a much larger scale.

Growing in containers makes considerable demands on the growing medium for air, water and nutrients because rooting is severely restricted. Compare the volume of containers such as flower pots, hanging baskets and grow bags with the volume that roots explore in the soil (see p. 140). When restricting plants like this, we undertake to ensure that all their water and nutrient requirements are provided through this relatively small volume. Consequently, growing in a restricted root volume brings two main challenges:

▶ providing a large amount of nutrient in a small amount of water without 'scorching' the roots (see p. 120)
▶ meeting the plant's water requirements while maintaining good gaseous exchange (see p. 147). There needs to be large pore spaces that do not collapse over the time the plant is in the container.

Limitations of using soils in containers

Soil is an inappropriate material to use in containers because most types lack the stability needed to

ensure that the crumb structure needed withstands the constant wetting (see p. 151). After watering a pot full of almost any soil, it is not long before it is only half full of soil: the soil collapses as the larger pores are reduced in size and small pores predominate. This severely reduces the aeration and drainage in the root environment. Consequently, alternative growing media (non-soils) are used in containers. These are generally called composts, but also plant substrates, plant growing media, or just 'mixes' or 'media'.

Composts

Materials alone or in combination are prepared and mixed to achieve a rooting environment that is free from pests and disease organisms and has constant air (oxygen) supplies, easily available water and suitable bulk density for the plant to be grown.

While lightweight mixes are usually advantageous, 'heavier' composts are sometimes formulated to give 'pot stability' for taller specimens. This should be achieved not by compressing the lightweight compost but by incorporating denser materials such as sand or grit. Quick-growing plants are normally the aim and loosely filling containers with the correct compost formulation, consolidated with a presser board and settling it with applications of water, will achieve this (see p. 130). The addition of nutrients has to take into account not only the plant requirements but also the nutrient characteristics of the components used. Most require the addition of lime (see p. 177) and all the major nutrients. Loamless composts require trace element supplements.

> **Compost** is growing medium used for growing plants in containers.

Over the years gardeners have added a wide variety of materials such as leaf mould, pine needles, spent hops, old mortar, crushed bricks, composted animal and plant residues, peat, sand and grit to selected soils to produce a compost with suitable physical properties. Typically there was a different one for each type of plant and many head gardeners guarded the details of their successful mixes. Significant developments were made as a result of the work done in the 1930s by Lawrence and Newell at the John Innes Institute. Before they undertook their primrose trial, they set out to eliminate all variation except the cultivars to be investigated. This included eliminating the variable results from the composts then being used. They demonstrated the importance of 'sterile' (pest- and disease-free), **stable** and **uniform** ingredients. They incorporated the developments in plant nutrition which had identified the problems associated with unbalanced nutrient supply, resulting in plants being either too 'hard' and slow growing or too 'soft' and disease prone. The range of loam-based composts that resulted from this work established the methods of achieving uniform production and reliable results with a single potting mixture suitable for a wide range of plant species.

Loam-based composts

Loam composts, typified by John Innes (JI) composts, are based on **loam sterilized** to eliminate the water-borne fungi (see damping off, p. 254) and insect pests. The loam should have sufficient clay and organic matter present to give good structural stability; the original John Innes specification identifies 'turfy clay loam'. Peat and sand (grit) are added to improve the physical conditions: the peat giving a high water-holding capacity and the coarse sand ensuring free drainage and therefore good aeration. There are two main John Innes composts: one for seed sowing and cuttings, the other for potting.

John Innes seed compost consists of two parts loam, one part peat and one part sand by volume. Well-drained 'turfy clay loam' low in nutrients with a pH between 5.8 and 6.5, undecomposed peat graded 3–10 mm with a pH between 3.5 and 5.0, and lime-free sand graded 1–3 mm should be used. Furthermore, 1,200 g of superphosphate and 600 g of calcium carbonate are added to each cubic metre of compost.

John Innes potting (JIP) composts consist of seven parts by volume 'turfy clay loam', three parts peat and two parts sand. To allow for the changing nutritional requirements of a growing plant, the nutrient level is adjusted by adding appropriate quantities of JI base fertilizer, which consists of two parts by volume hoof and horn, two parts superphosphate and one part potassium sulphate. To prepare JIP 1, 3 kg JI base fertilizer and 600 g of calcium carbonate are added to one cubic metre of compost. To prepare JIP 2 and JIP 3, double and treble fertilizer levels are used. This provides a comprehensive series of composts for growing in containers:

▶ JIP 1 has low nutrient levels but good phosphate levels for pricking out or potting up seedlings or cuttings.
▶ JIP 2 with the higher nutrient level is suitable for potting plants from JIP 1, potting up most houseplants, for use in hanging baskets, window

boxes and most plants being grown in medium-sized containers.
▶ JIP 3 nutrient content makes it suitable for final potting for the heavy feeding plants and for mature foliage plants.

Ericaceous alternative. While the standard JI composts are suitable for a wide range of species, some modification is required for some specialized plants. For example, calcifuge plants such as *Rhododendron*, *Camellias* and some heathers should be grown in a JI(S) mix in which sulphur is used instead of calcium carbonate to provide a sufficiently acid medium.

> **Calcifuge** plants are those adapted to grow on acid soils below pH 5.5.

Loam-based composts are well proven and are relatively easy to manage because of the water-absorbing and nutrient-retention properties of the clay present. This makes them a good choice for amateurs, those growing valuable specimens and for tall plants where 'pot stability' is important. Their main disadvantage has always been the difficulty in obtaining suitable quality loam ('turfy clay loam') as well as the high costs associated with sterilizing. Furthermore, the loam must be stored dry before use and the composts are heavy and difficult to handle in large quantities. Many loam-based composts now made have relatively low clay content and consequently exhibit few of its advantages. They have been superseded in horticulture generally by cheaper and cleaner alternatives.

Loamless composts

Loamless ('soilless') composts can provide the advantages of a uniform growing medium, but with components that are lighter, cleaner to handle, cheaper to prepare and which do not need to be sterilized (unless being used more than once). However, without loam the control of nutrients, including micronutrients, is more critical. Most of the components have low nutrient levels which manufacturers and growers who make their own compost are able to exploit because they can add nutrients accurately for their intended purpose.

Peat has, until recently, been the basis of most loamless composts. Peats are derived from partially decomposed plants and their characteristics depend on the plant species and the conditions in which

181

Table 15.1 Alternatives to peat

Organic materials	Inorganic materials
Coir	
Garden compost	Perlite
Green waste	Polystyrene
Leaf mould	Rockwool
Recycled landfill	Sand/grit
Straw	Vermiculite
Vermicomposts	
Wood chips/fibre	
Woodwastes	

they are formed but any one type is usually very uniform, such as **sphagnum moss peat**. This is less decomposed than other types of peat and has the ideal open structure for plants in containers.

Great efforts are being made to find **alternatives to peat** in order to preserve the wetland habitats where peat is harvested in Britain and Ireland and beyond. This peatland is also a significant carbon store that releases greenhouse gases (carbon dioxide) once it is made suitable for plants. The use of peat is not sustainable because it develops so slowly (see p. 7). A list of some of these alternative materials is given in Table 15.1.

A very wide range of loamless composts is available for both the amateur and the professional grower, but they broadly fit in to the following main categories:

▶ **Multipurpose composts**, as their name suggests, have been formulated for a wide range of purposes. The components are sufficiently fine to ensure successful growing from seed or cuttings while having sufficient air-filled porosity to grow larger plants on in a range of containers. These 'all-purpose' composts have a balance of nutrients that meets the needs of most plants grown in containers. There are sufficient nutrients to get the plants started, but additional feeding is usually needed after about a month. There is usually advice available as to what and how to add more nutrients ('feed') as the season progresses. Lime has been added to these, so these composts are not suitable for calcifuges ('lime-hating', ericaceous plants).
Most are now 'peat free' or 'reduced peat' which means that they are made up of a variety of ingredients. There are many on the market and they vary in performance and cost. Research continues and improvements are being made all the time. They are reviewed quite frequently in the consumer magazines and the gardening press

with descriptions of those on the market, their performance across a range of plants and cost.

▶ **Seed and cutting composts** are more precisely adjusted for use with seeds, particularly fine seed, and cuttings that benefit from having finer, more closely graded components and lower soluble nutrient content, although phosphate levels are maintained (see p. 166). Lime is added, so these composts are not suitable for calcifuges ('lime-hating', ericaceous plants).

▶ **Container composts** tend to be coarser, less closely graded and while usually suitable as a potting compost, it is appropriate for the larger plants being grown in larger containers. Lime is added so again these composts are not suitable for calcifuges ('lime-hating', ericaceous plants).

▶ **Ericaceous mixes** are specifically designed for calcifuges ('lime-hating', ericaceous plants).

Plant containers

There is an enormous range of containers used to meet the many different requirements of growing plants (see Figure 11.3).

Plastic containers predominate, with a variety of shapes and sizes. Most are very functional, but there are many decorative containers available with advantages in terms of lightness and cost. The black ones tend to heat up more than the standard terracotta-coloured ones and the contents of white plastic pots can be as much as 4°C lower than in other colours. Pots of white or light green plastic can transmit sufficient light to affect root growth adversely and encourage algal growth.

Clay pots are porous and water is lost from the walls by evaporation. Consequently, clay pots dry out more rapidly than plastic ones, especially in the winter and, although air does not enter through the walls, this can help to improve air-filled porosity. The higher evaporation rate also keeps the clay pots slightly cooler, which can be beneficial in hot conditions.

Biodegradable containers such as those made from paper have become popular because they can be planted directly. Some materials decompose more rapidly than others and there can be a temporary lock-up of nitrogen, but most such containers are now manufactured with added available nitrogen. It is essential that these containers are soaked and the surrounding soil is kept moist after planting, or the roots will fail to escape from the dry wall.

Hydroponics

Hydroponics (water culture) involves the growing of plants in water. The term includes the growing in solid rooting medium watered with a complete nutrient solution, which is more accurately called 'aggregate culture'. Plants can be grown in nutrient solutions with no solid material so long as the roots receive oxygen and suitable anchorage and support is provided.

> **Hydroponics** is the cultivation of plants in nutrient solution without soils.

Active roots require a constant supply of oxygen, but oxygen only moves slowly through water. This can be resolved by pumping air through the water in which the plants are grown (the same as the aeration of fish tanks), a method that has been used in experimental situations. Supplying oxygen to the roots is usually achieved on a large scale by growing in thin films of water, as created in the nutrient film technique (NFT) or a variation on the much older aggregate culture methods.

The advantages of hydroponic systems include:

▶ water – there is a constant supply of available water to the roots.
▶ nutrition – accurate control can be achieved and hence better growth and yield.
▶ conservation of water – evaporation is greatly reduced and loss of water and nutrients through drainage is minimal in recirculating systems.
▶ reduced costs – these can apply depending on the system as a result of labour reduction, reduced growing medium costs and/or quicker turn-around time between crops in protected culture.

Limitations include high initial costs of construction and the controls of the more elaborate automated systems. There are also different skills required for maintaining the correct pH and nutrient levels in a medium, water, that has no 'buffering capacity', that is, small changes in the water can quickly bring about a need for immediate correction – for example, pH can fall rapidly overnight to harmful levels (compare with changes that take years in soils before liming is needed). Other disadvantages apply to the different methods adopted.

Nutrient film technique (NFT)

This is a method of growing plants in a shallow stream of nutrient solution continuously circulated along plastic troughs or gullies. The method is commercially possible because of the development of relatively cheap non-phytotoxic plastics to form the troughs, pipes and tanks (Figure 15.3). There is no solid rooting medium: a mat of roots develops in the nutrient solution and in the moist atmosphere above it. The nutrient solution is lifted by a pump to feed the gullies directly or via a header tank. The ideal flow rate through the gullies appears to be about 4 litres per minute. The gullies are commonly made of disposable black/white polythene set on a graded soil

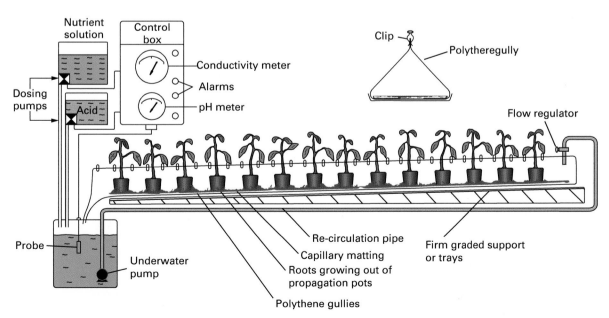

Figure 15.3 Nutrient Film Technique. The nutrient solution is pumped up to the top of the gullies. The solution passes down the gullies in a thin film created by capillary matting laid in the bottom which spreads the stream of water. The nutrient and pH levels in the catchment tank are monitored and adjusted as appropriate

or on adjustable trays. There must be an even slope with a minimum gradient of 1 in 100 with a flat bottom usually lined with capillary matting to ensure a thin film across the trough; areas of deeper liquid stagnate and adversely affect root growth.

The nutrient solution can be prepared on site from basic ingredients or proprietary mixes. It is essential that allowance be made for the local water quality, particularly with regard to the micronutrients such as boron or zinc that can become concentrated to toxic levels in the circulating solution. The nutrient level is monitored with a conductivity meter (measuring salt concentration levels) and by careful observation of the plants. Monitoring pH and maintaining at the correct level is also very important. Nutrient and pH control is achieved using, as appropriate, a nutrient mix: nitric acid or phosphoric acid to lower pH and, where water supplies are too acid, potassium hydroxide to raise pH. Great care and safety precautions are necessary when handling the concentrated acids during preparation.

The commercial NFT installations have automatic control equipment in which conductivity and pH-meters are linked to dosage pumps. The high- and low-level points also trigger visual or audible alarms in case of dosage pump failure. Dependence on the equipment may necessitate the grower installing fail-safe devices, a second lift pump and a standby generator. A variation on this method is to grow crops such as lettuce in gullies on suitably graded greenhouse floors (Figure 15.4a) or vertically to maximize floor area (Figure 15.4c). A characteristic of plants grown in water culture is their lack of root hairs (Figure 15.4b).

Aggregate culture

In aggregate culture, the nutrient solution is broken up into water films by an essentially inert solid medium such as coarse sand or gravel. More commonly today are materials such as perlite, rockwool or expanded clay aggregates (Figure 15.5). Rockwool, commonly used as an insulation material, is derived from rock which has been crushed, melted and spun into fibres. This provides a lightweight, absorbent, inert and sterile rooting medium usually supplied as polythene-wrapped slabs and cubes. Similarly, perlite is derived from rock which is crushed and the particles expanded to 20 times their original size to produce white, lightweight aggregate. These are porous with a rough surface so perlite has good water-holding capacity.

Typical systems make use of rockwool as polythene-wrapped slabs (Figure 15.6a), perlite in 'bolsters' (Figure 15.6b) or aggregate clay granules in gullies. These sit on a polythene-covered floor graded across the row. Polyurethane sheets are often placed underneath them to help create even slopes and insulate them from the cooler soil below. The plants are drip fed with a complete nutrient solution at the top, with the surplus running out through slits towards the bottom on the opposite side. Rockwool slabs are a very successful way of growing and lend themselves to a modular system. This method is widely used for a range of commercial crops such as tomatoes, cucumbers, peppers, melons, lettuce, carnations, roses, orchids and strawberries in protected culture.

When this method was first developed, the NFT system was copied. However, the recirculation of

(a) (b) (c)

Figure 15.4 Hydroponic lettuce growing: (a) NFT in shallow troughs on the floor; (b) close-up of roots grown in gullies filled with aggregate clay granules; (c) gullies arranged vertically to maximize cropping area

Figure 15.5 Hydroponic root media: (a) rockwool modules which are designed to receive the small plants and then at the appropriate stage the block is moved into or on to rockwool slabs; (b) aggregate clay granules are used to fill gullies and other containers

Figure 15.6 Hydroponic growing systems: (a) rockwool as polythene-wrapped slabs; (b) perlite in 'bolsters'

water was too difficult to maintain where the quality of water was poor. It was found that the surplus nutrient solution was most easily managed by allowing it to run to waste into the soil. However, this open system presents environmental problems and increasingly a closed system has had to be adopted. It is now becoming more usual to run the waste to a storage sump via collection gullies or pipes. Some of this can be used to irrigate outdoor crops if nearby. To recirculate the water it is necessary to have equipment to remove the excess fertilizer or accept a gradual deterioration of the nutrient solution. It is then flushed out to a sump when it becomes unacceptable.

There was also a risk of a build-up of water-borne pathogens and trace elements. Sources of infection such as *Pythium* (see p. 254) are minimized by isolation from soil and using clean water; the risk of recirculating pathogens is addressed by sterilizing the water.

Rockwool is not biodegradable, so the vast quantity now produced has created a serious disposal problem.

The slabs can be used successfully several times, if sterilized on each occasion, but eventually they lose their structure. Tearing them up and incorporating in composts or soils can deal with a limited amount. Far more can now be recycled in the production of new slabs. Increasingly, the mineral aggregate is being replaced by biodegradable materials processed to provide a suitable root environment for hydroponics, such as coir and green waste.

Expanded clay aggregates such as Leca or Hortag are made from processed clay which is heated and then expanded in a rotary kiln. After finishing at a very high temperature, sterile, tough, lightweight, honeycombed balls with a smooth surface finish are created. Granules 4–8 mm in diameter, giving a **capillary rise** of about 100 mm can be used to create a good root environment. They are non-toxic, environmentally inert and are used in hydroponic growing systems. They have an attractive dry surface making them useful in decorative containers (Figure 15.7).

Figure 15.7 Plant container with aggregated clay granule growing medium

> **Capillary rise** of water is its movement against gravity within thin tubes as with water moving up into paper towels.

The use of this material has proved beneficial for interior display containers (planters). Besides being lightweight and providing an attractive finish, trace element deficiencies occur less frequently when clay aggregates are used. Water can be supplied from a reservoir and the nutrients from an ion exchange 'battery' (Figure 15.8). More commonly, water and nutrition can be provided through fertigation (irrigation with nutrients in the water).

Alternatively, there are planters that store water in a sealed double wall equipped with a sensor that detects the drying in the root ball. When too dry, the sensor triggers the release of air into the water chamber causing water to seep out through a controlled water inlet. Water moves to the root ball by capillary action where the sensor detects when water supply needs to be shut off.

Green walls

It has been a long-standing part of gardening to cover walls with foliage or flowering plants to add height to the garden display, cover up an unattractive area or increase the areas available for wildlife (nectar for bees, insects for birds, nesting sites). This has been done by planting with climbers that can cling to walls (*Hedera* spp, *Parthenocissus tricuspidata*) or by providing trellis or wire work to tie in wall shrubs or fan-trained fruit trees (Figure 15.9). The covering can have other benefits, including protection of the wall

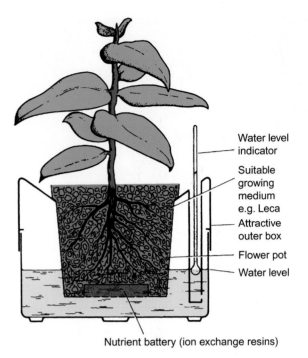

Figure 15.8 Planter with water reservoir. A water-level indicator is frequently incorporated and in some systems the nutrients are supplied from ion-exchange resins making them an attractive choice for interior landscapers

Figure 15.9 *Parthenocissus tricuspidata* on wall

from rain and providing some insulation with some claiming reduction in noise and improved air quality.

The new **green wall** technology increases the possibilities by having a framework that can be

Figure 15.10 Interior green wall

Figure 15.11 Commercial green wall

secured to a wall to provide platforms for grow bags for plants and a drip fertigation system (providing nutrient in the water). Alternatively there are ones with cells that can be filled with compost but others are hydroponic systems. These green walls can be purchased as kits for the domestic garden usually in the form of panels that can be linked to cover the area required. Interior green walls are supplied with appropriate lighting (Figure 15.10) and there are major installations on large buildings (Figure 15.11).

The plants can be grown from plugs or seed. Plants are selected from a wide range of decorative plants appropriate for the situation (sunny sites, shady/north facing walls, exposed sites). What appeals to many gardeners is the possibility of growing salads, herbs and fruit (e.g. strawberries), similar to the commercial production of crops such as lettuces (see Figure 15.4c).

Further reading

Bragg, N. (1998) *Grower Handbook 1 – Growing Media*. Grower Books.

Handreck, K.A. and Black, N.D. (2002) *Growing Media for Ornamentals and Turf*. University of New South Wales Press.

Pryce, S. (1991) *The Peat Alternatives Manual*. Friends of the Earth.

Smith, D. (1998. *Grower Manual 2 – Growing in Rockwool*. Grower Books.

Winterborne, J. (2005) *Hydroponics – Indoor Horticulture*. Pukka Press.

15

Please visit the companion website for further information:
www.routledge.com/cw/adams

Plant health maintenance

Figure 16.1 Fleece cover for a vegetable plot used against pests

This chapter includes the following topics:

- Physical control
- Cultural control
- Biological control
- Chemical control (herbicides, molluscicides, insecticides, fungicides)
- Selection of appropriate plants for particular situations
- Plant selection for resistance

Principles of Horticulture. 978-0-415-85908-0 © C. R. Adams, K.M. Bamford, J.E. Brook and M.P. Early.
Published by Taylor & Francis. All rights reserved.

Introduction

A general overview is given in this chapter for the various control measure available against weeds, pests and diseases. In the three following chapters, important weeds, pests and diseases are described with an emphasis on symptoms, damage and life cycles, and with brief comments on controls relevant to the particular organism causing the problem.

In this chapter, the human and environmental safety aspects of different control measures are given in some detail. Control measures that are mentioned briefly in the narrative of the following three chapters are covered here in more detail.

General comments on plant health maintenance

Gardeners should consider the whole range of control options (physical, cultural, biological, chemical and resistance) when confronted by a weed, pest or disease problem. It should be borne in mind that for **weeds**, cultural controls such as hoeing are an important alternative to the use of herbicides. With the small **pests**, physical controls such as deterrents and traps, and biological controls are increasingly being used, rather than insecticides. **Disease** control should, where possible, involve species and cultivars with plant resistance.

While the environmentally preferable non-chemical methods mentioned above may appeal to gardeners, there will be occasions when chemical methods may seem the only option. Control of **blight** on potato in a wet summer would be one example.

Physical control

> **Physical control** is a material, mechanical or hand control where the weed, pest or disease is directly blocked or **destroyed.**

Benefits of physical control:

▶ Once established, they often remain for a long time, e.g. wire or wood fencing used against rabbits and deer.
▶ They usually need little maintenance, e.g. plastic sheets placed on decorative borders to reduce weeds.

Limitations of physical control:

▶ Some physical methods are expensive to set up, e.g. fencing used against rabbits.

▶ Soil sterilization may lead to a worse disease situation if infected soil/plants are introduced.

Safe practice and environmental effects. In physical control, some hazards are:

▶ Unsafe use of cultivation equipment such as ploughs, rotavators, flame throwers and steam sterilization equipment, used to control weeds, pests and diseases.
▶ Unsafe removal of infected trees.
▶ Unsafe burning of infected plant material.

Natural balances:

▶ Of the physical controls described below, the only method that may affect natural balances is **partial soil sterilization**. If a newly sterilized soil is planted with some plants infected with a disease such as 'damping off' or *Fusarium* wilt, the spread of the disease may be more rapid than in a non-sterilized soil. This is because naturally occurring bacterial and fungal competitors to the disease have been killed off by the heat process.

Examples of physical controls

Barriers

Plastic sheets laid on the ground (and plants planted through slits in the plastic with ornamentals or soft fruit) are a very good way of reducing weed populations. **Biodegradable** forms of plastic used for this purpose are becoming popular.

Risks: when the plastic sheets are no longer needed, they should be disposed of in the correct location at the local council rubbish centre. They should not be burnt, producing toxic fumes.

Horticultural fleece placed over growing crops (see Figure 16.1) helps prevent the entry of pests such as cabbage root fly and cabbage white butterfly into rotation bed crops.

Fences using sturdy wire-mesh sunk into the ground to a depth of about 30 cm deter rabbits and small deer from digging under the fence. Fine-mesh **screens** placed over ventilation fans help to prevent the entry of pests such as fungus gnats or aphids from outside a greenhouse. Pots placed on small stands in **water-filled trays** are freed from flightless species such as red spider mite and adult vine weevil.

Traps

Pheromone traps containing a specific synthetic chemical similar to the attractant odour of the female moth are commonly used by gardeners to lure male codling and tortrix moths onto a sticky surface, thus

Figure 16.2 (a) A copper tape prevents slugs reaching Hostas in a pot; (b) glue band controls female winter moth on top fruit; (c) yellow sticky trap for controlling flying pests in greenhouses; (d) lettuce planted to attract slugs away from tomatoes

enabling an accurate assessment of their numbers and a more effective control. Comparable traps are available against plum moth, pea moth, leek moth and raspberry beetle.

Risks: none.

Rodent traps containing anticoagulant baits such as **difenacoum** are available. More recently, rat or mouse traps using a high-voltage charge are being used.

Risks: great care is needed to avoid access to such bait chemicals in the trap or in the store by children and pets.

Glue bands (Figure 16.2b) wrapped around an apple or pear trunk prevents the flightless winter moth female from crawling from the soil to lay her eggs. An adhesive **copper** tape may be placed around a pot to prevent slugs reaching plants such as *Hostas* (Figure 16.2a). Allotment owners sometimes use

Figure 16.3 Marigold placed in a greenhouse to deter insect pests

empty plastic milk **containers** to hold slug pellets that attract slugs, but prevent pets reaching the pellets (see Figure 18.6d).

Sticky traps (Figure 16.2c) are often hung in greenhouses. Their yellow colour attracts the flighted stages of many pests such as greenfly, whitefly, thrips and leaf miner adults.

Growing **sacrificial crops**, such as a row of lettuce between two row of tomatoes, can be used to attract slugs (Figure 16.2d), which are controlled in a trap nearby.

Deterrents

The odour from **onions** inter-planted with carrots may deter the carrot fly from attacking its host crop. **Marigolds** planted in amongst crops deter whitefly and aphids (Figure 16.3). **Ammonia** or **citronella** extracts present in commercially available products are sprayed on to bushes and trees to deter foxes, badgers and rabbits. An extract from a *Yucca* species is used as a commercial spray repellent against slug attacks.

Cultural control

> **Cultural control** is a procedure, or manipulation of the growing environment, that results in weed, pest or disease control. Gardeners, in their everyday activities, may remove or reduce damaging organisms in many different ways and thus protect the crop.

Benefits of cultural controls:

▶ They fit in with daily routines, e.g. regular feeding of plants.
▶ They have a long-lasting effects, e.g. removal of alternate host weed species such as shepherd's purse.

Limitations of cultural control:

▶ May be time-consuming to carry out.
▶ May lack the rapid control that is seen with pesticides.

Safe practice and environmental effects. While most controls described below need no serious comment in this section, **vegetative propagated** material such as bulbs and tubers (see p. 94), which are so commonly used by gardeners, bring with them the danger of spreading pests and diseases such as white rot of onions. Such material needs careful examination before being planted.

Natural balances. Two aspects of cultural control may be highlighted in terms of natural balances:

▶ **Rotation** of crops causes a beneficial change in the fungal and bacterial population in the soil. Soil pests such as potato **cyst nematode** may be captured by soil fungi and roots containing diseases such as **club root** may be rotted by soil bacteria and fungi at a time when the host crop (potato and cabbage in these two instances) is absent.
▶ **Organic** soils are claimed to contain more 'beneficial' fungal and bacterial species which maintain these natural balances.

Examples of cultural control

Soil fertility

While the correct content and balance of major and minor nutrients in the soil is recognized as vitally important for optimum crop yield and quality, it should be remembered that **plant resistance** to pests and diseases is also affected by nutrient levels in the plant. **Excessive nitrogen** levels, causing soft tissue growth, encourage the increase of peach-potato aphid, grey mould, *Fusarium* patch on turf and fireblight. Adequate levels of **potassium** help to control fungal diseases – for example, *Fusarium* wilt on plants such as peas, tomatoes and carnation, and tomato mosaic virus. Fertility provided by well-composted material usually provides nutrients to the plant at the correct concentrations.

Club root disease of brassicas is less damaging in alkaline soil with a **pH** (see p. 174) greater than 6, and lime may be incorporated before planting these crops to achieve this aim. Gardeners apply **mulches** (e.g. composted bark, grass cuttings or straw), to bare soil in order to control annual weeds by excluding the source of light. Black polythene **sheeting** is used in soft fruit to achieve a similar objective.

Risks: the use of increased levels of fertilizers to grow bigger crops and dense turf may (particularly in **sandy** soils) lead to nutrients leaching into drains and streams that may cause an increase in algae (**algal bloom**) with poor water quality and death of fish in streams and ponds.

Rotavating

Rotavating of soils enable a physical improvement in **soil structure** (see p. 147) as a preparation for the growing of crops. The improved drainage and tilth may reduce damping-off diseases, disturb annual and perennial weeds (e.g. chickweed and docks) and expose soil pests (e.g. leather jackets and cutworms) to the eager beaks of birds. Repeated rotavation may be necessary to reduce the underground rhizomes or taproots of perennial weeds such as couch grass. **Hoeing** annual weeds is an effective method, provided the roots are fully exposed and the soil is dry enough to prevent root re-establishment.

Crop rotation

Some important soil-borne pests and diseases attack specific crops. For example, potato cyst nematode (see p. 247) attack on potatoes and tomatoes, and club root (see p. 256) on members of the cabbage family. As these problems are soil borne, they are relatively slow in their spread, but are difficult to control, largely because they have a life-cycle stage that survives for several years in the ground. By the simple method of planting crops in different plots each year, this problem survival stage is excluded from the sensitive crop for a number of years, during which time the pest or disease numbers will slowly decline (Figure 16.4).

A gardener often creates five or six plots (sometimes bounded by wooden boarding) to isolate each plot and thus achieve successful rotation. Crops

Figure 16.4 Rotation plots used to reduce soil pests and diseases

belonging to the same plant family fit into the rotation system, because they have the same sensitivity to the particular pest or disease. Potatoes, tomatoes, peppers and aubergines are all members of the Solanaceae family. Cabbages, cauliflowers, Brussels sprouts, Chinese cabbage (and wallflowers) belong to the Brassicaceae (see p. 56). When considering a rotation plan, it is generally recommended that the gardener restricts member species of the same family to the same plot in the same growing season. A commonly used **four-year rotation** would be as follows: year 1, root crops such as carrots, beetroot and parsnips; year 2, brassicas; year 3, potatoes, celery, leeks and onions; year 4, peas and beans.

Rotation is **not** effective against pests and diseases which are unspecific, such as grey mould (*Botrytis*), which may attack a wide variety of plants. Rotation is also not likely to be effective against rapidly spreading airborne species such as greenfly and potato blight.

Risks: the **sclerotium** stage (see p. 252) of **white rot** disease (*Sclerotium cepivorum*) on onion and related crop such as leeks, garlic, chives and shallots is an especially difficult disease to deal with. It is able to survive in the soil for 15 years or more. It can be seen that a very long rotation period would be necessary to remove this serious disease. A four-year rotation is not normally sufficient.

Planting and harvesting times

Some pests emerge from their overwintering stage at about the same time each year (e.g. **cabbage root fly** in late April). By planting early to establish tolerant brassica plants before the pest emerges, a useful cultural control is achieved. Similarly, the deliberate planting of early potato cultivars enables harvesting before the maturation of **potato cyst nematode** cysts (see p. 247), so that damage to the crop and the release of the nematode eggs are avoided. Annual weeds may be induced to germinate in a prepared seedbed by irrigation. After weeds have been controlled with a herbicide such as **glyphosate**, a crop may then be sown into the undisturbed bed or **stale seedbed**, with less chance of further weed germination.

Vegetative propagation material

Such material (see also p. 94) is used in many areas of horticulture, as bulbs (e.g. tulips and onions), tubers (e.g. dahlias and potatoes), runners (e.g. strawberries), cuttings (e.g. chrysanthemums and many trees and shrubs) and graft scions in trees. The increase in nematodes, viruses, fungi and bacteria by vegetative propagation is a particular problem, since the organisms are inside the plant tissues, and since the plant tissues are sensitive to any drastic control measures. **Inspection** of introduced material may greatly reduce the risk of this problem. Soft, puffy narcissus bulbs, chrysanthemum cuttings with an internal rot, whitefly or red spider mite on stock plants, and virus on nursery stock, are all symptoms that would suggest either careful sorting or rejection of the stocks.

Clean stock schemes

Quality of vegetative material is monitored in Britain and Ireland by the government **Plant Health Propagation Scheme (PHPS)** and **Seed Potato Classification Scheme (SPCS)**. In this way, high-quality potatoes, bulbs and fruit plants (with low levels of virus diseases and nematodes) are available to the gardener from soft fruit, top fruit, potato and bulb industries.

Hygienic growing

Below are a few examples of the hundreds of 'common sense' activities that reduce pest and disease attack in greenhouses and gardens. During the crop, the grower should aim to provide optimum conditions for growth. Water content of soil should be adequate for growth (see **field capacity**, p. 151), but not be so excessive that root diseases (e.g. damping off in pot plants, club root of cabbage and brown root rot of conifers) are actively encouraged.

Water sources

Covering and regular cleaning of water tanks to prevent the breeding of **damping-off fungi** in rotting organic matter may be important in their control. Seed trays and pots should be thoroughly washed to remove all traces of compost that might harbour damping-off disease.

Companion planting

An increasingly common practice in the garden is the deliberate establishment of two or more plant species close together with the intention of deriving some horticultural benefit from their association. Such a situation may seem at first sight to encourage competition rather than mutual benefit. Supporters of companion planting consider that plant and animal species in the natural world show much evidence of mutual cooperation. Some experimental results have given support to the practice, but much of the evidence remains unproven. It should be stated, however, that while most commercial horticulturist producers in Western Europe grow blocks of a single species, millions of subsistence growers worldwide are using two or three different crop species inter-planted as a regular practice.

Two of many recommended companion-planting groupings are:

▶ potato, broad bean, sweet corn and onion
▶ carrot, leek, broad bean and broccoli.

Alternate hosts

Alternate hosts harbouring pests and diseases should be removed where possible. A few examples of weed alternate hosts are given here:

▶ Club root infects shepherd's purse (*Capsella bursa-pastoris*), and white blister 'rust' may be seen on charlock (*Sinapis arvensis*).
▶ Black bean aphid infests fat hen (*Chenopodium album*).
▶ In the greenhouse, glasshouse red spider mite may be found on chickweed.
▶ Speedwells may harbour stem eelworm.

Removal of infected plant material

In fruit tree species such as apple, routine **pruning** operations may remove serious pests such as fruit tree red spider mite eggs, and diseases such as canker and powdery mildew. Pruning should also aim to reduce the density of shoots in the centre of the tree. The resulting reduction in humidity provides a microenvironment that is less favourable to these diseases. **Tree stumps** harbouring serious underground diseases such as honey fungus should be removed manually or by means of a mechanical stump grinder. Making a feature of an infected

Figure 16.5 Tree stumps can be a long-term source of honey fungus

stump by placing a bird table on a tree stump is not a recommended activity in gardening (see Figure 16.5).

An environmental point

Removal of dead plant material from gardens can be considered from the opposite point of view. Many **beneficial species**, such as violet ground beetle, centipedes and ichneumon wasps, can use the **dead hollow stems** of garden perennials as overwintering refuges. Hedgehogs may spend the winter within piles of dead branches. Gardeners may wish to consider a **balance** between tidiness and their desire to achieve natural pest control.

Biological control

Biological control is the use of natural enemies to reduce the damage caused by a pest (or disease).

Benefits of biological control:

▶ Non-toxic to humans, wildlife and pets.
▶ Numbers of predators and parasites increase naturally.
▶ A balanced population of predators and parasites is reached.
▶ No build-up of resistant pests and diseases.

Limitations of biological control:

▶ Needs careful introduction and knowledge of life cycles.
▶ Can easily be affected by pesticides.

Safe practice and environmental impact. The main problems with biological control are:

▶ Unsuccessful application of biological control organisms can lead to a severe pest problem (e.g. late introduction of *Encarsia* wasp used against glasshouse whitefly).
▶ The incorrect introduction of a biological control organism can subsequently kill desirable or beneficial organisms in the environment (e.g. accidental introduction of the New Zealand flatworm (*Arthurdendyus triangulates*) in the 1960s has reduced earthworm numbers in some northern areas of Great Britain).

Risks can be **minimized** by the following:

▶ Understanding both the pest's and predator/ parasite's life cycles in order to achieve reliable control carefully.
▶ Choosing the best predator or parasite for the problem pest or disease concerned.
▶ Taking care that environmentally useful animal species are not subject to the attacks of the predators and parasites, such as the earthworms mentioned above.

Natural balances. In most horticultural situations, there are important examples of **natural balance** between species:

▶ With pests, their naturally occurring **predators and parasites** are an important form of crop protection, as described in this section of the chapter.
▶ With diseases, naturally occurring predators and parasites are less well understood, but the **nutritional** condition of the plant and the resulting naturally occurring bacterial and fungal populations on leaf, stem and root surfaces (see **organic growing**, p. 10) often help to slow a disease's progress.
▶ The garden represents a complex situation. There may be plant species present from every continent (see Chapter 2), and any of these plant species may be accompanied by a specific pest from its country of origin. Plant species that have been

195

established in Britain and Ireland for many years (e.g. apple) often have beneficial predators and parasites (e.g. *Aphelinus mali*, a wasp parasite on apple woolly aphid), introduced accidentally or deliberately from their country of origin, that reduce pest numbers. It is quite likely, however, that for the more **recently imported** plant species, there may not be appropriate predators or parasites to control a recently introduced pest occurring on the plant species in Britain and Ireland.

▶ Some horticultural practices can **disturb natural balances**. In a natural habitat such as a woodland, a **climax population** of plants and animals develops (see Chapter 3). Here, a complex balance exists between indigenous pests and their **predators/parasites**. The **food webs** include several types of predator/parasite found on each plant species that control (but do not eliminate) the pests. This development of food webs is not achieved to such an extent in most gardens, since a natural succession of wild plant species mentioned above is not desirable to gardeners who are aiming for optimum production of edible crops or for a pleasing aesthetic layout of decorative plants free from weeds.

▶ Regular **movement or removal** of cultivated plants without particular thought to the natural balance between predator/parasites and pests will make pest attacks more likely in the garden/ nursery situation. For example, if a gardener stops growing the 'poached-egg plant' (*Limnanthes douglasii*), the number of useful hoverflies (feeding on its pollen) may be reduced.

▶ The **removal** of the rotting hollow stems of herbaceous perennials and branches of decaying wood, which are common sheltering sites for parasitic wasps, predatory beetles and centipedes, may reduce the potential for control of pests.

▶ The lack of good **soil structure** (see p. 147) resulting from poor cultivation or inadequate incorporation of organic matter in a garden may hinder the movement of useful predatory animals such as centipedes in their search for underground soil pests.

▶ A poor physical preparation of soil and lack of attention to **pH and nutrient** levels in soil may result in poor soil microbial action (see p. 151).

▶ The **repeated planting** of crops or ornamentals into the same area of soil often leads to serious attacks of persistent soil-borne pests or diseases. Notable examples are club root disease on brassicas (see p. 256) and potato cyst nematode pest on potatoes (see rotation, p. 192). A comparable unbalanced situation is found when young trees and shrubs (such as roses) are planted into a soil previously occupied by an old specimen of the same plant species, with the resulting problem called '**replant disease**' often caused by a high level of nematodes (see p. 246) and *Pythium* fungus.

▶ The unconsidered use of **pesticides** may result in a rapid decrease in predators and parasites and may considerably delay their appearance and build-up in the following growing seasons.

Examples of biological control species

There are two sources of 'natural enemies' to pests (and occasionally diseases): the **indigenous** (i.e. they are locally present, in wild plant communities in Britain and Ireland) species and the **exotic** ones (from other countries). Garden pests may be controlled by **predators** that eat the pest, or by **parasites** that lay eggs within the pest (see also **food chains**, p. 38). These beneficial organisms are to be encouraged, and in some cases deliberately introduced. A range of important organisms useful in horticulture is now described.

Indigenous predators and parasites

Wild birds can contribute greatly to the control of horticultural pests. A pair of **blue tits** (*Cyanistes caeruleus*) can consume 10,000 caterpillars and a million aphids in a 12 month period. They will also eat scale insects, which are otherwise quite difficult to control. The installation of tit boxes (Figure 16.6a) is a worthwhile activity.

Hedgehogs (*Erinaceus europeus*) belong to the insectivore group of mammals. Although their preferred diet is beetles, caterpillars and earthworms (up to 200 g per day), they will also eat slugs. Sometimes saucers of water or half-strength milk are placed in the garden for them to drink during dry summer periods. Care must be taken that they are not exposed to dead slugs which had previously consumed slug bait containing methiocarb or metaldehyde, as these slugs will be toxic to the hedgehog. Placing slug pellets in containers that prevent hedgehog entry is strongly recommended (see p. 229). Hedgehogs are encouraged to enter gardens by means of small holes cut into the base of a fence panel. Wooden hedgehog-shelters are commercially available for placing in quiet corners of large gardens. Heaps of logs and piles of leaf litter in a quiet location are suitable for their daytime and overwintering retreat. Care should be taken in winter that hibernating hedgehogs are not burnt in bonfires.

Figure 16.6 (a) A blue tit box; (b) a blue tit feeder

16

Gardeners having ponds should remember that hedgehogs quite commonly fall into the water at night, and will avoid drowning if a part of the pond wall has a slope or a small ramp provided for their escape.

Frogs and toads commonly leave their ponds in damp weather and may contribute greatly to the control of slugs and ground-living insect pests. The **common frog** (*Rana temporaria*) is smooth-skinned, about 7 cm in length, and greenish-brown or yellow in colour. It is seen most commonly from March to October. The frog's egg mass is laid in spring as a large round clump, usually in shallow water. The species' numbers have decreased in the countryside in recent years, and it is most commonly seen nowadays in garden ponds. The introduced green and brown **marsh frog** (*Pelophylax ridibundus*), found mainly in Kent and East Sussex, is large (up to 17 cm in length), has dark blotches on its body and often has a yellow stripe down its back. The **common toad** (*Bufo bufo*) has a grey or brown, warty appearance. The female may reach 9 cm in length, while the smaller male is more commonly 6 cm. The toad's egg mass is laid in spring in long strings. This species may be active all year round in the UK when the weather is mild. It prefers deep ponds. The introduced **midwife toad**, which

can reach 5 cm in length, is found most commonly in Bedfordshire, South Yorkshire and Devon. The male holds on to the egg mass on his back (hence the common name). This species makes a characteristic high-pitched piping sound. These amphibians commonly leave their ponds in damp weather and may contribute greatly to the control of slugs and ground-living insect pests.

Lacewings (e.g. *Chrysopa carnea*) are pale green insects, 1.5 cm, which fold their transparent wings over their bodies when at rest. Several hundred eggs are laid per year, each on the end of fine stalks, on the underside of leaves. They are useful horticultural predators, their hairy larvae (Figure 16.7b) eating aphids and mite pests, often reaching the prey in leaf folds where ladybirds cannot reach. They also are now used commercially in greenhouses.

The 40 British species of **ladybird** beetle (Figure 16.8) are a welcome sight to the gardener. Almost all are predatory. The red **seven-spot ladybird** (*Coccinella 7-punctata*) emerges from the soil in spring, mates and lays about 1,000 elongated yellow eggs on the leaves of a range of weeds such as nettles, and crops such as beans, throughout the growing season. Both the emerging slate-grey and yellow larvae and the adults

Figure 16.7 (a) Lacewing adult; (b) lacewing larva control aphids (courtesy of Bioline Ltd)

feed on a range of aphid species. Wooden ladybird shelters and towers are now available to encourage the overwintering of these useful predators. A worrying development in the last few years has been the rapid spread and increase of the **harlequin ladybird** from South East Asia. This species is larger (6–8 mm long) and rounder than the seven-spot species (4–5 mm). It has a wider food range than other ladybird species, consuming other ladybirds' eggs and larvae, and eggs and caterpillars of moths. Furthermore, it is able to give humans a slight but irritating bite.

Hoverflies (Figure 16.9), superficially resembling wasps, are commonly seen darting or hovering above flowers in summer. Many of the 250 British species (e.g. *Syrphus ribesii*) lay eggs in the midst of aphid colonies, and their legless light green-coloured grubs (resembling small green maggots) consume large numbers of aphids. The flowers of some garden plants are especially useful in providing pollen for the hoverfly adults and therefore encouraging aphid control in the garden. Summer flowering examples are poached-egg plant (*Limnanthes douglasii*), baby-blue-eyes (*Nemophila menziesii*) and Californian poppy (*Romneya coulteri*). Later summer and autumn examples are *Phacelia tanacetifolium* and ice plant (*Sedum spectabile*).

Other beneficial insects include the common wasp, parasitic wasps, anthocorid bugs and ground beetles. Predatory mites and parasitic nematode worms are also important in controlling pest numbers.

Insect pests may also be parasitized by specialized **parasitic fungi**. The numerous species of web-forming and hunting **spiders** (Figure 16.9d) are very important in the reduction of many types of insects.

Occasionally, weeds are controlled biologically. The **cinnabar moth** caterpillar (*Tyria jacobaeae*) may remove the foliage of groundsel and ragwort. A **rust** (*Puccinia lagenophora*) is commonly seen infecting groundsel, but unfortunately also attacks cinerarias.

Increased attention is being given by horticulturists to the careful **selection of pesticides** (if they are needed) to avoid unnecessary destruction of the predators and parasites described above (see also p. 203).

Exotic (introduced) predators and parasites

In greenhouses and polythene tunnels, **high temperatures** (often all year round) and subtropical species of plants bring with them **exotic** pests and diseases from other countries and they increase very rapidly, and may become resistant to pesticides.

Biological control of exotic pests requires exotic **predators and parasites**; and so, the health of the major greenhouse crops in Britain and Ireland is due in large measure to these introduced predators and parasites. Two organisms, a South American mite that eats all stages of the glasshouse red spider mite, and a tiny South-East Asian wasp that parasitizes the glasshouse whitefly, have been used for many years and are briefly described below.

Phytoseiulus persimilis (Figure 16.10a) is a 1 mm globular, deep orange, predatory tropical mite used in greenhouse production to control glasshouse red spider mite (**two-spotted mite**, see also p. 245). The predator's short egg–adult development period (seven days), laying potential (50 eggs per life cycle) and appetite (five pest adults eaten per day) explain its extremely efficient action.

Encarsia formosa (Figure 16.10b) is a tiny (2 mm) wasp, which lays an egg into the glasshouse whitefly third and fourth **scale stage** (see also p. 236), causing it to turn black and eventually to release another wasp. It requires temperatures above 22°C to be effective.

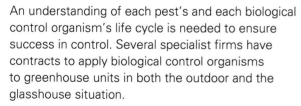

Figure 16.8 (a) Ladybird adult; (b) ladybird eggs; (c) ladybird larva; (d) ladybird pupa; (e) ground beetle is a predator on soil borne pests

An understanding of each pest's and each biological control organism's life cycle is needed to ensure success in control. Several specialist firms have contracts to apply biological control organisms to greenhouse units in both the outdoor and the glasshouse situation.

There are more than fifty biological control species (available from **specialist companies**) that are used against garden and glasshouse pests such as aphid, caterpillars, flea beetle, glasshouse whitefly, leaf miners, mealy bugs, two-spotted mite, fungus gnat, thrips, vine weevil and slugs. These have been selected for their effectiveness against their chosen pest species. Also, they have been tested to ensure they do not interfere with the natural balances in the garden/local habitat.

Chemical control

> **Chemical control** is the use of a chemical substance intended to prevent or kill a destructive weed, pest or disease.

The number of pesticides available to gardeners has decreased in recent years, reflecting the need to use only products that are safe to humans, pets and wildlife.

Benefits of chemical control:

▶ Chemical control produces rapid control.
▶ Products are easily accessible.

199

Figure 16.9 (a) Hoverfly feeding on pollen; (b) common wasp; (c) ladybird predator on mealy bugs (courtesy of Bioline Ltd); (d) garden-web spider

Limitations of chemical control:

▶ Products can be dangerous to humans, animals and plants.
▶ Products can cause resistant strains of pests, diseases and weeds to develop.

Safe practice and environmental impact. In chemical control, the **hazards** include the following possible outcomes:

> A **hazard** is something with the potential to do harm.

▶ **Acute** poisoning of humans, pets, farm animals, bees and wild animals.
▶ **Accumulation** of pesticides that lead to toxic levels in humans, pets, farm animals.
▶ Cancer-inducing effects in human.
▶ Damage to cultivated and wild plants, especially by herbicides.
▶ Contamination of streams and dams.

▶ Development of strains of rodents, insects, mites and fungi **resistant** to pesticides.

Risks can be **minimized** by adhering carefully to the following:

> **Risk** can be measured by the chances of something happening and the level of consequence if it did.

▶ Restricting chemical applications to only those situations that justify such a control measure; in many instances, other control measures may be preferable and less harmful.
▶ Choosing the least harmful chemical to effectively control the problem organism.
▶ Reading the instructions on the product label.
▶ Choosing the correct clothing, where necessary.
▶ Measuring the correct amount of concentrate and water (where relevant) calculating (where appropriate) the amount of pesticide and water

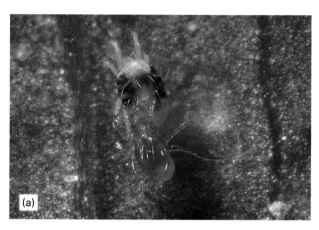

Figure 16.10 (a) *Phytoseiulus* eating a two-spotted mite; (b) *Encarsia* wasp laying an egg into a whitefly 'scale' (courtesy of Bioline Ltd)

necessary for application to the crop area in question

▶ Mixing the two, avoiding spillage on to skin, clothing and the surrounding area.

▶ Applying the product so that the same area is not covered more than once, at any one time.

▶ Applying the product under suitable dry, wind-free weather conditions.

▶ Applying the product so that other humans, beneficial animals, waterways and adjacent plantings are avoided.

▶ Avoiding spray drift, especially with herbicides.

▶ Carefully storing the pesticides in a secure, safe, dry place away from children and pets.

Other information on **safety and environmental impact** and on **natural balances** is given separately below in each of the four main sections: on herbicides (p. 201), molluscicides (p. 202), insecticides (p. 203) and fungicides (p. 214).

> The word **'pesticide'** is used here to cover all crop protection chemicals, including herbicides (for weeds), molluscicides (for slugs and snails), insecticides (for insects), acaricides (for mites), nematicides (for nematodes) and fungicides (for fungi).

Each container of pesticide available for garden use contains several ingredients. The **active ingredient's** role is to kill the weed, pest or disease. The other constituents help in the product's storage.

Herbicides (weedkillers)

Benefits and limitations of herbicidal control: see general points for 'chemical control' on p. 199.

Safe practice and environmental impact of herbicidal control: in most garden situations, the almost complete removal of weeds such as those described in Chapter 17 is considered a useful or necessary activity.

▶ Care needs to be taken when mixing and applying herbicides. A concentrated active ingredient in the trade product represents a particular risk if adequate protective clothing is not worn. Products for gardeners are chosen for their low level of toxicity but gardeners still need to be careful.

Natural balances:

▶ Some weeds are important in maintaining butterfly species. For example, the leaves of **stinging nettle** (*Urtica dioica*) are the main food source for caterpillars of the comma, peacock, red admiral and painted lady butterflies, some of Britain and Ireland's most beautiful insects (Figure 16.11). Removal of all nettle plants in an area is likely to reduce these insects.

▶ Care needs to be taken when spraying herbicides near ponds. **Glyphosate**, for example, is toxic to fish, especially when water temperatures are high.

▶ Some herbicides may have some toxicity to beneficial insects (e.g. **clopyralid** lawn weedkiller on ladybirds).

▶ Careless spraying of the wrong herbicides may cause damage or death of desirable plants. For example, **glyphosate**, a total weedkiller, sprayed on a lawn, will completely kill off the grass.

▶ Even the vapour of a herbicide may be damaging. For example, extremely small amounts of lawn herbicide containing **2,4-D** can spoil the growth of tomato plants and prevent fruiting. The vapour originating from a container of concentrate containing 2,4-D thoughtlessly stored inside a

Figure 16.11 Comma butterfly. Its caterpillars feed on nettle leaves

greenhouse may well be sufficient to cause similar damage to tomatoes.

Restoring and maintaining natural balances:

- Avoid killing plants which maintain beneficial predators, parasites and pollinating species such as hoverflies, butterflies (Figure 16.11) and bees.
- Avoid spray drift onto useful and desirable plants.

Types of herbicide action

Herbicides can be classified into **three** groups according to their action against weeds:

- **Contact herbicides** enter the leaf or stem, and then kill the tissues of susceptible plants in the particular area where they have entered. An example is **diquat** used by private gardeners to control annual weeds on waste ground.
- **Translocated herbicides** enter the leaf, stem or roots and then move via the vascular system (see p. 79) to reach all parts of the plant. This property is particularly useful in controlling **perennial** weeds with their extensive underground root systems (e.g. dock), rhizomes (e.g. couch) or stolons (e.g. yarrow), which are otherwise hard to reach for control. **Glyphosate** is an example of a **total** translocated weedkiller.
- **Selective herbicides** are able to kill the chosen weed species through the leaf (or root), but leave the surrounding garden plants **unaffected**. The best example in gardening is the active ingredient **2,4-D** that is used in lawn herbicides to kill off a range of broadleaved weeds while leaving the grass unaffected. It should be emphasized that a **careful reading** of the instructions on the herbicide packet is necessary. For example, spraying this active ingredient immediately after

cutting the grass can lead to the grass becoming scorched. It is equally important to remember that any spray being blown onto neighbouring garden plants will damage (or kill) the plants concerned. Lawn clippings from a lawn recently sprayed with **2,4 D** should not be used as a mulch on flower beds.

- **Residual herbicides** containing active ingredients such as **diflufenican** remain chemically stable and active over a period of months at the soil surface or in the soil. They are usually in a mixed formulation containing a total translocated ingredient such as glyphosate. These products are used for total control, usually in situations such as gravel paths, away from direct contact with garden plants.

Slug killers (molluscicides)

Benefits and limitations of molluscicide control: see general points for 'chemical control' on p. 199.

Safe practice and environmental impact of slug killers:

- **Metaldehyde**, while being very effective against slugs (and snails), is reported to be be toxic to small mammals, birds and some predatory insects (see p. 197).
- **Ferric phosphate** is considered to be less dangerous to garden wildlife than **copper** ingredients (that may build up in soil and affect earthworm numbers).
- **Methiocarb**, previously a common slug killer, is considered by many gardeners to be unsuitable environmentally as it is toxic to many forms of wildlife.

Minimization of risks:

- While slugs are undoubtedly a constant major pest, the use of nematode parasites (see p. 230) and barriers to slugs (see p. 229) provide alternatives to chemical control.
- Use of resistant cultivars, and encouragement of predators such as hedgehogs, ducks, frogs and ground-living beetles may considerably lessen slug damage for private gardeners.
- Use ingredients such as **ferric phosphate** rather than **metaldehyde** or **methiocarb** where possible to reduce environmental problems.
- Place pellets inside containers that allow access to slugs, but prevent the entry of mammals and birds (see p. 229).
- Grow less susceptible species or cultivars. For example, 'Pentland Dell' potato cultivar is one of the least affected by slugs.

Natural balances: Britain and Ireland has a climate that may bring damp conditions to the garden or horticultural unit at any time of year. Many plant cultivars (such as lettuces, potatoes and *Hosta*) are highly bred to have succulent tissues that unfortunately favour slugs. Slugs exploit these situations and are rated by gardeners as the **worst** garden pest.

Examples of molluscicide active ingredients:

▶ **Ferric phosphate** is an inorganic salt, formulated in pellets. It is relatively non-toxic to mammals and birds. This ingredient is acceptable to organic growers.

▶ **Metaldehyde** causes the slug to produce excessive amounts of mucus, and consequently to become dry and die. Metaldehyde is applied as pellets (or sometimes as a spray formulation). There is some danger to children, pets and wild animals.

Insecticides and acaricides (used against insect and mite pests)

Benefits and limitations of insecticidal control: see general points for chemical control on p. 199.

Safe practice and environmental impact of insecticidal and acaricidal control: while most acutely toxic and persistent insecticides are no longer approved for garden use in Britain, there are still important points to be remembered:

▶ Care needs to be taken when mixing and applying insecticides. A **concentrated** active ingredient within the trade product represents a particular risk if adequate protective clothing (see p. 201) is not worn. Products for gardeners are chosen for their low level of toxicity to humans but gardeners still need to be careful.

▶ Most of the insecticide ingredients available to the private gardener have some toxicity to **beneficial** insects and mites living alongside pest species. Alternative strategies (see physical, cultural and biological control in this chapter) such as deterrents, fleeces, traps and encouraging beneficial predators and parasites should be considered before taking the option to use an insecticidal spray.

Natural balances:

▶ The use of insecticides/acaricides can seriously affect natural balances. While the increase in insect and mite numbers is balanced in nature by **beneficial** predators and parasites such as ladybirds and parasitic wasps (see biological control, p. 196), careless spraying of an insecticide such as **deltamethrin** may kill off these useful species.

▶ When using selected biological control species such as *Encarsia* wasp against glasshouse whitefly (see p. 198) in **greenhouses**, the selection of appropriate insecticides/acaricides such as **fatty acids** for control of other pests (see integrated control, p. 201) reduces *Encarsia* deaths.

▶ Insecticide products are dangerous to animals in **ponds** (fish, snails and insects), and care is needed not to spray near ponds.

Restoring and maintaining natural balances:

▶ Read the pesticide product label carefully to check whether the chosen active ingredient kills biological control and pollinating species. Internet sources will give more detailed information.

▶ Be particularly careful when spraying in glasshouses using introduced biological control species, or near ponds.

Entry point for insecticides into insects

The insects and mites have **three** main points of weakness for attack by pesticides: their **waxy exoskeletons** (see p. 232) may be penetrated by wax-dissolving contact chemicals; their abdominal **spiracles** allow fumigant chemicals to enter tracheae or are blocked by 'sticky' pesticide formulations; and their **digestive systems**, in coping with the large food quantities required for growth, may take in stomach poisons.

Examples of insecticide active ingredients

Three groups of insecticides are described.

▶ **Natural plant extracts**. Such products contain natural products such as **alginates/polysaccharides** and act by blocking the breathing holes (spiracles) of pests such as aphids, thrips and mites. They have been given clearance for use by organic growers. A smoke formulation based on extracts from **garlic** enables the gardener to fumigate the glasshouse against pests such as glasshouse whitefly, aphids and two-spotted mite, without needing to remove the glasshouse plants.

▶ **Deltamethrin** belongs to the pyrethroid group. It has both cuticle and stomach action. It is effective against caterpillars, and outdoors. It is residual for a period of up to three weeks. It may reduce the effectiveness of biological control by killing off useful predators and parasites.

▶ Potassium salts of **fatty acids** work by contact action, dissolving the cuticle of pests such as aphids, whitefly, spider mites, mealy bugs and scale insects.

Nematicides

No active ingredients are at present available to gardeners for nematode control.

Fungicides

Benefits and limitations of fungicidal control: see general points for chemical control on p. 199.

Safe practice and environmental impact of fungicidal control: fungicides used in gardening do not represent the danger of human and environmental toxicity that may be found in insecticides.

▶ However, care needs to be taken when mixing and applying fungicides. The **concentrated** trade product represents a particular risk. Products for gardeners are chosen for their low level of toxicity to humans (see toxicity aspects of pesticides, p. 200), but gardeners still need to be careful.

Natural balances:

▶ While the fungicide ingredients available to the private gardener do not have a high toxicity to beneficial animals that is seen with the insecticides and acaricides, it should be remembered that fungi are found throughout gardens, helping to break down dead plant material to form humus. There are also many fungi acting in **beneficial** ways, as biological control agents of pests, against disease-causing fungi on the leaf surface (phyllosphere), around roots (rhizosphere) and in composting (see pp. 158–161).

▶ Disease controls with broad-spectrum ingredients such as copper, or systemic chemicals such as **myclobutanil** may reduce the levels of these useful fungi, but strong evidence to support this claim is not yet available.

Restoring natural balances:

▶ Provide plants with optimal conditions of soil fertility (see p. 140) and microclimate that will reduce the likelihood of fungal (and bacterial) infections, but will encourage beneficial bacterial activity.

▶ Provide plants (particularly seedlings) with disease-free soil and composts. Avoid the introduction of infected plants (see p. 194).

▶ Choose cultivars with a proven record of **plant resistance** where possible (see p. 205).

▶ Read the pesticide product **label** carefully to check whether the chosen active ingredient kills biological control and pollinating species. Internet sources will give more detailed information.

▶ Be particularly careful when spraying in glasshouses using introduced **biological control species**, or near **ponds**.

Action of fungicides

Fungicides must act against the disease but not seriously interfere with plant activity. Fungicides may act either in a **protectant** way on the plant surface, or in a **systemic** way inside the plant.

Examples of fungicide active ingredients

▶ **Inorganic chemicals** contain no carbon. Two chemicals are available to gardeners. Copper salts mixed with slaked lime (**Bordeaux mixture**) form a microscopically thin protective barrier to fungi such as potato blight when sprayed on the leaf. Another copper formulation (**Cheshunt mixture**) has a more 'tender' action on young roots of seedlings and young plants while controlling damping-off disease. Fine-grained (colloidal) **sulphur**, applied as a spray, controls powdery mildews and apple scab.

▶ **Myclobutanil** belongs to the conazole group. It is protectant and systemic, on powdery mildews, black spot of rose and apple scab.

Formulations

Active ingredients are mixed with other ingredients to increase the efficiency and ease of application, prolong the period of effectiveness, or reduce the damaging effects on plants and humans. The whole product (**formulation**) in its bottle or packet is given a trade name, which often differs from the name of the active ingredient. The main formulations are liquids, wettable powders, dusts and baits.

Plant damage

The commonest damage is from herbicide sprays. Care should be taken not to spray in windy weather or to spray too close to garden plants. Insecticides and fungicide sprays may harm plants, particularly in **hot weather** when the leaves may be scorched. Plants growing in **greenhouses** are more susceptible because their leaf cuticle is thinner than that of plants growing at cooler temperatures. Careful examination of the pesticide (particularly herbicide) packet **label** often prevents this form of damage occurring.

Product label

The 'statutory area' on the label present on each packet or bottle of pesticide must provide the following details:

▶ The fact that the product is for garden use.

▶ The plant species, crop or situation where treatment is permitted and the maximum dose or concentration.

- The maximum number of treatments.
- The latest time of application, or harvest interval (days between application and harvest).
- Any specific restrictions, such as clothing required and temperature at which application should be made (the nature of the protective clothing stated on the label commonly reflects the LD_{50} status of the ingredient).
- A reminder to read all other safety precautions on the label and directions for use on labels of pesticide containers intended for gardener use.
- A **blue logo** is now included, where relevant, showing a picture of a child and a dog, to indicate that there is danger from the product if eaten by children, pets or wildlife (e.g. **metaldehyde** slug pellets).

Selection of plants

The gardener often has **two** important decisions when considering the most suitable species/cultivar to grow. The first is the **choice of the plant** appropriate for the garden location. The second is the use of a suitable **plant resistance** to diseases and pests.

Plant species suitable for specific garden locations

The garden can be seen as a pattern of small habitats (see p. 38), each presenting particular soil and microclimate conditions that are favourable to particular **plant species**. Among the many decisions a gardener has to make, choosing the most suitable plant species for each location is one of the most important. Failure to choose sensibly may result in weak plant growth and susceptibility to diseases (and pests). More detailed information can be found in reference books on garden species.

Benefits of correct species choice:

- Optimum plant growth is achieved.
- Healthy plants are usually less prone to diseases and pests.

Locations in the garden

Table 16.1 lists some of the common garden locations, and some plant species suited to each location. In its more suitable place, a plant species is less likely to be affected by pests, diseases and physiological disorders because its general health (including its level of nutrients) will be suitable for balanced growth.

Table 16.1 Suitable garden plants for different locations (see also p. 21 for hardiness ratings)

Garden location	Suitable garden species
Sunny	*Ceanothus thyrsiflorus* (shrub), *Lonicera japonica* (climber), *Lamium maculatum* (perennial), *Petunia* x hybrid (annual)
Dry shade	*Gaulthera shallon* (shrub), *Cissus striata* (climber), *Tellima grandifolia* (perennial)
Moist shade	*Skimmia japonica* (shrub), *Passiflora coccinea* (climber), *Anemone* x hybrids (perennial)
Sandy soil	*Cytisus scoparia* (shrub), *Tropaeolum tricolorum* (climber), *Erymgium tripartitum* (perennial), *Antirrhinum* cvs (annual)
Heavy soil	*Cornus alba* (shrub), *Rosa filipes* (climber), *Filipendula ulmaria* (perennial)
Alkaline soil	*Berberis darwinii* (shrub), *Wisteria sinensis* (climber), *Gypsophila paniculata* (perennial), *Ageratum* cvs (annual)
Acid soil	*Pieris japonica* (shrub), *Berberidopsis corallina* (climber), *Uvularia grandiflora* (bulb)
Protected site (by a wall)	*Buddleia crispa*, *Garry elliptica* (shrubs), *Solanum crispum* (climber), *Salvia involucrate* (perennial)
Windy site	*Euphorbia characias* (small shrub), *Schisandra rubrifolia* (climber), *Geranium sanguineum* (perennial), *Limnanthes douglasii* (annual)

Plant selection for resistance

Benefit of resistance control:

- The selected resistant plant/cultivar produces appropriate chemicals that work **inside** the plant to combat a pest or disease without any further intervention being necessary from the gardener.
- In this way, resistance may greatly reduce the need for chemical control.

Limitations of resistance control:

- New strains (pathotypes) of disease or pest may develop that overcome the plant's resistance.

Natural balances: since plant resistance works against the pest or disease from inside the plant, there are not likely to be any changes in the balance of food chain species (see p. 38) when resistant cultivars are chosen. The indirect effect will be that, in the case of pest resistance, a pest's predator and parasite numbers may decrease in that locality.

Examples: a few examples of resistant cultivars are given here. 'Pentland Dell' is less prone to slug attack than most potato cultivars. The tomato cultivar 'Primato' has resistance to mosaic virus, *Fusarium* wilt and *Verticillium* wilt. The lettuce 'Beatrice' resists both downy mildew and lettuce root aphid.

Further reading

British Crop Protection Council (2014) *UK Pesticide Guide*. BCPC.

Brown, L.V. (2008) *Applied Principles of Horticulture*. Butterworth-Heinemann.

Buczacki, S. and Harris, K. (2005) *Pests, Diseases and Disorders of Garden Plants*. Collins.

French, J. (2007) *Natural Control of Garden Pests*. Aird Books.

Greenwood, P. and Halstead, H. (2009) *Pests and Diseases*. Dorling Kindersley.

Helyer, N. (2003) *A Colour Handbook of Biological Control in Plant Protection*. Manson Publishing.

Hessayon, D.G. (2009) *Pest and Weed Expert*. Transworld Publishers.

RHS (2013) Weed killers for home gardeners. Online: www.rhs.org.uk/media/pdfs/advice/WeedkillersForGardeners. RHS Advisory Service.

RHS (2014) Pesticides for home gardeners. Online: www.rhs.org.uk/media/pdfs/advice/pesticides. RHS Advisory Service.

RHS (2013) Fungicides for home gardeners. Online: www.rhs.org.uk/media/pdfs/advice/fungicides. RHS Advisory Service.

Thompson, K. (2009) *The Book of Weeds*. Dorling Kindersley.

Please visit the companion website for further information:
www.routledge.com/cw/adams

CHAPTER 17
Level 2

Garden weeds

Figure 17.1 Ground elder growing under a path

This chapter includes the following topics:

- Damage
- Weed identification
- Weed biology
- Ephemeral weeds
- Annual weeds
- Perennial weeds

Principles of Horticulture. 978-0-415-85908-0 © C.R. Adams, K.M. Bamford, J.E. Brook and M.P. Early.
Published by Taylor & Francis. All rights reserved.

Introduction

This chapter gives an overview of weeds and their importance in private gardens. A representative range of weed species is described to cover most horticultural locations. Non-chemical and chemical controls are discussed.

Control measures and their impact on humans and the environment are dealt with in more detail in Chapter 16.

> A **weed** is a plant of any kind that is growing in the wrong place.

Damage

Problems caused by weeds may be categorized as follows:

- **Reduction of crop productivity** occurs because of competition between the weed and the plant for water, nutrients and light. The cultivated plants are deprived of these major requirements and poor growth results. The extent of this competition is largely unpredictable. The major effects are seen when **light** is excluded from the garden plants as they are crowded out by the weeds. Similarly, the availability of **nutrients** and **water** becomes restricted when the weeds out-compete the garden plants.

- **Ephemeral** weeds crop productivity is further reduced when large numbers of seeds such as hairy bittercress (*Cardamine hirsuta*) and shepherd's purse (*Capsella bursa-pastoris*), chickweed (*Stellaria media*) and groundsel (*Senecio vulgaris*) may be **introduced** into a cultivated soil with poor-quality composts or farmyard manure.

- **Annual** weeds, such as field speedwell (*Veronica persica*) and annual meadow grass (*Poa annua*), may also be introduced in this way.

- **Perennial** weed seeds are not normally introduced in such large amounts in poor compost. However, their underground organs in such species as creeping buttercup (*Ranunculus repens*), ground elder (*Aegopodium podograria*), couch (*Agropyron repens*) and creeping thistle (*Cirsium arvense*) may survive poor composting (see compost temperature p. 161).

- **Uncontrolled growth** of many weeds will inevitably produce serious plant losses (Figure 17.2).

- **Reduced visual appeal of plants in a garden**. This is another way that weeds affect the garden. The conscientious gardener may consider that any plant spoiling the appearance of plants in pots, borders, paths or lawns should be removed, even though

Figure 17.2 Chickweed and sow thistle crowding out a cabbage crop

the growth of garden plants themselves may not be badly affected. Bindweed growing up shrubs and ground elder sprouting from beneath concrete garden paths are two examples of this kind of effect.

- **Alternate hosts of pests and diseases**. Horticultural pests and diseases are quite commonly found on weeds. Sow thistle and chickweed often support whitefly and red spider mite in greenhouses. Groundsel is infected by a rust disease that attacks cinerarias (see also p. 194).

- **Other weed effects**. Weeds may affect **drainage** by preventing the flow of water along ditches (e.g. by chickweed). Weeds such as redshank that have strong stringy stems may clog mowing **machines** (Figure 17.3). Shiny dark-coloured **poisonous** fruits of the climbing black nightshade may be confused with blackcurrant fruits.

Weeds in different garden locations

Four main garden situations are listed below, together with some weeds that are commonly found in them.

- In **recently cultivated soil**, the seeds of ephemeral and annual weeds such as chickweed and common speedwell, which have been brought to the soil surface, often germinate in large numbers. A common perennial weed problem in this location is couch grass.

Figure 17.3 Redshank can clog lawn mowers

Figure 17.4 Creeping cinquefoil produces roots at its internodes

- In the **herbaceous perennial border**, annual weeds are often relatively few (crowded out by the garden plants and relatively easily removed by hoeing). However, perennial weed species such as dandelions, perennial sow thistle, creeping cinquefoil (Figure 17.4) and horsetail grow alongside the garden plants. These must be carefully removed with a garden fork. Herbicidal sprays are likely to kill garden plants if great care is not taken.
- In **woody perennial plantings** (shrubs and trees), perennial weeds range from the shorter species such as broadleaved dock, ground elder and couch, to the climbing species such as small bindweed.
- In **lawns**, common weed species are perennials such as slender speedwell, creeping buttercup, yarrow, dandelion and broadleaved dock. In coarser grass, creeping thistle can establish and spread.

Weed identification

As with any problem in horticulture, recognition and identification are essential before any reliable control measures can be attempted. The weed **seedling** causes little damage to a crop but will quickly grow to be the damaging adult plant, bearing seeds that will spread. The seedling stage is relatively easy to control, whether by physical or by chemical methods. Identification of this stage is therefore important and with a little practice the gardener or grower may learn to recognize the important weeds using such plant features as cotyledon and leaf shape, colour and hairiness of the cotyledons and first true leaves (Figure 17.5).

Mature weeds may be identified using an illustrated **flora book**, which shows details of leaf and flower characters.

Weed biology

> An **ephemeral** weed is a weed that has several life cycles in a growing season.

> An **annual** weed is a weed that completes its life cycle in a growing season.

> A **perennial** weed is a weed that lives through several growing seasons.

The range of weed species includes flowering plants, ferns, mosses, liverworts and algae. These species display one or more special features of their life cycle, which enable them to compete as successful weeds against the crop and cause problems for the horticulturist.

- **Ephemeral weeds** such as groundsel and chickweed produce seeds through much of the year and seeds often germinate more quickly than crop seeds and thus emerge from the soil to crowd out the developing plants. Their seeds germinate

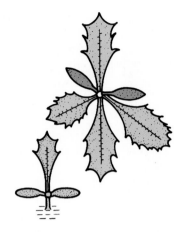

Chickweed (×1.5)
Bright green. Cotyledons have a light-coloured tip and a prominent mid-vein. True leaves have long hairs on their petioles

Groundsel (×1.5)
Cotyledons are narrow and purple underneath. True leaves have step-like teeth

Large field speedwell (×1.5)
Cotyledons like the 'spade' on playing cards. True leaves hairy, notched and opposite

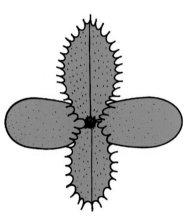

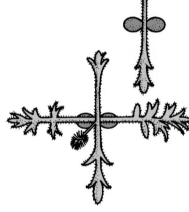

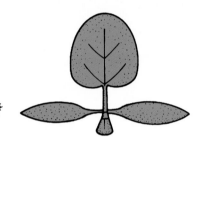

Creeping thistle (×1.5)
Cotyledons large and fleshy. True leaves have prickly margins

Yarrow (×1.5)
Small broad cotyledons. True leaves hairy and with pointed lateral lobes

Broad-leaved dock (×1.5)
Cotyledons narrow. First leaves often crimson, rounded with small lobes at the bottom

Figure 17.5 Seedlings of common weeds. Notice the difference between cotyledons and true leaves (courtesy of Blackwell Scientific Publications)

throughout the year. Their roots are often quite shallow.

▶ **Annual weeds** such as speedwells, annual meadow grass and fat hen are similar to the ephemerals in their all-year-round seed production. Their seeds take longer to ripen than those of ephemerals. They may live for a year or more. They often develop deeper roots than ephemerals.

▶ **Perennial weeds** such as ground elder, creeping buttercup, couch grass, yarrow and docks have long-lived root system. Each species has an underground organ that is difficult to control. Ground elder and couch have long lateral rhizomes; docks have deep swollen roots; and creeping thistle has long lateral roots. While seed production may be high, it is the spreading **underground organs**

that present the main problems to gardeners, enabling the weed to emerge quickly from the soil in spring, often from considerable depths if they have been ploughed in. The chopping-up of underground organs by spades and rotavators may cause these species to increase in cultivated soils.

Spread of weeds

Weeds may be spread in a number of ways:

▶ Fruits such as those in hairy bitter cress (and Himalayan balsam) discharge seeds explosively to a considerable distance.

▶ Seeds of many species from the Asteraceae (Compositae) family, such as groundsel, thistles and dandelion, are carried along in the wind by seed 'parachutes'.

Figure 17.6 Cleavers plant that produces 'sticky-bud' fruit

- Seeds of chickweed and dandelion may be spread by the moving water in ditches.
- The fruit of cleavers weed (Figure 17.6) stick to clothes and hair of humans and animals in a manner similar to 'velcro'. Chickweed seed is held in a similar way. Groundsel and annual meadow grass seeds become sticky in damp conditions and are able to stick to boots and machinery wheels.
- Some of the seeds of groundsel, annual meadow grass, yarrow and dock survive digestion in the stomachs of birds. Chickweed and annual meadow grass seed are also able to survive rabbit digestive systems.
- Cut stems of slender speedwell are moved by grass mowers.
- Ants carry around the seeds of speedwell.
- The underground horizontal roots, stolons and rhizomes of perennial weeds such as thistle, yarrow and couch, respectively, slowly spread the weed from its point of origin.
- Ploughs and rotavators move around cut underground fragments of weeds such as thistles, yarrow, dandelion and couch.
- Commercial seed stocks can be contaminated with seeds of weeds such as speedwells and couch.

Other aspects of weed biology

- **Soil conditions** may favour certain weeds. Sheep's sorrel (*Rumex acetosella*) prefers acid conditions. Mosses are found in badly drained soils. Knapweed (*Centaurea scabiosa*) competes well in dry soils. Common sorrel (*Rumex acetosa*) survives well on phosphate-deficient land. Yorkshire fog grass (*Holcus lanatus*) invades poorly

fertilized turf. Nettles and chickweed prefer highly fertile soils.
- **The growth habit** of a weed may influence its success. Chickweed, creeping buttercup, slender speedwell and creeping cinquefoil produce horizontal (prostrate) stems bearing numerous leaves that prevent light reaching emergent crop seedlings or turf. Groundsel and fat hen have an upright habit that competes less for light in the early period of weed growth. Perennial weeds such as bindweed, cleavers and nightshades are able to grow alongside and climb up woody plants, such as cane fruit and border shrubs, making control difficult.
- **Seed production** may be high in certain species. A scentless mayweed plant (perennial) may produce 300,000 seeds, fat hen (annual) 70,000 and groundsel (annual) 1,000 seeds in a growing season.
- **Dormancy** (see also p. 68) is seen in many weed species. In this way, weed seed germination commonly continues over a period of four or five years after seed dispersal, presenting the grower with a continual problem. Groundsel is something of an exception, since many of its seeds germinate in the first year.
- **Perennial weeds** with swollen underground organs provide the greatest problems to the horticulturist. This is especially so in long-term crops such as soft fruit and turf because foliage-acting and residual herbicides may have little effect on these underground organs.
- **Fragmentation** of above-ground parts may be important. A lawnmower used on turf which has slender speedwell weed growing in it cuts and spreads the delicate stems that establish (like cuttings) in other parts of the lawn in **damp** conditions.
- **Greenhouse** production generally suffers less from weed problems because composts and border soils are regularly sterilized, but weeds such as sow thistle, chickweed (see Figure 17.8), groundsel and hairy bitter cress may become established.

Ephemeral weeds

Hairy bitter cress (*Cardamine hirsuta*)

This is an annual species of the Brassicaceae family.

Damage and location. It is common throughout Britain and Ireland. It is often seen in gardens and is particularly common on bare ground, in greenhouses and at the side of paths. The compost

Figure 17.7 Hairy bitter cress can quickly invade bare ground and greenhouses

of container grown plants in garden centres may also be covered by this species (and this is one means by which the weed may be introduced into gardens) (Figure 17.7). It is recorded up to an altitude of 1,500 m. It is a common weed of gardens, greenhouses, paths, railways and waste ground.

Life cycle. Hairy bitter cress grows as an annual (or biennial) plant. It flowers throughout the year but peaks from March to August. It is self-pollinated. Seed is most commonly released in May and June (less often in September–October). A large plant may release several thousand seeds. Weeding may encourage this dispersal. High temperatures dry the dispersed seed and induce the ripening process that allows germination. Very few unripe seeds germinate after dispersal. The peak time of germination is between July and August and between November and December. The seedlings, being frost-tolerant, survive all but the severest of winters. The species is able to complete its life cycle in as little as five weeks. In fertile soils, the life cycle may take longer. The dormant seeds of this species in the soil can lead to a relatively persistent 'seed-bank' that emerges over several years.

Spread. This is by means of the 'explosive' discharge mechanism which can result in the seeds travelling up to a metre away from the mother plant. Also, the seeds become sticky when wet and can be spread on tools and clothing.

Control. As with all annual weeds, it is important to destroy the plant before seed can be produced. Seedlings should be removed by **cultural** methods such as cultivations and hoeing, among ornamentals or vegetables, preferably when soils are dry. In this

way, flowering and releasing seed is prevented. It should also be remembered that pieces of the stem are able to re-establish in suitably moist conditions. In container plants, weed seedlings should be removed before seed production and dispersal begin. Covering border soil or bare soils with a 15 cm deep mulch of bark or compost will be effective in suppressing the emergence of weed. There are several **chemical** control options for this weed. For bare soil and the side of paths, a choice of two contact chemicals, **fatty acids** and **pelargonic acid**, are available. Where weed infestations are high, the translocated chemical **glyphosate** may be used. It is advised that the weed is controlled by this chemical at the flowering stage in order to have maximum effect. Extreme care should be taken not to allow spray to reach garden annuals, or any leaf growth on perennials as this chemical is not selective and is capable of killing all plant species. Glyphosate may be sprayed around the base of woody plants so long as there are no emerging suckers visible. Glyphosate is almost immediately neutralized when in contact with soils, and therefore normally presents no risk to garden species being planted into soil that have recently been treated by this chemical.

Chickweed (*Stellaria media*) and mouse-eared chickweed (*Cerastium fontanum*)

These belong to the plant family Caryophyllaceae.

Damage and location. The first species is found in many horticultural situations as a weed of bare soil, among herbaceous and woody perennials, and in vegetables, soft fruit and greenhouse plantings. It has a wide distribution throughout Britain and Ireland, growing on land up to altitudes of 700 m, and is most important on nutrient-rich, heavy soils.

Life cycle. The seedling cotyledons are pointed with a light-coloured tip, while the true leaves have hairy petioles (Figure 17.8). The adult plant has a characteristic lush appearance and grows in a prostrate manner over the surface of the soil; in some cases it covers an area of 0.1 m², its leafy stems crowding out young plants as it increases in size. Small white, five-petalled flowers are produced throughout the year, the flowering response being indifferent to day length. The flowers are self-fertile.

An average of 2,500 disc-like seeds (1 mm in diameter) may result from the oblong fruit capsules produced by one plant. Since the first seed may be dispersed within six weeks of the plant germinating, and the plant continues to produce seed for several months, it can be seen just how prolific the species is. The large numbers of seed (up to 14 million/

Figure 17.8 (a) Chickweed plant; (b) seedling

ha) are most commonly found in the top 7 cm of the soil where, under conditions of light, fluctuating temperatures and nitrate ions, they may overcome the dormancy mechanism and germinate to form the seedling. Many seeds, however, survive up to the second, third and occasionally fourth years. Figure 17.11 shows that germination can occur at any time of the year, with April and September as peak periods. Chickweed is an alternate host for many aphid-transmitted viruses (e.g. cucumber mosaic), and the stem and bulb nematode.

Spread. The seeds are normally released as the fruit capsule opens during dry weather. They survive digestion by animals and birds and may thus be dispersed over large distances. Irrigation water may carry them along channels and ditches.

Mouse-eared chickweed is a close relative of chickweed, but is a perennial, with oval leaves and many leaf hairs that give the leaves a slightly fluffy appearance (Figure 17.9). It often grows in turf, lying close to the soil surface and escaping lawnmower blades. It prefers damp, poorly drained ground. It

Figure 17.9 Mouse-eared chickweed in turf

213

may be infected by cucumber mosaic virus carried by peach-potato aphid (see p. 230).

Control. Chickweeds are controlled by a combination of methods. **Physical** controls include partial sterilization of soil in greenhouses, while the **cultural control** of hoeing in the spring and autumn periods prevents the developing seedling from developing and flowering. Mulching is effective against germinating weeds. Gardeners can use the non-selective, non-persistent herbicide **glyphosate** for control in such situations as ornamental beds containing woody perennials, and in cane fruit, but care is needed to avoid spraying foliage of garden plants.

Mouse-eared chickweed in turf is reduced by mowing the grass close to the ground, by the **cultural** control method of raking out the weed's horizontal stems in spring, and by seeding over bare areas in spring or autumn with grass varieties that are suitable for the growing situation. Herbicides containing a combination of active ingredients such as **clopyralid, 2,4-D, MCPA** may be effectively used against mouse-eared chickweed (as a spot spray or over the whole area) when it is growing in turf.

Shepherds's purse (*Capsella bursa-pastoris*)

Damage and location. This is an annual (and biennial) weed (Figure 17.10) that is commonly found in gardens, and other cultivated and wayside locations, and is found growing on most soil types, sometimes in large numbers. It has been recorded in up to 1,750 ft in Britain and Ireland.

Life cycle. Flowering and seed production occur throughout the year, most commonly in May to October. Plants can produce several thousand seeds. Three generations of weed may occur per year. Most seeds germinate after a 'maturation period' of two years, but this period is much shorter in some instances, especially when seeds are on the soil surface and exposed to daylight. Turning-over soil can result in dormant seeds (from a deeper level) germinating at the soil surface. Shepherd's purse is a host of two important diseases, **club root** and **white blister 'rust'**, and the pest **mealy cabbage aphid**.

Spread. The weed can be introduced, as seed, into the garden through garden compost, or on new plants purchased from nurseries or garden centres. It can spread from neighbouring gardens.

Control. Several **cultural** methods are available for control. Young plants can be pulled out by hand. Seedlings can be hoed when the soil surface is dry. Avoid disturbing the soil so that dormant seeds are

Figure 17.10 Shepherd's purse plant showing typical small white flowers and heart-shaped fruits

not brought to the surface. **Mulch** around garden perennials, using a 5 cm thick layer of composted bark or garden compost to inhibit the weeds' germination and subsequent growth. Spraying young weeds with a translocated weedkiller such as **glyphosate** will be effective; but great care must be taken that the spray does not reach garden plants.

Groundsel (*Senecio vulgaris*)

This belongs to the Asteraceae (Compositae) plant family.

Damage and location. This is a very common and important weed, particularly on heavy soil. Its high level of seed production and the ability of its seed to germinate soon after dispersal from the plant lead to dense mats of the weed. It grows on both rich and poor soils up to almost 600 m in altitude. It can be a problem in the garden, in soft fruit and in herbaceous and woody perennials, but is most commonly serious after soil cultivation in vegetable plots.

Life cycle. The seedling cotyledons are narrow, purple underneath, and the first true leaves have

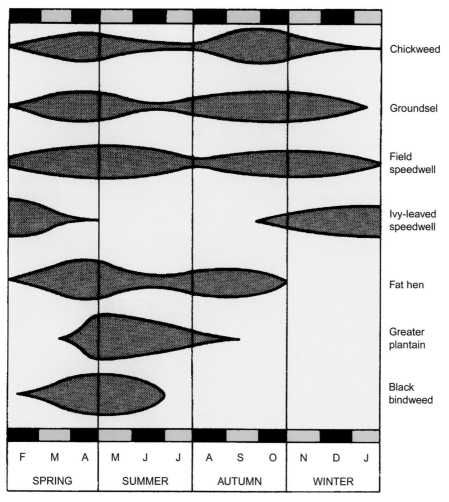

Figure 17.11 Annual and perennial weed periods of seed germination throughout the year (reproduced by permission of Blackwell Scientific Publications)

17

Figure 17.12 (a) Groundsel plant; (b) seedling

step-like teeth (Figure 17.12). The adult plant has an upright habit and produces as many as 25 yellow, small-petalled flower heads. Flowering occurs in all seasons of the year. Each flower produces about 45 column-shaped seeds, 2 mm in length, densely packed in the fruit head. As can be seen in Figure 17.11, the seeds may germinate at any time of the year, with early May and September as peak periods. Since there may be more than three generations of groundsel per year (the autumn plants surviving the winter) and each generation may give rise to 1,000 seeds, it is clear why groundsel is one of the most successful colonizers of cultivated ground. Its role as a symptomless carrier of the wilt fungus *Verticillium* increases its importance in certain crops such as tomatoes and hops.

Spread. The seeds bear a mass of fine hairs. These hairs, in dry weather, can parachute seeds along on air currents for many metres. In wet weather the seeds become sticky and may be carried on the feet of animals, including humans. The seeds survive digestion by birds and thus can be transported in this way.

Control. A combination of control methods may be necessary for successful control. **Cultural** control is by hoeing or cultivation, particularly in spring and autumn to prevent developing seedlings from flowering. Care should be taken not to allow uprooted flowering groundsel plants to release viable seed. The gardener can use the herbicide **glyphosate** for control in uncultivated areas, and in such situations as ornamental beds containing woody perennials, and in cane fruit. Great care is needed to avoid spraying foliage of garden plants.

Annual weeds

There are at least 50 **successful** annual weed species found in gardens. This book can cover only a few examples that illustrate the main points of life cycle, spread and control. Two species, annual meadow grass and speedwell, are described below.

Annual meadow grass (*Poa annua*)

This belongs to the plant family Poaceae (Graminae).

Damage and location. This species is a fairly small annual (or short-term perennial) found in herbaceous and woody perennial borders, on sports grass surfaces, on paths and in vegetable plots (Figure 17.13). It is quite often seen on golf greens where its flowering heads and its yellowish green foliage make it conspicuous. It does not thrive on acid soils or those low in phosphates. Despite its relatively small size,

Figure 17.13 Annual meadow grass adult plant

it often emerges in sufficient quantities to smother crop seedlings. Its seed may be present as an impurity in grass seed bought at the garden centre. (Special selections of seed containing this species may sometimes be deliberately used by seed companies as a component of turf seed mixtures.)

Life cycle. Flowers can occur at any time of year and are usually self-pollinated. About 2,000 seeds per plant are produced from April to September. Plants will flower and seed even when mown regularly, because the weed is able to flower at a plant height of less than 1 cm. Seeds germinate from February to November, with the main peaks in early spring and in autumn. Some seed will germinate soon after their release; others can remain viable in soil for at least four years.

This weed species can be the alternate host of a number of nematode species such as root knot nematode (*Meloidogyne minor*), which is increasingly being recognized as a problem on fine turf (especially on creeping bent, *Agrostis stolonifera*).

Spread. This weed has no obvious dispersal mechanism. Most seeds fall around the parent plant and become incorporated into the soil, but heavy winds may spread the seeds a few metres. Seeds may be carried around on boots and wheels of machinery. Worms may bring seeds to the soil surface in worm casts.

Control. This is achieved by a variety of methods. In flower beds and vegetable plots, the **cultural** action of hoeing normally controls the weed, especially when it is in the young stage. Deep digging-in of seedlings and young plants is also usually effective. Mulching with material such as wood bark is effective against germinating weeds in flower beds and fruit areas.

For turf, the general advice is to avoid patches of bare turf, to avoid over-liming and to lift any weed flower heads flattened by mowing so that the mower can be more effective. Mowing in different directions helps to achieve the same result. Not surprisingly, there is **no herbicidal** control for a grass weed on a lawn, or on sports surfaces. For control in bare ground, ornamental beds containing woody perennials and cane fruit, the amateur gardener can use the non-selective, non-residual herbicide **glyphosate**, but care is needed to avoid spraying foliage of garden plants.

Speedwells: large field speedwell (*Veronica persica*) and slender speedwell (*V. filiformis*)

These belong to the plant family Scrophulariaceae.

Damage and location. The first species, the large field speedwell (*V. persica*), is an important annual weed in vegetable plots, crowding out young crop plants and reducing growth of more mature stages (Figure 17.14). The second species, a slender, mat-forming, creeping perennial speedwell (*V. filiformis*), once considered a desirable rock garden plant introduced from Turkey, has become a common turf problem (Figure 17.15) .

Life cycle. The seedling cotyledons are oval, while the true leaves are opposite, notched approximately triangular and hairy (Figure 17.14) in both species. The leaves of *V. filiformis* are about half the size of *V. persica*.

Veronica persica produces up to 300 bright blue flowers, 1 cm wide, per plant. The flowers are self-fertile and occur throughout the year, but mainly between February and November. The adult plant produces an average of 2,000 light-brown, boat-shaped seeds (2 mm across). The seeds of this

Figure 17.14 (a) Large field speedwell; (b) seedling

species germinate below soil level all year round, but most commonly from March to May (see Figure 17.11), the winter period being necessary to break dormancy. Seeds may remain viable for more than two years.

Figure 17.15 Slender speedwell plant has small leaves and is often seen growing in turf

Figure 17.16 Creeping buttercup in turf spreads by means of runners

Veronica filiformis produces small self-sterile purple-blue flowers between March and May, and spreads by means of prostrate stems which root at their nodes to invade fine and coarse turf, especially in damp areas. Segments of this weed cut by lawnmowers easily root and thus increase the species. Seeds are not important in its spread.

Spread. Seeds of *V. persica* falling to the ground may be dispersed by ants. Seed of this species can be spread as contaminants of crop seed. *V. filiformis* does not produce seed. Its slow spread is mainly by means of lawnmower activity.

Control. Field speedwell (*V. persica*) is controlled by a combination of methods. The cultural action of hoeing or mechanical cultivation, particularly in spring, prevents developing seedlings from growing to mature plants and producing their many seeds.

The gardener can use the herbicide **glyphosate** for control in bare soil, and in such situations as ornamental beds containing woody perennials, and in cane fruit but care is needed to avoid spraying foliage of garden plants.

The slender speedwell (*V. filiformis*) represents a different problem for control. **Cultural** controls such as regular close mowing and spiking of turf

remove the high humidity necessary for this weed's establishment and development.

Gardeners may have difficulty controlling this weed with the range of turf weedkillers available. Herbicides containing **fluroxypyr** will have some effect. They can control the weed in a turf seedbed using a total, contact chemical such as **glyphosate**, a few weeks **before** sowing the turf seed. This '**stale seed bed**' method leaves the turf to establish, relatively undisturbed by weeds. Great care is needed to avoid spraying foliage of garden plants.

Perennial weeds

Seven species – creeping buttercup, ground elder, couch, creeping thistle, yarrow, dandelion and broadleaved dock – are described below to demonstrate the different features of their biology (particularly the parts of the plant that allow the plant to **survive** from year to year) that make them successful perennial weeds. The flowering period of these weeds is mainly between June and October but the main problem for gardeners and growers is the plants' ability to survive and reproduce vegetatively (see also p. 133).

The horsetails, mosses and liverworts that are unrelated to flowering plants are covered in a separate section (p. 94).

Creeping buttercup (*Ranunculus repens*)

A member of the plant family Ranunculaceae.

Damage and location. This is a common weed in turf (Figure 17.16), especially when the soil is a heavy clay or the location is prone to water logging. The species may grow at altitudes up to 1700 m. It is most common in neutral pH conditions but may occur on slightly acid or lime soils, especially when bare patches have been left by machinery, sports footwear or rabbit damage.

Life cycle. Creeping buttercup may grow up to 60 cm in height at flowering time. Horizontally growing **runners** develop above ground in May and develop roots along their length, representing the main problem to turf grass. Each new rooting point gives rise to a new plantlet during the summer months, so that after a few years of uncontrolled growth, the buttercup colony occupies a large circular area of turf. The low growing point of this weed keeps it out of reach of most lawnmowers. Its long, fibrous roots reach deep down into the soil, giving it resistance to dry weather, and represent a problem for herbicidal control.

Spread. The two-year-old plant produces bright yellow flowers from May to August, and an average of 700 seeds are produced by each plant. The plant dies after seed production, but leaves behind developing daughter plants. Seeds may fall to the ground and survive for seven years in the soil, or may be eaten by birds and thus spread.

Control. **Cultural** controls include weeds being carefully uprooted with a small fork (ensuring that the growing point and main roots are removed) while raking up the weed's runners before cutting the grass helps to reduce spread. Autumn spiking and drainage of lawns lead to less weed establishment. Selective weedkillers such as those containing a mixture of **2,4-D**, **dicamba** and **fluroxypyr** penetrate down to the weed's roots without killing turf. In bare ground, buttercups may be controlled by **glyphosate** herbicide.

Ground elder (*Aegopodium podagraria*)

This belongs to the Apiaceae family. Ground elder is so-named because it has a leaf shape superficially resembling that of the elderberry shrub. It is, however, not related to this species.

Damage and location. This weed is perennial, and can rapidly form a dense matt of foliage that crowds out plants in beds and borders (see Figure 17.1).

Life cycle. Ground elder produces white underground **rhizomes** (see p. 92) under the soil but close to the surface. These invade gardens by creeping under fences from neighbouring garden and wasteland. The weed may also be present in the compost of introduced pot plants that are then grown in the garden.

Control. **Cultural controls** include carefully removing rhizomes with a garden fork, as they occur close to the surface of the soil. However, eradicating it completely needs constant attention as the small pieces of rhizome remaining in the soil can develop into a new plant.

One rather prolonged but effective control measure involves the temporary removal and potting of shrubs and herbaceous border plant from the weed-affected areas. This procedure is then followed by placing over the area a large sheet of thick, opaque **black plastic** that starves the weed of light and moisture. It is recommended that this sheet is left in position for a year to achieve control. It may be necessary to repeat this procedure if the weed emerges above ground again.

When this weed is found in **lawns** and growing independently (i.e. not linked to another ground elder plant by a rhizome), regular weekly mowing of the grass with a low blade cut will gradually starve the weed of its stored food reserves.

Chemical controls include:

▶ **Glyphosate** is a translocated herbicide that is effective against ground elder. The chemical moves down from the leaves to the underground roots and rhizomes, thus killing the whole plant. Three important points should be noted: glyphosate kills all types of plants – any sensitive garden plants need to be protected at the time of spraying; use a sheet of polythene in order to avoid any spray drift reaching them; gel formulations that are wiped onto the weed will be less likely to cause a problem.

▶ Mid–summer sprays achieve the best results as the weed has considerable leaf at this time. In this way, the chemical is deposited over a large area of foliage, and there will be more internal movement of the chemical down to the roots and rhizomes. However, a second spray may be applied in late summer, if the weed has not yet died.

▶ Spraying in the evening is more effective because the chemical is absorbed by the leaves more effectively at this time.

Couch grass (*Agropyron repens*)

Belongs to the plant family Poaceae (Graminae).

Damage and location. This grass is thought to have been introduced to Britain and Ireland 2,000 years ago by the Romans as a medicinal plant. It is sometimes called 'twitch', and is a widely distributed and important weed found at altitudes up to 500 m. It quite often takes over bare plots and those growing ornamentals, vegetable or fruit.

17

Figure 17.17 Couch grass plant showing rhizomes

Life cycle. The dull-green plant is often confused, in the vegetative stage, with the creeping bent (*Agrostis stolonifera*). However, the small 'ears' (ligules) at the base of each leaf are a distinguishing feature of couch. The plant may reach a metre in height and often grows in clumps. Seeds (9 mm long) are produced only after cross-fertilization between different strains of the species, and the importance of the seed stage, therefore, varies from field to field. The seed may survive deep in the soil for up to 10 years.

Spread. Couch seeds may be carried in grass seed batches over long distances. From May to October, stimulated by the high light intensity, overwintered plants produce horizontal rhizomes (Figure 17.17) just under the soil; these white rhizomes may spread 15 cm per year in heavy soils, 30 cm in sandy soils. They bear scale leaves on the underground nodes that remain suppressed during the growing period and do not produce new stems. But, in autumn, rhizomes attached to the mother plant often grow above ground to produce new plants that survive the winter. If the rhizome is cut by cultivations such as digging or ploughing, fragments containing a node and several centimetres of rhizome are able to grow into new plants. The rapid growth of couch plants creates severe competition for light, water and nutrients in any infested crop.

Control. This is achieved by a combination of **cultural** and **chemical** methods. In bare soil, deep digging or ploughing (especially in heavy land) exposes the rhizomes to drying. Further control by rotavating the weed when it reaches the one- or two-leaf stage disturbs the plant at its weakest point, and repeated rotavating will eventually cut up couch rhizomes into such small fragments that they are unable to propagate.

The gardener can use the herbicide **glyphosate** for control in such situations as bare soil and ornamental beds containing woody perennials, and in cane fruit, but **not** in growing crops. Great care is needed to avoid spraying foliage of garden plants.

Creeping thistle (*Cirsium arvense*)

This belongs to the plant family Asteraceae (Compositae).

Damage and location. This species is a common weed in grass and perennial crops (e.g. apples), where it forms dense clumps of foliage, often several metres across.

Life cycle. The seedling cotyledons are broad and smooth, the true leaves spiky (Figure 17.18). The mature plant is readily recognizable by its dark green spiny foliage growing up to a metre in height. It is found in all areas, even at altitudes of 750 m, and on saline soil. The species is dioecious (see p. 100), the male plant producing spherical and the female slightly elongated purple flower heads from July to September. Only when the two sexes of plants are within about 100 m of each other does fertilization occur in sufficient quantities to produce large numbers of brown, shiny fruit, 4 mm long. The seeds may germinate beneath the soil surface in the same year as their production, or in the following spring, particularly when soil temperatures reach 20°C. The resulting seedlings develop into a plant with a taproot that commonly reaches 3 m down into the soil.

Spread. Seeds are wind-borne using a parachute of long hairs. The mature plant produces **lateral roots** which grow out horizontally about 0.3 m below the soil surface and may spread the plant as much as 6 m in one season, causing a large dense patch of thistles. Along their length are produced adventitious buds that, each spring, grow up as stems. Under permanent grassland, the roots may remain dormant for many years. Soil disturbance, such as ploughing, breaks up the roots and may result in a worse thistle problem.

Figure 17.18 Creeping thistle plant showing lateral roots

Figure 17.19 Yarrow plant showing stolons at the base of the stem

Control. The seedling of this weed is not normally targeted, although an attentive gardener may use a hoe against this easily identified stage. Cutting down plants at the flower bud stage when sugars are being transferred from the roots upwards is a **cultural** control measure that helps weaken the plants. The main cultural control, however, is the removal of main roots and lateral roots by deep digging.

Gardeners are able to use herbicide products which contain a mixture of **dicamba**, **MCPA** and **mecoprop-P** (all of these are translocated down to the roots) on turf. **Glyphosate** can be used for control in such situations as bare soil and ornamental beds containing woody perennials, and in cane fruit, but **not** in growing crops. Great care is needed to avoid spraying foliage of garden plants.

Yarrow (*Achillea millifolium*)

This belongs to the plant family Asteraceae (Compositae).

Damage and location. This strongly scented perennial, with its spreading flowering head (Figure 17.19), is a common hedgerow plant found on most soils at altitudes up to 1,200 m. Its persistence, together with its resistance to herbicides and drought in grassland, makes it a serious turf weed.

Life cycle. The seedling leaves are hairy and elongated with sharp teeth (Figure 17.5). The mature plant has dissected pinnate leaves (see p. 89) produced throughout the year on wiry, woolly stems, which commonly reach 45 cm in height, and which from May to September produce flat-topped white to pink flower heads. Each plant may produce 3,000 small, flat seeds annually. The seeds germinate on arrival at the soil surface.

Spread. Seeds are dispersed by birds. When not in flower, this species produces below-ground and above-ground **stolons** (see Figure 17.19) which can grow up to 20 cm long per year. In autumn, rooting from the nodes occurs and new stems appear.

Control. Control of this weed may prove difficult. Routine scarification of turf does not easily remove the roots. Gardeners may use products containing **2,4-D** and **mecoprop** against this weed.

Figure 17.20 Dandelion plant growing in turf

Dandelion (*Taraxacum officinale*)

This belongs to the plant family Asteraceae (Compositae).

Damage and location. This species is a perennial with a stout **taproot**. It is a weed in lawns (Figure 17.20), orchards, and on paths. Several similar species such as mouse-ear hawkweed (*Hieracium pilosella*) and smooth hawk's beard (*Crepis capillaris*) present problems similar to dandelion in turf.

Life cycle. Seedlings emerge mainly in March and April. Flowers are produced from May to October. An average of 6,000 seeds is produced by each plant. Most seeds survive only one year in the soil, but a few may survive up to five years. Mature plants can survive for 10 years.

Spread. Seeds are wind dispersed by means of tiny 'parachutes' and may travel several hundred metres. They are also able to spread in the moving water found in ditches and by animals through their digestive systems. The plant may survive and regenerate from roots, after they have been chopped up by spades or rotavators.

Control. Physical removal of the deep root by a sharp trowel is recommended, but this leaves bare gaps in turf for invasion by other weeds. Gardeners use products containing the two translocated ingredients **2,4-D** and **dicamba** that can reach and kill the stout root of the dandelion. **Glyphosate** can be used for control in such situations as bare soil and ornamental beds containing woody perennials, and in cane fruit, but **not** in growing crops. Great care is needed to avoid spraying foliage of garden plants.

Broadleaved dock (*Rumex obtusifolius*)

This belongs to the plant family Polygonaceae.

Damage and location. This is a common perennial weed in vegetable plots, turf and bare soil.

Figure 17.21 Broad-leaved dock showing a swollen taproot

Life cycle. The seedling cotyledons are narrow (Figure 17.5). The seedling's first true leaves are broad and often crimson-coloured. The mature plant is readily identified by its long (up to 25 cm) shiny green leaves (Figure 17.21), known to many as a cure for 'nettle rash'. The plant may grow 1 m tall, producing a conspicuous branched inflorescence of small green flowers from June to October. The seed represents an important stage in this perennial weed's life cycle, surviving many years in the soil and most commonly germinating in spring. Like most *Rumex* spp., the seedling develops a stout, branched taproot, which may penetrate the soil down to 1 m in the mature plant, but most commonly reaches 25 cm. Segments of the taproot, chopped by digging and rotavating, are capable of producing new plants.

Spread. The numerous plate-like fruits (3 mm long) may fall to the ground or be dispersed by seed-eating birds such as finches. They are sometimes found in commercial batches of seed.

Control. High levels of seed production, a tough taproot and a resistance to most herbicides present a problem in the control of this weed.

Figure 17.22 (a) Horsetails have an extensive rhizome system; (b) horsetail emerging from soil

In turf, cultural control involving removal of the deep root by a sharp trowel is recommended, but this leaves bare gaps in turf for invasion by other weeds. A product containing **dicamba**, **MCPA** and **mecoprop-P** is effective, especially against young dock plants. **Glyphosate** can be used for control in such situations as bare soil and ornamental beds containing woody perennials, and in cane fruit, but **not** in green-stemmed crops. Great care is needed to avoid spraying foliage of garden plants.

Horsetails

Horsetails (that belong to the order Equisetales) are related to ferns. They have a more complex structure than mosses and liverworts (see p. 52). One species, the **field horsetail** (*Equisetum arvense*), may be a serious weed in turf, soft fruit and in nursery stock (Figure 17.22), and especially in damp soil situations. Its roots produce persistent **rhizomes** (see p. 92).

Spread. Underground rhizomes may be dispersed and be the agents of spread when the weed is dug up.

Control. A **cultural** control involves regular cutting down of the weed and maintaining good levels of soil fertility (to allow the garden plant to compete). Sprays of herbicides containing **glyphosate** may be successful on bare ground, especially if the foliage is quite young and has been lightly **bruised**, to allow better spray entry through the thick **waxy epidermis**. Great care is needed to avoid spraying foliage of garden plants.

Mosses

Moses (that belong to the division Bryophyta) include three species that may be important weeds in gardens. The small **cushion-forming moss** (*Bryum* spp.) grows on sand capillary benches in greenhouses, and in acid turf that has been closely mown. **Feathery moss** (*Hypnum* spp.) is common on less closely mown, unscarified turf, or on the compost surface in pot plants (Figure 17.23). A third type of moss (*Polytrichum* spp.), which looks quite different from the other two species, is erect and has a rosette of leaves. It is found in dry acid conditions around golf greens. **Cultural** methods such as improved drainage, aeration, liming, application of fertilizer and removal of shade usually achieve good results in turf. Control with contact scorching **chemicals** (e.g. **ferrous sulphate**), applied in spring or summer may give temporary results. Moss on **sand benches** becomes less of a problem if the sand is regularly washed.

Liverworts

Liverworts (in the Hepatophyta division, fairly closely related to the mosses, see p. 52) have several weed

Figure 17.23 (a) Hypnum moss in turf; (b) pot plant with Hypnum moss and Pellia liverwort; (c) *Polytrichum* moss that may be serious in turf

species in the *Pellia* genus that are recognized by their thick, flat leaves growing on the surface of pot plant compost (Figure 17.23). These organisms increase only when the soil and compost surface is excessively wet, or when nutrients are so low as to limit plant growth. Products containing **pelargonic acid** are effective against these weeds on pot plants that have **woody** stems (but not pot plants with green, non-woody stems as they may be killed by this chemical).

Further reading

British Crop Protection Council (2014) *UK Pesticide Guide*. BCPC.

Brown, L.V. (2008) *Applied Principles of Horticulture*. 3rd edn. Butterworth Heinemann.

Blamey, M., Fitter, R. and Fitter, A. (2013) *Wild Flowers of Britain and Ireland*. Bloomsbury.

Hessayon, D.G. (2009) *Pest and Weed Expert*. Transworld Publishers.

RHS (2013) Weed killers for home gardeners. Online: www.rhs.org.uk/media/pdfs/advice/WeedkillersForGardeners. RHS Advisory Service.

Thompson, K. (2009) *The Book of Weeds*. Dorling Kindersley.

Please visit the companion website for further information:
www.routledge.com/cw/adams

CHAPTER 18
Level 2

Garden pests

Figure 18.1 Rabbit eating garden plants

This chapter includes the following topics:

- Mammal pests
- Bird pests
- Slugs and snails
- Insects
- Insect pests
- Mite pests
- Other arthropod pests
- Nematode pests

Introduction

Why do we need to control pests? Most gardeners will have seen aphids on their roses and broad beans and wanted to achieve some level of control before the pest becomes unsightly, or damages the plants. This chapter describes some of the pest species seen in gardens, from the mammals such as rabbits down to the tiny nematodes. In addition to descriptions of damage and control, the **link between life cycle and controls** is highlighted in an attempt to make the control more meaningful. Control measures and their impact on humans and the environment are dealt with in more detail in Chapter 16.

Mammal pests

There are relatively few mammal pest species in Britain and Ireland, but they can cause serious damage to plants in the garden and horticultural units. The rabbit, grey squirrel and mole are described below.

> A **pest** is a mammal, bird, insect, mite, mollusc or nematode that is damaging to plants.

The rabbit (*Oryctolagus cuniculus*)

The rabbit (Figure 18.1) is common in most countries of central and southern Europe. It was introduced to Britain and Ireland around the eleventh century by the Normans, and became an established pest in the nineteenth century. In some countries where it was introduced, such as Australia, it has become an extremely damaging pest.

Damage. The rabbit may consume 0.5 kg of plant food per day. Young turf and cereal crops are the worst affected, particularly when in the seedling stage, and quite large areas may be destroyed. The resulting bare areas of fine turf on golf courses may allow lawn weeds such as yarrow (see p. 221) to become established. Rabbits may move from cereal crops to horticultural holdings. Stems of top fruit (such as apples) and ornamental shrubs and trees may have the cambium, phloem and xylem tissues interrupted (**ring-barked**), particularly in early spring when other food is scarce. Vegetables such as carrots, lettuce, annual bedding plants and herbaceous perennials are common targets for this pest (Figure 18.2).

Life cycle. The rabbit's high reproductive ability enables it to maintain large numbers, even when continued control methods are being used. The female (doe), weighing about 1 kg, can reproduce within a year of its birth, and may have three to five litters of three to six young ones

Figure 18.2 Rabbit damage on carrots

in one year, commonly in the months of February to July. The young are blind and naked at birth, but emerge from the underground nest after only a few weeks to find their own food. Large burrow systems (warrens) penetrating as deep as 3 m in sandy soils, may contain as many as 100 rabbits. Escape or bolt holes running off from the main burrow system may allow the rabbit to escape from predators such as weasels.

Control. Rabbit control is, by law, the responsibility of the land owner. Preventive measures are effective. Three physical control measures may be mentioned. Brick walls, or wire mesh **fencing** with the base dug 30 cm underground and with this base facing outwards at an angle, represent an effective barrier to the pest. Sturdy sheet **plastic guards** coiled round the base of exposed young trees prevent ring-barking. Repellent chemicals such as **aluminium ammonium sulphate**, sprayed on bedding displays and young trees, deter the pest. Rabbits are deterred by some garden plants. Six resistant examples of the 150 listed species are *Agapanthus*, dahlia, snowdrop, *Clematis*, *Berberis* and *Weigela*.

The grey squirrel (*Sciurus carolinensis*)

The grey squirrel was introduced into Britain and Ireland from North America in the late nineteenth century, at a time when the native red squirrel population was being reduced by disease. The grey squirrel (Figure 18.3) has become dominant in many areas, but the red squirrel survives, particularly in central and southern Scotland, the Lake District, central and north Wales, and the Isle of Wight.

Damage. The horticultural damage caused by grey squirrels varies with each season. In spring, germinating bulbs may be eaten, and the bark of

Figure 18.3 Grey squirrel

platforms (dreys) high up in trees. The female may become pregnant at an early age (six months). As the squirrels have few natural enemies, and they live high above ground, control is difficult in most areas. In some northern parts of Britain and Ireland, however, pine martens (a tree-inhabiting relative of the weasel) are important predators of grey (and red) squirrels.

Control. During the months of April to July, when most damage is seen, **cage traps** containing desirable food (e.g. maize seed), reduce the squirrel population to less damaging levels. Spring traps placed in natural or artificial tunnels achieve rapid results at this time of year if placed where the squirrel is commonly found. It is suggested that the caged squirrel is enticed from the trap into a strong sack and killed with a blunt instrument. It should be noted that releasing captured squirrels back into the wild is illegal. **Ultrasonic deterrent** devices are claimed to have a repelling action against squirrels up to a distance of 30 m. **Netting** may be placed over frames covering sensitive crops such as raspberries and strawberries to reduce the squirrel's foraging. The use of anti-coagulant poisons such as **warfarin** is **not** permitted for gardeners. Trained operators need to be employed to carry out this procedure.

The mole (*Talpa europea*)

The mole is found in all parts of mainland Britain but is not present in Ireland.

Damage. This dark grey, 15 cm long mammal (Figure 18.4), weighing only about 90 g, uses its shovel-shaped feet to create an underground system 5–20 cm deep and up to 0.25 ha in extent. The tunnel contents are excavated into mole hills. The resulting root disturbance to turf and crops causes wilting, and

many tree species stripped off (**ring-barking**). In summer, pears, plums, raspberries and peas may suffer. Autumn provides an alternative food source in the form of wild plant seeds, but apples and potatoes may be damaged at this time. In winter, little damage is done. Gardens and horticultural units located near wooded areas can be badly damaged by this agile and inventive pest.

Life cycle. Squirrels most commonly produce two litters of three young from March to June, in twig

18

Figure 18.4 (a) Mole (source: Wikimedia Commons); (b) mole hills in grassland

may result in serious losses. It should be noted that the plants themselves are not eaten by the mole.

Life cycle. Throughout the year, in its dark environment, this solitary animal moves, actively searching for earthworms, slugs, millipedes and insects. About five hours of activity are followed by about three hours of rest. Only in spring do males and females meet. In June, one litter of two to seven young are born in a grass-lined underground nest, often located underneath a dense thicket. Young moles often move above ground to find a new territory, reaching maturity at about four months. Moles live for about four years.

Control. Natural predators of the mole include tawny owls, weasels and foxes. The main control method is trapping, best carried out between October and April, when tunnelling is closer to the surface. Gardeners can use **pincer** or **half-barrel traps** placed in fresh tunnels and inserted carefully so as not to greatly change the tunnel diameter. The soil must be replaced so that the mole sees no light from its position in the tunnel. The animal enters the trap, is caught and starves to death. On areas such as lawns, the drenching of a soluble product containing **aluminium ammonium sulphate** (that acts as an irritant to the mole, when reaching the tunnel) is often used. Smoke formulations can be inserted inside mole tunnels: the resulting vapour deposits a chemical such as **castor oil** on the inside of the tunnel and this keeps earthworms away from the tunnel, causing the mole to move away from the area. **Sonar deterrent** devices (including solar-powered models), when inserted into soil, deliver a frequency of sound intolerable to the mole that causes the animal to move away from the disturbance.

Deer

Deer (notably roe deer and muntjac) are increasingly become important pests as more woodland areas are developed in Britain and Ireland.

Bird pests

The wood pigeon (*Columba palumbus*)

Damage. This attractive-looking, 40 cm long, blue-grey pigeon with white under-wing bars is known to horticulturists as a serious pest on most outdoor edible crops (Figure 18.5). In spring, seeds and seedlings of crops such as brassicas, beans and germinating turf may be systematically eaten. In summer, cereals, clover and soft fruit receive its

Figure 18.5 Wood pigeon

attention; in autumn, tree fruits may be taken in large quantities; while in winter, cereals and brassicas are often seriously attacked, the latter when snowfall prevents the consumption of other food. The wood pigeon is attracted to high-protein foods such as seeds when they are available.

Life cycle. Wood pigeons lay several clutches of two eggs per year from March to September. The chicks from August/September clutches are the most likely to survive. Eggs are laid on a nest of twigs situated deep inside the tree, and hatch after about 18 days, the young remaining in the nest for 20–30 days. Predators such as jays and magpies eat many eggs, but the main population-control factor is the lack of food in winter. Numbers in Britain and Ireland are boosted a little by migrating Scandinavian pigeons in April, but the large majority of birds are resident in the Britain and Ireland and do not migrate around the country.

Control. The wood pigeon spends much of its time feeding on wild plants and only a small proportion of its time is spent eating crops. Control of the whole population therefore seems ethically unsound and is both costly and impracticable. **Physical** control involves the protection of particular areas by means of disturbing or scaring devices. In gardens, plastic or stainless steel strips bearing spikes may be placed in pigeon vantage points on buildings and walls and reduce their damage to garden plants. **Ultrasonic scarers** are effective over a small area. **Water-scare** equipment, switched on by the pest's presence, is becoming more popular with gardeners. Five-centimetre mesh **netting** placed over an area prevents the pigeon's entry to such targets as young plants and soft fruit. Gel or liquid formulations of **aluminium ammonium sulphate** applied to plant areas or brushed on to pigeon vantage places such as roof ledges act as an irritant and deter the pest. Life-size-**model predators** such as owls or falcons may be purchased and placed in a suitably prominent place in the garden.

Bullfinch

The bullfinch is another bird pest. It can seriously affect soft and tree fruit yields by nipping out fruiting buds overwinter.

Slugs and snails

Slugs and snails belong to the phylum Mollusca.

Damage. The slug, unlike the snail, lacks a shell and this permits its movement under the soil in search of its food source: seedlings, roots, tubers and bulbs. It feeds by means of a file-like tongue (**radula**), which cuts through the plant material held by the soft mouth. It can also scoop out cavities in affected plants (Figure

Figure 18.6 (a) Slug damage on carrot; (b) black *Arion* slug; (c) garden snail; (d) slug trap made from a plastic milk bottle; (e) carnivorous *Testacella slug* that eats earthworms and compost worms

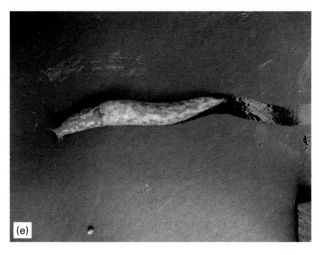

18.6). In moist, warm weather it often causes above-ground damage to leaves of bedding plants, new turf, lettuce and Brussels sprouts. Snails, while not being as serious pests as slugs, nevertheless are capable of causing considerable damage in private gardens to young plants, to particularly 'snail-attractive species' such as hostas, and to crops in greenhouses such as tomatoes and petunias.

Life cycle. Slugs and snails are **hermaphrodite** (bearing in their bodies both male and female organs). They mate in spring and summer, and lay clusters of up to 50 round, white eggs in rotting vegetation, the warmth from which protects this sensitive stage during cold periods. Slugs range in size from the keeled slug (*Milax*), 3 cm long, to the black garden slug (*Arion*), which reaches 10 cm in length. Slugs move slowly by means of an undulating foot, the slime trails from which may indicate their presence. The three species of mottled slug (*Testacella*), occurring mainly in southern Britain, are quite common, but rarely seen, living predominantly underground. The mottled slugs are carnivorous, feeding on earthworms and compost worms, sometimes on other slugs (Figure 18.6).

Control. Predators such as centipedes, ground beetles and glow-worm larvae are natural controlling species against slugs and snails. Digging of garden plots exposes slugs to birds such as thrushes and reduces slug egg numbers. There are some examples of slug-resistant cultivars. While most lilies are prone to slug attack, *Lilium hennyi* and *Lilium tigninum* are rarely affected. *Hosta species*, which compete with lettuce as the slug's favourite food, may be grown with more confidence if blue-leaved cultivars are chosen. **Coarse sand** placed round the base of pots growing plants such as hostas helps prevent leaf damage. **Copper strips** placed around pots (Figure 16.2a) or on the edge of glasshouse benches deter snails. A parasitic **nematode** (*Phasmarhabditis hermaphrodita*), drenched into soil in the warmer months, is increasingly being used to limit slug numbers. A spray containing extracts from a *Yucca* species is now available for small areas of control. The most effective methods are the two ingredients, **ferric phosphate**, a pellet that is increasing in popularity with amateur gardeners, and **metaldehyde** used as small, coloured pellets (which include food attractants such as bran and sugar). Metaldehyde is also available as a liquid formulation. Pellets containing **methiocarb** (that acts as a stomach poison) are progressively less used now by gardeners, for reasons given below. (Use of metaldehyde and methiocarb in gardens has recently been claimed to be a major contribution to the decline in the numbers of birds such as **song thrushes** that eat the poisoned slugs. Hedgehog, frog and insect predator numbers may also be similarly affected. A simple device, such as that seen in Figure 18.6d, helps gardeners to prevent the entry and poisoning of mammals and birds. As can be seen, it uses a modified plastic milk carton (containing the slug pellets) which retains the dead slugs within the container, and prevents entry of mammals and birds. Trade product are available that give similar results.)

Insects

Belonging to the large group of **Arthropoda**, which include also the woodlice, mites, millipedes and symphilids (Table 18.1), insects are horticulturally the most important arthropod group, both as pests and as beneficial soil animals.

Insect structure and biology

The body of the adult insect is made up of segments and is divided into three main parts: the head, thorax and abdomen (Figure 18.7). The **head** bears three pairs of moving mouthparts. There are two main methods of **feeding** (see Figure 18.8). Caterpillars, sawfly larvae and beetles have **biting** mouthparts. In the aphids and their relatives, the mandibles and maxillae are fused to form a delicate tubular **stylet** that sucks up liquids from the plant phloem tissues. Thrips adults and larvae tear and **rasp** plant tissues.

Insects remain aware of their environment by means of compound eyes which are sensitive to movement (of predators) and to colour (of flowers). The **thorax** bears three pairs of legs, and in most insects, two pairs of wings. The **abdomen** bears breathing holes (spiracles) along its length, which lead internally to a breathing system of tracheae. The blood is colourless, circulates digested food and has no breathing function. The digestive system, in addition to its food-absorbing role, removes waste cell products from the body by means of fine, hair-like growths located near the end of the gut.

Since the animal has an **external skeleton** made of tough chitin, it must shed and replace its 'skin' (**cuticle**, see Figure 18.7) periodically by a process called **ecdysis**, in order to increase in size.

The **two main groups** of insect develop from egg to adult in different ways. In the first group (**Exopterygta**), which includes the aphids, thrips and earwigs, the egg hatches to form a first stage (instar) called a **nymph**, which resembles the adult in all but size, wing development and possession of sexual organs. Successive nymph instars more closely resemble the adult. Two to seven instars (growth stages) occur before the adult emerges (Figure 18.9).

Table 18.1 Arthropod groups found in horticulture

Group	Key features of group	Habitat	Damage
Woodlice (*Crustacea*)	Grey, seven pairs of legs, up to 2 mm in length	Damp organic soils	Eat roots and lower leaves
Millipedes (*Diplopoda*)	Brown, many pairs of legs, slow moving	Most soils	Occasionally eat underground tubers and seed
Centipedes (*Chilopoda*)	Brown, many pairs of legs, very active with strong jaws	Most soils	Beneficial
Symphilids (*Symphyla*)	White, 12 pairs of legs, up to 8 mm in length	Glasshouse soils	Eat fine roots
Mites (*Acarina*)	Variable colour, usually four pairs of legs (e.g. red spider mites)	Soils and plant tissues	Mottle or distort leaves, buds, flowers and bulbs; soil species are beneficial
Insects (*Insecta*)	Usually six pairs of legs, two pairs of wings		
Springtails (Collembola)	White to brown, 3–10 mm in length	Soils and decaying humus	Eat fine roots; some beneficial
Aphid group (*Hemiptera*)	Variable colour, sucking mouthparts, produce honeydew (e.g. greenfly)	All habitats	Discolour leaves and stems; prevent flower pollination; transmit viruses
Moths and butterflies (*Lepidoptera*)	Large wings; larva with three pairs of legs, and four pairs of false legs and biting mouthparts (e.g. cabbage white butterfly)	Mainly leaves and flowers	Defoliate leaves (stems and roots)
Flies (*Diptera*)	One pair of wings, larvae legless (e.g. leatherjacket)	All habitats	Leaf mining, eat roots
Beetles (*Coleoptera*)	Horny front pair of wings which meet down centre; well-developed mouthparts in adult and larva (e.g. wireworm)	Mainly in the soil	Eat roots and tubers (and fruit)
Sawflies (*Hymenoptera*)	Adult like a queen ant; larvae have three pairs of legs, and more than four pairs of false legs (e.g. rose-leaf curling sawfly)	Mainly leaves and flowers	Defoliation
Thrips (*Thysanoptera*)	Yellow and brown, very small, wriggle their bodies (e.g. onion thrips)	Leaves and flowers	Cause spotting of leaves and petals
Earwigs (*Dermaptera*)	Brown, with pincers at rear of body	Flowers and soil	Eat flowers

18

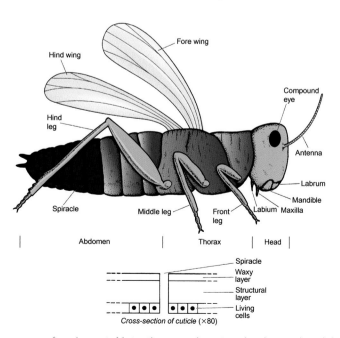

Figure 18.7 External appearance of an insect. Note the mouthpart, spiracles and cuticle – the three main entry points for insecticides

231

This development method is called **incomplete metamorphosis**. In contrast, in the second group of insects (**Endopterygota**), including the moths, butterflies, flies, beetles and sawflies, the grubs (larvae) undergo a remarkable change (**complete metamorphosis**) within the pupa. The egg hatches to form a first **instar**, called a **larva**, which usually differs greatly in shape from the adult. For example, the larva (caterpillar) of the cabbage white bears little resemblance to the adult butterfly.

Some damaging larval stages are shown in Figure 18.10 and these can be compared with the often more familiar adult stage.

The method of **overwintering** differs between insect groups. The aphids survive mainly as the eggs, while most moths, butterflies and flies survive as the pupa. The speed of increase of insects varies greatly between groups. Aphids may take as little as 20 days to complete a life cycle in summer, often resulting in vast numbers in the period June to September. However, the wireworm, the larva of the click beetle, usually takes four years to complete its life cycle.

Insect groups are classified into their appropriate order (Table 18.1) according to their general appearance and life-cycle stages.

Insect pests

There follows now a selection of insect pests in which each species' particular features of life cycle are given. While comments on control are mentioned here, the reader should also refer to Chapter 16 for details of specific types of control (biological, chemical, cultural and so on) and for explanations of terms used.

Aphids and their relatives (*order Hemiptera*)

This important group of insects has the egg–nymph–adult life cycle (see Exopterygota, p. 228) and sucking mouthparts.

Peach-potato aphid (*Myzus persicae*)

Damage. The species commonly occurs in greenhouses and outdoors in hot summers. The adult and nymph stages of this pest may cause three types of damage. First, using the sucking stylet (see Figure 18.8), it may inject a digestive juice into the plant phloem, which, in the sensitive tissues of buds may lead to distortion. Second, having sucked up sugary phloem contents, the aphid excretes a sticky substance called honeydew that may block leaf stomata and reduce photosynthesis, particularly when dark-coloured fungi (**sooty moulds**) grow over the honeydew. Third, the aphid stylet may suck up and then transmit viruses such as severe mosaic virus and leaf roll virus (see p. 262) on potato.

Life cycle. This aphid varies in colour from light green to orange, measures 3 mm in length (see Figure 18.11a), and has a complex life cycle, shown in Figure 18.12, alternating between the winter host (peach) and the many summer hosts such as potato and bedding plants. In spring and summer,

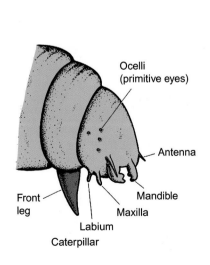

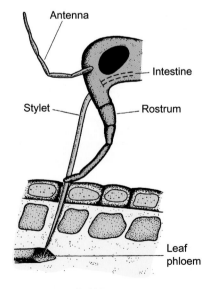

Figure 18.8 Mouthparts of the caterpillar and aphid. Note the different methods of obtaining nutrients. The aphid selectively sucks up dilute sugar solution from the phloem tissue

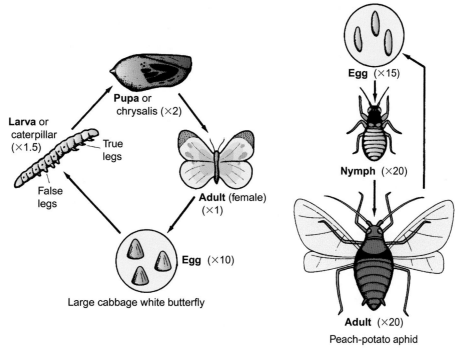

Figure 18.9 Life cycle stages of a butterfly and aphid pest. Note that the four stages of the butterfly life cycle are very different in appearance. The nymph and adult of the aphid are similar

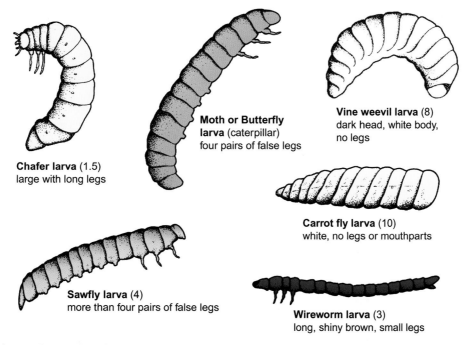

Figure 18.10 Insect larvae that damage crops

the emerging females give birth to nymphs directly without any egg stage (a process called **vivipary**), and without fertilization by a male (a process called **parthenogenesis**). Only in autumn, in response to decreasing day length and outdoor temperatures, are both males and females produced. These have wings and fly to the winter host, the peach. Here, the female is fertilized and lays thick-walled black

eggs. In greenhouses, the aphid may survive the winter as the nymph and adult stage on plants such as *Begonia* and chrysanthemum, or on weeds such as fat hen.

Spread. Occurs mainly in early summer by winged females.

Control. This aphid can be controlled in several ways. In outdoor crops, several organisms such as ladybirds,

18

233

Figure 18.11 (a) Peach-potato aphid; (b) poached-egg plant, pollen source for hoverflies; (c) hoverfly on *Nepeta* sp.

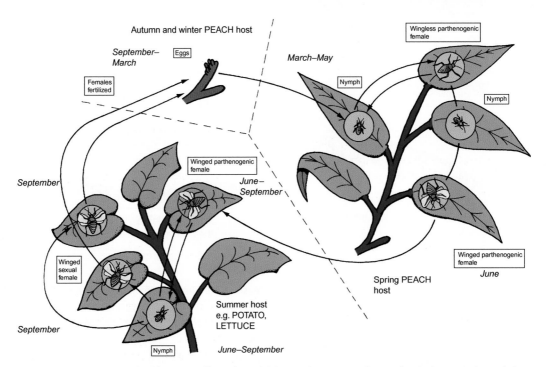

Figure 18.12 Peach-potato aphid life cycle. Female aphids produce nymphs on both the peach and the summer host. Males are produced only in autumn. Eggs survive the winter. In greenhouses, the pest may remain active throughout the year

Figure 18.13 (a) Black bean aphid; (b) spindle tree (*Euonymus europaeus*), winter host of black bean aphid

lacewings, hoverflies (see Figures 16.7 and 16.8), and parasitic fungi (see also p. 198), naturally found in the environment, may reduce pest numbers. For the greenhouse, small parasitic wasps such as *Aphidius matricariae* and *Aphelinus abdominalis* are available for purchase by gardeners. **Abamectin**, **fatty acids** and **pyrethrins** are three insecticide active ingredients options for spraying.

Black bean aphid (*Aphis fabae*)

This 2 mm long black aphid (Figure 18.13) is commonly referred to as 'blackfly', although it is not a fly species. Most individuals are black, but some are dark olive green. The species on bean should **not** be confused with another common black aphid (*Myzus cerasi*) which is found on cherry and ornamental *Prunus* cultivars.

Damage. The most commonly damaged plant species are broad beans, but runner beans, French beans, peas and red beet can be affected. The pest sucks the sugary phloem liquid from the leaves and flowers, often leading to a **sooty mould** covering the plant. Plant growth is affected and seed production in beans can be seriously reduced, with small misshapen pods being produced.

Life cycle. This species survives the winter mainly as the egg stage on the small shrub *Euonymus europaeus* (Figure 18.13). In March/April, female nymphs emerge which, at maturity, fly to the summer host plants such as beans and beet. Female adults may lay as many as five young per day, which themselves are ready to produce more young within 14 days. In warm winters, the pest may survive on field beans. Several weed species such as docks, poppies and fat hen can act as **alternate hosts** (see p. 198) in the summer period. In autumn, male winged aphids develop and, along with the winged females, fly to the *Euonymus* or spindle tree host where the overwintering eggs are laid.

Spread. Is achieved in spring by winged female aphids.

Control. This aphid is attacked by several predators and parasites such as ladybirds, lacewings, hoverflies, tiny parasitic wasps and parasitic fungi (see biological control, p. 196), naturally found in the environment. The gardener can choose between insecticide products containing different classes of chemical: **pyrethrins**, **natural fatty acids** or **abamectin**.

Other aphids

There are many other horticulturally important aphid species. The **cabbage aphid** (*Brevicoryne brassicae*) affects cabbages and other brasssicas. The **apple woolly aphid** (*Eriosoma lanigerum*) causes galls on apple stems (see Figure 18.14). The **rose aphid** (*Macrosiphum rosae*) attacks young shoots of rose (see Figure 18.14).

Glasshouse whitefly (*Trialeurodes vaporariorum*)

This small insect (Figure 18.15a), looking like a tiny white moth (with wings held horizontally when at rest), was originally introduced from the tropics but now causes serious problems on a range of glasshouse food and flower crops. It should not be confused with the very similar, but slightly larger, **cabbage whitefly** on brassicas (Figure 18.15b). (Another greenhouse pest, **tobacco whitefly**, is mentioned briefly after the glasshouse whitefly section.)

Damage. All stages after the egg have sucking stylets, which extract a sugary liquid from the phloem, often causing large amounts of honeydew and sooty moulds

18

Figure 18.14 (a) Rose aphid; (b) apple woolly aphid; (c) aphids become swollen and brown when parasitized by a tiny *Aphidius* wasp

on the leaf surface. Plants that are seriously attacked include fuchsias, cucumbers, chrysanthemums and pelargoniums. Chickweed or sowthistle weeds in greenhouses may harbour the pest over winter in all stages of the pest's life cycle (see alternate hosts, p. 194).

Life cycle. The adult glasshouse whitefly is about 1 mm long, with white wings and white body. It is able to fly from plant to plant. The fertilized female lays about 200 tiny, white, rugby-ball-shaped eggs in a circular pattern on the under-surface of the leaf over a period of several weeks. After turning black, the eggs hatch to produce nymphs (crawlers), which soon become flat, immobile scales. The last scale stage to develop is thick walled and is called a 'pupa' (Figure 18.15a), from which the male or female adult emerges. Three days later, the emerging female starts to lay eggs again. The whole life cycle takes about 32 days in spring, and about 23 days in the summer. It can be seen that a 200-fold increase in numbers every month can quickly lead to a serious pest outbreak. The skins of the egg and pupa stages are relatively thick and these stages are more difficult to control with insecticides.

Spread. This is mainly by introduced plants, or by chance arrivals of adults through doors or vents.

Control. This is achieved in several ways. Gardeners should remove weeds in greenhouses, such as chickweed or sow thistle that harbour the pest. Careful inspection of the **lower** leaf surfaces of introduced plants achieves a similar result. There is a reliable form of **biological control**. This involves a minute exotic **wasp** (*Encarsia formosa*) which is available for purchase (see Figure 16.10b). It lays an egg inside the last scale stage of the whitefly. The developing whitefly is eaten away by the wasp grub and the scale turns black and soon releases the next generation of wasps (see p. 198 for more details). Control is usually most effective if applied when whitefly numbers are low. Some sprays used against other pests may kill the wasp, if not carefully chosen. Gardeners can use spray products containing specially formulated **fatty acids** to control young and adult pest stages. If chemical control is to be used, it is suggested that a serious infestation of this pest receives a weekly chemical spray until the pest is controlled. In this way, the more sensitive young scales and adult stages are targeted as they emerge from the thick-walled, resistant egg and scale stages, respectively.

(e)

Figure 18.15 (a) Adult glasshouse whitefly and pupa (courtesy of Bioline Ltd); (b) cabbage whitefly; (c) bud blast; (d) capsid damage on apple; (e) capsid damage on a leaf; note the holes caused by distorted leaf growth

Tobacco whitefly (*Bemisia tabaci*)

This is a different species from the common glasshouse whitefly, and has become quite common in Britain and Ireland since its introduction in the 1980s. It is similar in size (about 1 mm in length) but may be recognized by the more vertical ('tent-like') way it holds its wings, revealing its light yellow-coloured body. The female, unlike glasshouse whitefly, lays eggs in a random fashion. It may transmit viruses such as **tomato yellow leaf-curl virus**. This pest species is not controlled by the

Encarsia wasp parasite mentioned above. It is, however, parasitized by the exotic wasp *Eretmocerus mundus*, which is commercially available.

Leaf hoppers

These squat light brown or green insects are about 8 mm long and affect a wide variety of crops (e.g. potato, rose, *Primula* and calceolaria). The adults, as their name suggests, can avoid danger by hopping off a leaf. They are also able to fly from plant to plant. They live on the under-surface of leaves, causing a fine mottling of the upper surface. In strawberries, *Euscelis lineolatus* is a vector of green-petal disease. On rhododendron, *Graphocephala fennahi* carries the serious **bud blast** fungal disease that kills off the flower buds (Figure 18.15c). August and September sprays of products containing **fatty acids** prevent insect egg-laying inside buds of rhododendron.

Common green capsid (*Lygocoris pabulinus*)

This is a light green insect resembling a large (5 mm) active aphid. It occurs in small numbers on fruit trees, shrubs and flowers. Although it is rarely seen, it can cause considerable damage to bud, leaves and fruit by injecting toxic juices that cause distorted plant growth (see Figure 18.15d and e). Aphid sprays are sometimes used against this pest.

Thrips (order Thysanoptera)

Owing to their increased activity during warm humid summer weather, thrips are sometimes called 'thunder flies', and are known for their ability to get into people's hair in sultry summer weather and cause itching. They may be important pests in greenhouses.

Onion thrips (*Thrips tabaci*)

Their toxic salivary juices cause a plant reaction: silver streaks in onion leaves, straw-brown spots several millimetres in diameter on cucumber leaves and white streaks on carnation petals. The **pupa** stage occurring in the soil may cause difficulties in control. Gardeners are able to use a product for thrip control containing **natural plant extracts** that blocks the insect's breathing holes. The predatory bug *Orius laevigatus* is an effective biological control.

Western flower thrip (*Frankliniella occidentalis*)

This, sometimes found on greenhouse and outdoor flower and vegetable crops, is an introduced species. It has been shown to transmit some plant viruses.

Moths and butterflies

This insect group characteristically contains adults with four large wings and curled feeding tubes. The larva (**caterpillar**), with six small legs and eight false legs, is modified for a leaf-eating habit (see Figure 18.10). Some species are specialized for feeding inside fruit (**codling moth** on apple), underground (**cutworms**, see Figure 18.18), inside leaves (**oak leaf miner**, see Figure 18.19) or inside tree branches (**leopard moth**). The gardener may find large webbed caterpillar colonies of the **lackey moth** (*Malacosoma neustria*) on fruit trees and hawthorns, or the **juniper webber** (*Dichomeris marginella*) causing webs and defoliating junipers.

Large cabbage white butterfly (*Pieris brassicae*)

Damage. Leaves of cabbage, cauliflower, Brussels sprouts and other hosts such as wallflowers and the shepherd's purse weed are progressively eaten away. The defoliating damage of the larva may result in skeletonized leaves, with only the main veins showing.

Life cycle. This well-known pest on cruciferous plants emerges from the overwintering pupa (chrysalis) in April and May and, after mating, the females lay batches of 20–100 yellow eggs on the underside of leaves. Within a fortnight, groups of larvae emerge and soon moult to produce the later instars, which reach 25 mm in length and are yellow or green in colour, with clear black markings (Figure 18.16b). They have well-developed mandibles. The pupa stage occurs usually in June in a crevice or woody stem, and is held to its host by silk threads. A second generation of the adult emerges in July, giving rise to a more damaging infestation than the first. The second pupa stage overwinters. Care should be taken not to confuse the cabbage white larva with the large smooth green or brown larva of the cabbage moth, or the smaller light green larva of the diamond-backed moth, both of which may enter into the hearts of cabbages and cauliflowers, presenting greater problems for control.

Spread. The species is spread by the adults.

Control. There are several forms of control against the cabbage white butterfly. A naturally occurring small wasp (*Cotesia glomerata*) lays its eggs inside the caterpillar, and the yellow cocoons emerge (see Figure 18.16d). A virus disease may infect the pest,

Figure 18.16 (a) Cabbage white adult; (b) cabbage white caterpillar; (c) cabbage white pupa; (d) yellow pupae of *Cotesia glomerata* wasp that parasitizes cabbage white caterpillars

18

causing the larva to go grey and die. Birds such as tits and starlings eat the plump larvae. Private gardeners often cover brassicas plants with fine-mesh nets to prevent adults reaching the plants. When damage becomes severe, spray products containing **lambda-cyhalothrin** may be used.

Winter moth (*Operophthera brumata*)

This pest's timing of life-cycle stages is unusual.

Damage. These are pests, which may be serious on top fruit, such as apples, but also attack woody plants such as currants, roses and beech. The caterpillars eat away leaves in spring and early summer and often form other leaves into loose webs, reducing the plant's photosynthesis. They occasionally scar young apple fruit.

Life cycle. The pest emerges as the adult form from a soil-borne pupa in November and December;

hence the species' common name. The male is a greyish-brown moth, 2.5 cm across its wings, while the female is **wingless**, looking at first sight rather like a spider (Figure 18.17b). The female crawls up the tree and lays 100–200 light green eggs around a number of buds. The eggs hatch in spring at bud burst to produce green larvae with faint white stripes. These larvae move in a characteristic **looping** fashion and when fully grown, descend on silk threads at the end of May before pupating in the soil until early winter.

Spread. This is slow because the females do not have wings.

Control. A common control is a **glue band** (see Figure 16.2b), which is wound around the main trunk of the tree in October preventing the flightless female moth's progress up the tree. A spray of **deltamethrin** in spring may be used if necessary.

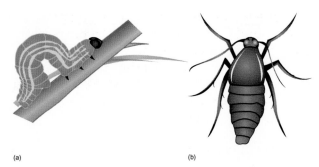

Figure 18.17 (a) Winter moth caterpillar moves in a looping fashion; (b) winter moth female looks rather like a spider

Cutworm (e.g. *Noctua pronuba*)

Damage. The caterpillars of the **yellow underwing moth**, unlike most other moth larvae, live in the **soil**, nipping off the stems of young plants and eating holes in succulent crops, such as bedding plants, lawns, potatoes, celery, turnips and conifer seedlings. The damage superficially resembles that caused by slugs.

Life cycle. (Figure 18.18) The adult moth, 2 cm across, with brown forewings and yellow or orange hind wings, emerges from the shiny, soil-borne, chestnut brown pupa from June to July, and lays about 1,000 eggs on the stems of a wide variety of weeds. The **young** caterpillars, having fed on weeds, descend to the soil and in the later stages cause the damage described above, eventually reaching about 3.5 cm in length. They are grey to grey-brown in colour, with black spots along the sides. Several other cutworm species, such as **heart and dart moth** (*Agrotis exclamationis*) and turnip moth (*Agrotis segetum*), may cause damage similar to that of the yellow underwing moth. In all three species, their typical caterpillar-shaped larvae should not be confused with the legless leatherjacket, which is also a common underground larva. The cutworm species normally have two life cycles per year, but in hot summers this may increase to three.

Spread. The larvae are able to crawl from plant to plant, but most spread is by the actively flying adults.

Control. Gardeners remove a good proportion of the cutworm larvae as they dig plots over, maybe with a robin perched nearby. Also, good weed control reduces cutworm damage. There are, currently, **no approved pesticides** available to the private gardener against this pest.

Leopard moth (*Zeuzera pyrina*)

The large white caterpillar of this species tunnels into branches and trunks of a wide range tree species, such as apple, ash, birch and lilac. The tunnelling may

Figure 18.18 Yellow underwing moth with cutworm and dark brown pupa

weaken the branches of trees, which in high winds commonly break.

Brown-tailed moth (*Euproctis chrysorrhoea*)

Although not yet common as a garden pest, this species is increasing in the countryside. The hairy brown caterpillars with white markings are seen in large communal 'silky tents' on hawthorns, blackthorns and brambles. The caterpillar's hairs can cause considerable **skin irritation** to humans.

Other moth pests

Recently reported moth pests in the UK are leek moth (*Acrolepiosis assectella*), holm oak leaf-mining moth (*Phyllonorycter messaniell*, Figure 18.19) and horse chestnut leaf miner (*Cameraria obridella*).

Flies

This group of insects typically have only a single pair of functioning wings. The hind wings are modified into little stubs which act as balancing organs. The larvae are legless and elongated, and their mouthparts, where present, are simple hooks. The larvae are the only stage causing plant damage in gardens.

Figure 18.19 Holm oak leaf miner (symptoms are caused by small moth caterpillars)

Chrysanthemum leaf miner (*Chromatomyza syngenesiae*)

Damage. This leaf miner symptom is caused by a small fly species (contrast the leaf miners in Figure 18.19), the larvae of which can do serious damage to horticultural crops by tunnelling through the leaf. This species is found on members of Asteraceae plant family, which includes chrysanthemum, cineraria and lettuce.

Life cycle. The flies emerge at any time of the year in greenhouses, but normally only between July and October outdoors. These adults, which measure about 2 mm in length and are grey-black with yellow underparts, fly around with short hopping movements. The female lays about 75 minute eggs singly inside the leaves, causing small white spot symptoms to appear on the upper leaf surface. The larval stage is greenish white in colour, and tunnels into the **pallisade mesophyll** (see p. 115) of the leaf, leaving behind the characteristic mines seen in Figure 18.20a. On reaching its final stage, the 3.5 mm long larva develops into a brown pupa within the mine. The adult soon emerges from the pupa. The total life cycle takes about three weeks during the summer months.

Spread. This pest is spread by the adult stage.

Control. Alternate host weeds such as groundsel and sow thistle should be controlled. Yellow sticky traps (see p. 191) remove many of the adult flies in greenhouses. Certain chrysanthemum cultivars show some resistance. Gardeners using products containing **abamectin** have some control of the pest, especially in the early growing season.

Other leaf miners

The occurrence of **South American leaf miner** (*Liriomyza huidobrensis*) and **American serpentine leaf miner** (*Liriomyza trifolii*), which are able to damage a **wide variety** of greenhouse plants, has, in recent years, created problems for horticulture.

Carrot fly (*Psila rosae*) is a widespread and serious pest on umbelliferous crops (carrots, celery and parsnips), where tunnelling by the grubs makes the roots useless (Figure 18.20b). The **cabbage root fly** (*Delia radicum brassicae*) causes similar damage in brassica crops.

Leatherjacket (*Tipula paludosa*) is an underground pest, the larva (see Figure 18.10) of the 'daddy long legs' (crane fly) seen in late summer. It is a natural inhabitant of grassland, causes most problems on turf, but also damages potatoes cabbages, lettuce and strawberries, especially after in wet springs and early summers.

Fungus gnat or sciarid fly (*Bradysia* spp.) have larvae that are small (3 mm), translucent with a black head. They feed on fine roots of greenhouse pot plants such as cyclamen, orchid and freesia, especially when the plants are overwatered. This species is also a pest in mushroom crops. Biological control by tiny nematodes (*Steinernema feltiae*) and a small beetle (*Atheta coriara*) are available to the gardener. The adults are also caught by yellow **sticky traps** (see page 191).

A recently introduced species **Hemerocallis gall midge** (*Contarinia quinqenotata*), which distorts flower buds of day lilies, is becoming increasingly common in southern Britain.

Beetles

This group of insects have adults with hard, horny front wings (elytrae) which, when folded, cover the delicate hind wings used for flight. The meeting point of these hard wing cases produces the characteristic straight line down the beetle's back over its abdomen. The thick skin (cuticle) of beetles enables many of them to live successfully underground. Most beetle species are **beneficial**, helping in the breakdown of humus (e.g. dung beetles), or feeding on pest species (see ground beetle, p. 199). A few beetles such as vine weevil, wireworm, chafer and, more recently, lily beetle can cause serious plant damage.

Figure 18.20 (a) Chrysanthemum leaf mines; (b) carrot fly damage

Vine weevil (*Otiorhyncus sulcatus*)

This species belongs to the beetle group, but as with all weevils, possesses a longer snout on its head than other beetles.

Damage. The larva stage is the most damaging, eating away roots of crops such as cyclamen and begonias in greenhouses, primulas, strawberries, young conifers and vines outdoors, and causing above-ground yellowing and wilting symptoms similar to root diseases such as *Verticillium* wilt. Digging around in the plant's root zone will, however, quickly show the unmistakable white grubs (Figure 18.21a). The black adults move around at night and are therefore rarely seen by the gardener. They may eat out neat holes in leaves or round the edges of the foliage of plants such as rhododendron, raspberry and grapes, and many herbaceous perennials.

Life cycle. The adult is 9 mm long, black in colour, with a rough textured cuticle (Figure 18.21b). The forewings are fused together, the pest being incapable of flight. No males are known. The female lays eggs (mainly in August and September) in soil or compost, next to the roots of a preferred plant species. Over a period of a few years, she may lay 1,000 eggs as she visits many plants. The larvae are white and legless, with a characteristic chestnut-brown head. They reach 1 cm in length in December when they pupate in the soil before developing into the adult.

Spread. This is achieved by the female adult crawling around at night, or by the too frequent movement of pots containing plants with grubs eating their roots.

Control. Gardeners sometimes use traps of corrugated paper placed near infested crops. Inspection of plants at night by torchlight may reveal the feeding adult. The parasitic nematode *Steinemena carpocapsae* is used by incorporating it as a drench into compost or soil. The insecticide **thiacloprid** that was used as an effective drench against the larvae of this pest has recently been suspected of a possible involvement in bee deaths ('**colony collapse disorder**') in Europe.

Wireworm

Click beetles (*Agriotes lineatus*) are commonly found in grassland, but the wireworm larvae (see Figure 18.10) will bore through other plants such as potatoes and carrots. Some gardeners dig in green manure crops to lure wireworms away from underground roots and tubers. A nematode (*Heterorhabdis megadis*) can be purchased for biological control.

Garden chafer (*Phyllopertha horticola*)

This is increasingly proving a problem on turf, where the large white grubs eat the roots. Small yellow patches appear in the lawn, notably in summer when the grubs are becoming fully grown. While healthy, well-fertilized lawns often show minimal damage, the gardener may need to use a biological control, involving the tiny nematode (*Heterorhabditis bacteriophora*).

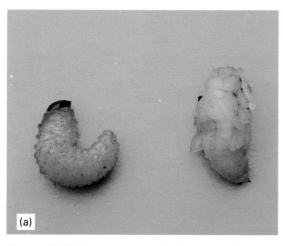

Figure 18.21 (a) vine weevil larva and pupa; (b) vine weevil adult; (c) adult vine weevil damage: on grape leaf

Lily beetle (*Lilioceris lilii*)

This small beetle has steadily increased in numbers since 1940 in the southern Britain and Ireland, recently reaching parts of northern England, central Scotland and Northern Ireland. The adult is bright red; its head and legs are black; and it is 8 mm in length (Figure 18.22b). The larvae often have a dark slimy appearance, and are 8–10 mm in length, orange in colour, with black heads. Adults and larvae defoliate a wide range of lilies and fritillaries. **Deltamethrin** sprays may be used. A parasitic wasp (*Lemophagus arrabundus*), present in mainland Europe, has not yet established in Britain and Ireland.

Other beetle pests

Springtime attack of **flea beetle** (*Phyllotreta* species, 3 mm long, black in colour, some with a yellow stripe) on leaves of young cruciferous plants (e.g. cabbages and stocks) may become serious. Horticultural fleece draped over young brassicas helps in control (see Figure 18.22a).

In **recent years** in the UK, three other increasingly common beetle problems have been reported. The **viburnum beetle** (*Pyrrhalta viburni*, 5 mm long, light brown in colour) is found on *Viburnum opulus*, *V. tinus* and *V. lantana*. The **rosemary leaf beetle** (*Chrysolina americana*, 8 mm long, metallic green with purple stripes) occurs on lavender, rosemary and thyme. The **asparagus beetle** (*Crioceris asparagi*, 7 mm long, metallic black body with four white spots) is becoming more common. Chemical sprays containing ingredients such as **pyrethrum** or **deltametrin** are sometimes used against these new pests.

Sawflies

Sawflies, together with bees, wasps and ants, are classified in the order *Hymenoptera*, which have adults with two pairs of translucent wings, and with the fore wings and hind wings being locked together by fine hooks. The slender waist-like first segments of the abdomen give these adults a characteristic appearance. Adult sawflies resemble flying ants. Sawfly larvae in many species resemble moth caterpillars (see Figure 18.10), but some such as **pear slugworm** (*Caliroa cerasi*) are black, have no visible legs and look rather slug like.

Gooseberry sawfly (*Nematus ribesii*) is an important pest on gooseberries, redcurrants and white currants, but not on blackcurrants. Extensive damage to foliage may be caused by the caterpillars, often leaving only the main leaf veins (Figure 18.23). The green caterpillars with black spots are easily recognized. Often three life cycles per year occur. Picking-off caterpillars in spring is recommended. Sprays containing **pyrethrum** are effective.

Large rose sawfly (*Arge pargana*) caterpillars, skeletonizing rose leaves are controlled by sprays containing **deltamethrin**.

New sawfly problem. Berberis sawfly (*Arge berberidis*) has recently spread to many parts of the UK. It defoliates the leaves of *Berberis* and *Mahonia*.

Ants. Two species, the **black ant** (*Lasius niger*) and the **red ant** (*Myrmica rubra*) may commonly be seen in UK gardens. They encourage aphids that produce honeydew (see p. 233). **Ant nests** located in lawn areas spoil the appearance of the turf. Removal of the ant nest avoids chemical control, but nests may re-establish fairly quickly. Pyrethrin products containing ingredients such as **permethrin** are commonly used as dusts to control ants.

Springtails (order Collembola) are wingless insects, 2 mm in length, that are very common in soils, and

18

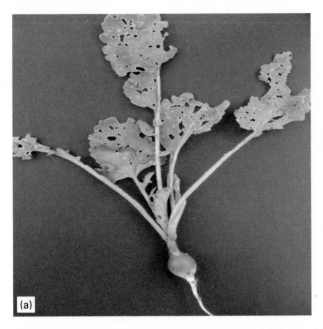

Figure 18.22 (a) Flea beetle damage on radish leaves; (b) Lily beetle

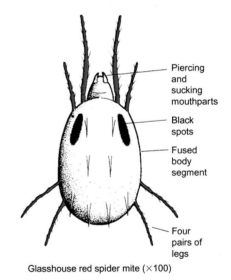

Glasshouse red spider mite (×100)

Figure 18.23 Gooseberry sawfly damage

Figure 18.24 Two-spotted mite. Note that this pest is very small (0.8 mm) and can easily escape a gardener's attention

normally help in the breakdown of soil **organic matter**. They may cause damage in greenhouses on cucumber roots.

Mite pests

The mites (*Acarina*) are grouped with ticks, spiders and scorpions in the *Arachnida*. Although similar to insects in many respects, they are distinguished from them by the possession of **four pairs of legs**, a **fused body** structure (no clear abdomen or thorax) and the **absence of wings** (Figure 18.24). Many of the tiny soil-inhabiting mites serve a useful purpose in breaking down plant debris. Several above-ground species are

serious pests on plants. The life cycle is composed of egg, larva, nymph and adult stages.

Two-spotted mite or glasshouse red spider mite (*Tetranychus urticae*)

Damage. The piercing mouthparts of the mites inject poisonous juices which cause localized death of leaf cells. This results in a **fine mottling** symptom on the leaf. In large numbers, the mites can kill off leaves and eventually whole plants. Numerous fine silk strands are produced in severe infestations (Figure 18.25b)

Figure 18.25 (a) Fine mottling symptoms caused by two-spotted mite on a palm leaf; (b) fine strands ('webbing') produced by the mites; (c) two-spotted mite being attacked by *Phytoseiulus* predator

Life cycle. This pest thrives best in high greenhouse temperatures. It lives mainly on the **under-surface** of leaves. It is 1 mm long, is yellowish in colour, with two black spots (Figure 18.25a) but become red in autumn The female lays about 100 tiny spherical eggs on the underside of the leaf. The life-cycle length varies markedly from 62 days at 10°C to six days at 35°C, when the pest's multiplication potential is extremely high. In autumn, the female (with eggs inside her) hibernates (diapauses). She often emerges in March or April. A second species (*Tetranychus cinnabarinus*) is orange-red in colour all year and does not hibernate.

Spread. This occurs when adults and nymphs crawl from plant to plant. This pest is often carried into greenhouses on pot plants, unnoticed because of its small size.

Control. This may be achieved in several ways. Gardeners should carefully check incoming plants for the presence of the mite, using a hand-lens if necessary. A predatory mite, *Phytoseiulus persimilis*, is available from specialist **biological control** companies and is introduced into cucumber and other greenhouse plants (see also p. 195 for more details). There are spray products containing **fatty acids** or **plant extracts** that control the mite on edible and non-edible plants. **Abamectin** is used on non-edible crops.

Gall mite of blackcurrant (*Cecidophyopsis ribis*)

This tiny mite lives inside the blackcurrant bud (Figure 18.26) causing distorted leaf and flower development, and carries blackcurrant reversion disease. 'Ben Hope' blackcurrant cultivar is resistant to the pest. Disease-free bushes should be planted. Pruning out infected stems is useful. A spray of colloidal **sulphur** in May and June reduces the dispersal of the mites as they emerge from infected buds.

Tarsonemid mite (*Tarsonemus pallidus*)

This mite causes distortion of developing leaves and flowers (resulting from the pest's injected toxins). These are the main symptom of this pest, particularly

Figure 18.26 Blackcurrant gall mite. Note the swollen spherical buds

on *Fuchsia*, *Pelargonium* and *Cyclamen*. Care should be taken to prevent introduction of infested plants and propagation material into glasshouses. No reliable control is available for the gardener.

Other mites

Three other horticulturally important mites require a mention. The **fruit tree red spider mite** (*Panonychus ulmi*) can cause serious leaf mottling of ornamental *Malus* and apple. The **conifer spinning mite** (*Oligonychus ununguis*) causes spruce leaves to yellow, and the mite spins a web of silk threads. **Bryobia mite** (e.g. *Bryobia rubrioculus*) is found on a wide range of trees. It may attack fruit trees, and more seriously cause damage to greenhouse crops such as cucumbers, if blown in from neighbouring trees. The two-spotted mite predator, *Phytoseiulus*, is not effective against bryobia mite.

Also, recently, there have been two newly reported mite problems in Britain and Ireland. Hazel shrubs are being attacked by **hazel big bud mite** (*Phytopus avellanae*). **Fuchsia gall mite** (*Aculops fuchsiae*) is causing young *Fuchsia* leaves to become red and distorted.

Other arthropod pests

In addition to insects and mites, the phylum Arthropoda contains three other horticulturally relevant groups, the *Crustacea* (woodlice), *Symphyla* (symphilids) and *Diplopoda* (millipedes), which can cause damage to plants (see also Table 18.1).

Woodlouse (*Armadillidium nasutum*) damage is confined mainly to greenhouse plants where stems and lower leaves of succulent plants such as cucumbers, but occasionally young outdoor transplants, may be nipped. This species (Figure 18.27a) has adapted to terrestrial life, but still requires damp conditions to survive. **Soil-placed glue traps** used for a range of soil pests are effective in control.

Symphilids (*Scutigerella immaculata*) resembling tiny white millipedes eat root hairs and may cause lettuce to mature without a heart, and also allow *Botrytis* infections. Their presence may be detected by placing suspect soil into a bucket of water to identify the pest floating on the water surface.

Millipedes are slow-moving creatures with many legs (two pairs to each body segment). Several species are useful in breaking down soil organic matter, but two pest species, the flat millipede (*Brachydesmus superus*) and a tropical species (*Oxidus gracilus*), can cause damage to roots of strawberries and cucumbers, respectively.

Centipedes superficially resemble millipedes, but are much more active. They help to control soil pests by searching for insects, mites and nematodes in the soil (Figure 18.27b).

Nematode pests

This group of organisms, also called **eelworms**, is found in almost every part of the living world. They range in size from the large animal parasites, such as *Ascaris* (about 20 cm long) in the guts of livestock to the tiny soil-inhabiting species (about 0.5 mm long). Non-parasitic species in soil are often beneficial, feeding on plant remains and soil bacteria, and helping in the formation of **humus** (see p. 158). The general structure of the nematode body is shown in Figure 18.29. A feature of the plant parasitic species is the **spear** in the mouth region, which is thrust into plant cells. Salivary enzymes are then injected into the plant and the plant juices sucked in by the nematode. Nematodes are very active animals, moving in a wriggling fashion in soil moisture films, most actively when the soil is at **field capacity** and more slowly as the soil either waterlogs or dries out. **Symptoms**

 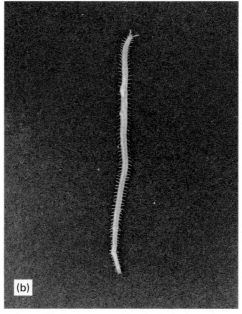

Figure 18.27 (a) Woodlouse, can be a pest in greenhouses; (b) Geophilus centipede, a useful predator on soil pests

caused by these tiny animals are sometimes confused with those caused by fungi or bacteria.

One horticulturally damaging species is described below. Four others are mentioned.

Potato cyst nematode (*Globodera rostochiensis* and *G. pallida*)

Damage. This serious pest is found in most soils that have grown potatoes. Leaves become yellow and plants become stunted (Figure 18.28) and occasionally killed. The distribution of damage in the plot is characteristically in patches. Tomatoes grown in greenhouses and outdoors may be similarly affected. The pest may be diagnosed in the field by the tiny, mature white or yellow, onion-shaped females (**cysts**) that are **attached to** the potato roots (a hand-lens is needed for this observation).

Life cycle. A proportion of the eggs in the soil hatch in spring, stimulated by chemicals that are produced by neighbouring potato roots. The larvae invade the roots, disturbing **translocation** in xylem and phloem tissues, and sucking up plant cell contents. When the adult male and female nematodes are fully developed, they wriggle to the outside of the root, and the now swollen female leaves only her head inserted in the plant tissues (Figure 18.29). After fertilization, the white female swells and becomes almost spherical, about 0.5 mm in size, and contains 200–600 eggs. As the potato crop reaches harvest, the female changes colour. In *G. rostochiensis* (the golden nematode), the change is from white to **yellow** and then to dark

Figure 18.28 Potatoes stunted by potato cyst nematode

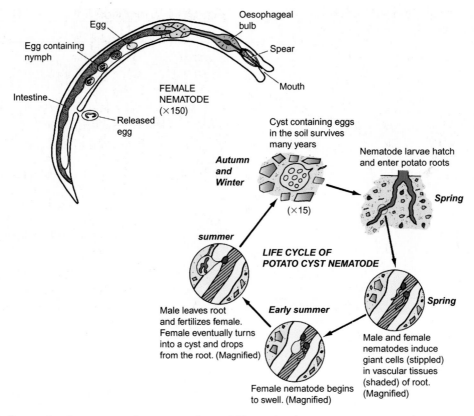

Figure 18.29 Generalized structure of a nematode, and life cycle of potato cyst nematode

brown, while in the other species, *G. pallida*, no yellow phase is seen. The significance of the species difference is mentioned later in the control section. Eventually, the female changes to a dark brown colour and falls from the root into the soil. This stage, which looks like a minute brown onion, is called the **cyst**, and the many eggs inside this protective shell may survive for 10 years or more in the soil.

Spread. This nematode spreads, as cysts, with the movement of infested soil, on boots and tyres. In peat-soil areas, the cysts often travel along with the wind-blown soil.

Control. Several forms of control are available against this pest. Since it attacks only potatoes and tomatoes, **rotation** is a reliable way of helping control. Growing potatoes one year in five will often greatly reduce this pest. **Early** cultivars of potatoes are lifted before most nematodes have reached the cyst stage, and thus escape serious damage. Some potato cultivars, such as 'Pentland Javelin' and 'Maris Piper', are **resistant** to golden nematode strains found in Britain and Ireland, but not to *G. pallida*. Since the golden nematode is dominant in the south of Britain, the use of resistant cultivars has proved more effective in this region. Growing a mustard crop to the flowering stage, and then incorporating it into the soil as a '**green manure**' encourages the increase in soil fungi that entangle

and parasitize the nematodes. No chemical forms of control are available to the gardener.

Other nematode pests

Stem and bulb nematode (*Ditylenchus dipsaci*) can damage onions (causing a soft puffy appearance called 'bloat'). Affected carrots have a dry mealy rot. Stems of beans are swollen and distorted. Narcissus bulbs show brown rings when cut across and their emerging leaves show slightly raised yellow streaks. Weed control of the chickweed alternate host is recommended. Rotation with resistant crops such as lettuce and brassicas reduces pest numbers. Use of planting material from infested soil should be avoided. All these cultural methods help to reduce this pest.

Chrysanthemum eelworm (*Aphelenchoides ritzemabosi*) causes a blotching and purpling of the leaves that spreads to become a dead brown, V-shaped area between the veins of the leaves. The lower leaves are worst affected because the eelworm invades initially from the soil. This nematode also attacks *Saintpaulia* and strawberries. For control, all chrysanthemum plant debris should be disposed of. Warm-water treatment of dormant chrysanthemum stools (e.g. at 46°C for five minutes) is effective for outdoor-grown plants.

Root knot eelworm (*Meloidogyne* spp.) is a worldwide important pest in tropical countries. It used to be a serious pest in British greenhouses, but the use of 'grow bags' has reduced its survival and spread. The pest causes large root **galls** up to 4 cm in size, on the roots of plants such as chrysanthemum, *Begonia*, cucumber and tomato, with resulting wilting and poor plant growth. **Partial-soil sterilization** (see p. 190) helps reduce the problem, especially if all root debris has been removed. Tomato rootstocks, such as '**KVNF**', grafted onto other cultivars resist the nematode.

Migratory plant nematodes

Unlike the nematodes described above, the migratory species feed only from the **outside** of the root. The dagger nematodes (e.g. *Xiphinema diversicaudatum*) and needle nematodes (e.g. *Longidorus elongatus*), which reach lengths of 0.4 and 1.0 cm, respectively, attack the young roots of crops such as rose, raspberry and strawberry, and cause stunted growth. In addition, these species transmit important viruses, **arabis mosaic** on strawberry and **tomato black ring** on ornamental cherries. The nematodes also may survive on the roots of a wide variety of weeds. The use of clean plant stock and good weed control helps keep these problems at bay.

Further reading

Alford, D.V. (2002) *Pests of Ornamental Trees, Shrubs and Flowers*. Wolfe Publishing.

British Crop Protection Council (2014) *UK Pesticide Guide*. BCPC.

Brown, L.V. (2008) *Applied Principles of Horticulture*. 3rd edn. Butterworth Heinemann.

Brown, S. (2005) *Sports Turf and Amenity Grassland Management*. Crowood Publications.

Buczacki, S. and Harris, K. (2005) *Pests, Diseases and Disorders of Garden Plants*. Collins.

French, J. (2007) *Natural Control of Garden Pests*. Aird Books.

Greenwood, P. and Halstead, H. (2009) *Pests and Diseases*. Dorling Kindersley.

Helyer, N. (2003) *A Colour Handbook of Biological Control in Plant Protection*. Manson Publishing.

Hessayon, D.G. (2009) *Pest and Weed Expert*. Transworld Publishers.

RHS (2011) Rabbit resistant plants. Online: apps.rhs.org.uk/advicesearch/profile.aspx?pid=209. RHS

RHS (2014) Pesticides for home gardeners. Online: www.rhs.org.uk/media/pdfs/advice/pesticides. RHS Advisory Service.

18

Please visit the companion website for further information:
www.routledge.com/cw/adams

CHAPTER 19
Level 2

Garden diseases and disorders

Figure 19.1 Honey fungus toadstools that have emerged from an infected tree base

This chapter includes the following topics:

- Fungal diseases
- Bacteria
- Viruses
- Plant disorders
- Symptoms of disease and plant disorders

Principles of Horticulture. 978-0-415-85908-0 © C.R. Adams, K.M. Bamford, J.E. Brook and M.P. Early. Published by Taylor & Francis. All rights reserved.

Introduction

The damp climate of Britain and Ireland favours a range of diseases. The great variety of garden plants and crop species leads to further opportunities. This chapter presents a range of these damaging problems, highlights life-cycle details that explain their success and gives a range of control measures available to the gardener. The distinct but sometimes confusing area of 'plant disorders' is covered at the end of the chapter.

A brief review of fungi, bacteria and viruses is followed by a description of some of the most important garden disease problems.

Structure and biology of fungi, bacteria and viruses

Fungi, commonly called moulds, cause serious losses in all areas of horticulture. Some details of their classification are given in Chapter 4.

A fungus is composed, in most species, of microscopic strands (hyphae), which may occur together in a loose structure (mycelium), form dense resting bodies (sclerotia, see Figure 19.3) or produce complex underground root-like strands (rhizomorphs, see Figure 19.15). The group of organisms causing such diseases as damping off (*Pythium*), potato blight (*Phytophthora*) and downy mildew (*Peronospora*) have recently been re-classified as a separate group called fungus-like 'Oomycetes', distinct from true fungi. Also, the organism causing club root (*Plasmodiophora*)

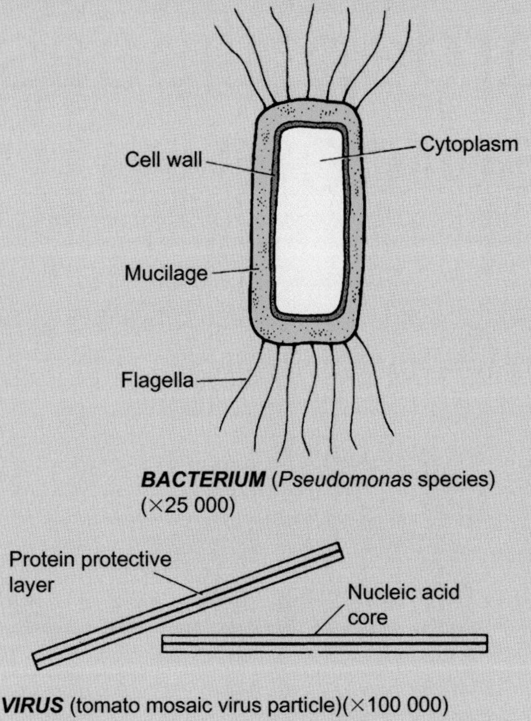

BACTERIUM (*Pseudomonas* species) (×25 000)

VIRUS (tomato mosaic virus particle)(×100 000)

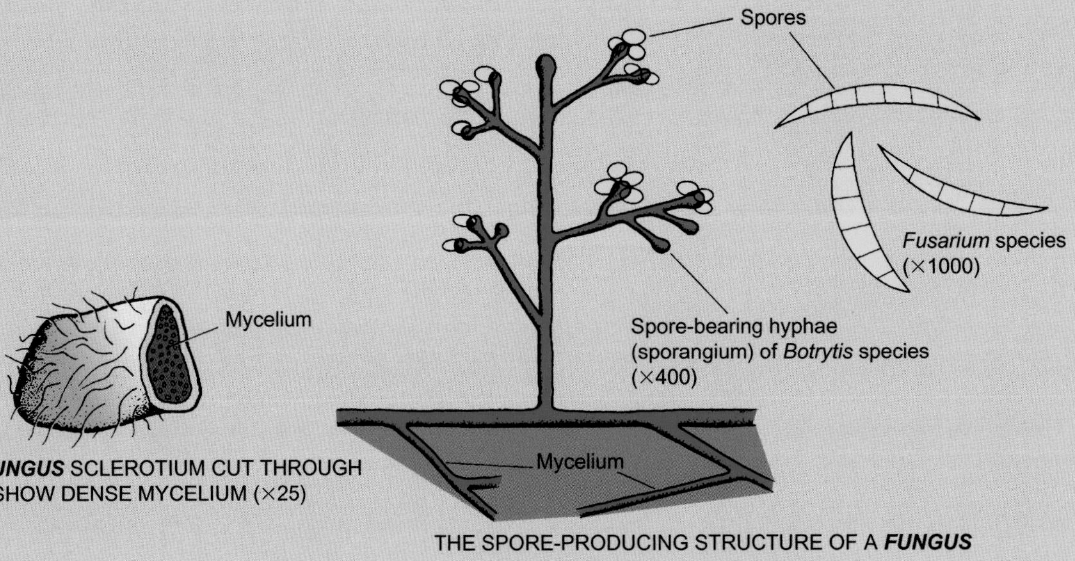

A FUNGUS SCLEROTIUM CUT THROUGH TO SHOW DENSE MYCELIUM (×25)

THE SPORE-PRODUCING STRUCTURE OF A FUNGUS

Figure 19.2 Microscopic structure of fungi, bacteria, and viruses

is quite different from fungi, producing a jelly-like structure (plasmodium) inside the cells of the host plant (see p. 76).

The hyphae in most fungal species are able, in the appropriate circumstances, to produce spores. Two soil-borne fungi, *Rhizoctonia (Thanatephorus) solani* (that causes a damping-off symptom, and black scurf disease on potatoes) and *Sclerotium (Stromatinia) cepivorum* (that causes white rot on onion, see Figure 19.3) are two notable exceptions, where the hyphae only very rarely produce spores. Wind-borne spores are generally very small (about 0.01 mm), not sticky and often borne by special hyphae protruding above the leaf surface (e.g. those of grey mould), so that they catch turbulent wind currents. Water or rain-borne spores are often sticky (e.g. those of damping off).

Minute asexual spores produced without fusion of two hyphae commonly occur in seasons favourable for disease increase – for example, humid weather for downy mildews and dry, hot weather for powdery mildews. Sexual spores, produced after hyphal fusion, commonly develop just before unfavourable conditions occur (e.g. a cold, damp autumn). They are produced as single spores in the oomycetes. In the powdery mildews and many other species in the Ascomycota (see p. 62), spores are produced in groups within a protective hyphal spore-case (like a tiny black spherical flask less than 1 mm in size) often observable to the naked eye. Different genera and species are identified by microscopic measurement of the shape and size of the spores, or the spore-bearing spore cases, or by the appearance of the hyphae (especially those special hyphae producing spores, see Figure 19.2).

Figure 19.3 White rot on onion. Note the black sclerotia which enable this disease to survive long periods in the soil

Gardeners without microscopes must use the symptoms ('easily visible features or signs that are characteristic of the disease') to identify the particular problem.

> A **plant disease** is an unhealthy condition in a plant caused by a fungus, bacterium or virus.

While disease-causing or parasitic fungi are the main concern of this chapter, it should be realized that in many parts of the garden and environment, there are useful saprophytic fungi that usefully break down organic material such as dead roots, leaves, stems and sometimes decaying tree stumps (see p. 62), and there are useful symbiotic fungi that may live in close association with the plant, e.g. mycorrhizal fungi in fine roots of conifers and many other garden species.

Infection

The spore of a leaf-infecting fungal parasite, after landing on the leaf in damp conditions, produces a germination tube, which being delicate and easily dried out, must enter through the cuticle or stomata within a few hours before dry, unfavourable conditions recur. Within the leaf, the hyphae grow, absorbing food until, within a period of a few weeks, they produce a further crop of spores (see Figure 19.5). Leaf diseases such as potato blight often increase very rapidly when conditions are favourable.

Roots may be infected by spores (e.g. in damping off), by hyphae (e.g. in wilt diseases), by sclerotia (e.g. in *Sclerotinia* rot) or by rhizomorphs (e.g. in honey fungus). Root diseases are generally less affected by short periods of unfavourable conditions and often increase at a slower, more constant rate; although in hydroponic systems (see p. 183), increase is likely to be much more rapid.

Phyllosphere

On the surface of leaves and stems, there lives a population of micro-organisms (mainly bacteria) which occupy a microhabitat commonly called the phyllosphere. These bacteria may be 'casual' or 'resident'. Casual organisms such as *Bacillus* spp. mainly arrive from soil, roots and water, and are more common on leaves closer to the ground. These species are capable of rapid

19

increase under favourable conditions, but then may decline. Resident organisms such as *Pseudomonas* spp. may be weakly parasitic on plants, but more commonly persist (often for considerable periods) without causing damage, and on a wide variety of plants.

There is increasing evidence that phyllosphere bacteria may reduce the infection of diseases such as powdery mildews, *Botrytis* diseases on lettuce and onion, and turf grass diseases. Practical disease control strategies by phyllosphere organisms have not been developed, but there remains the general principle that a healthy, well-nourished plant will be more likely to have organisms on the leaf surface available to reduce fungal infection.

There are also phyllosphere fungi such as *Alternaria*, *Cladosporium* and *Fusarium* that are commonly found on leaf surfaces and these may prove to be useful in combating fungal disease infection in a similar way to the bacteria described above.

Fungal diseases

Practical relevance of fungal classification

The classification of fungi referred to on p. 62 has some practical implications in understanding fungal disease life cycles and control. Species within a fungal 'division' often have similar methods of spread, and of survival. Their similarity of spore structure, of hypha structure and of biochemistry also means that they are often susceptible to control by the same (or similar) fungicide active ingredient. For example, the potato blight 'fungus' is closely related to the damping-off 'fungi' and the downy mildew 'fungi', and these three types of 'fungi' are effectively controlled by fungicide active ingredients containing copper salts. Because of the horticulturally practical aspects of fungal classification, the appropriate fungal division is given for each disease as it appears in the text (for example, the division 'Zygomycota' is mentioned against potato blight, damping off and downy mildew).

Potato blight (*Phytophthora infestans*)

This fungus is a member of the Zygomycota division of fungi (see p. 62).

Damage. This important disease is a constant threat to potato production, and caused the Irish potato famine in the nineteenth century. The first symptoms seen are yellowing of the foliage, which quickly goes black and then produces a white bloom on the under-surface of the leaf in damp weather. The stems may then go black, killing off the whole plant. The tubers may have dark surface spots that, internally, appear as a dry, deep red-brown rot. This 'fungus' may attack tomatoes, the most notable symptom being the dark brown blisters on the fruit (Figure 19.4).

Life cycle. The fungus survives the winter as mycelium and sexual spores in the tubers (Figure 19.5). The spring emergence of infected shoots results in the production of asexual spores.

Spread. The spores are spread by wind, land on potato leaves or stems and can, after infection, result in a further crop of spores within a few days under warm wet weather conditions. The disease can spread very

Figure 19.4 Potato blight: (a) leaf symptoms on potato; (b) fruit symptoms on tomato

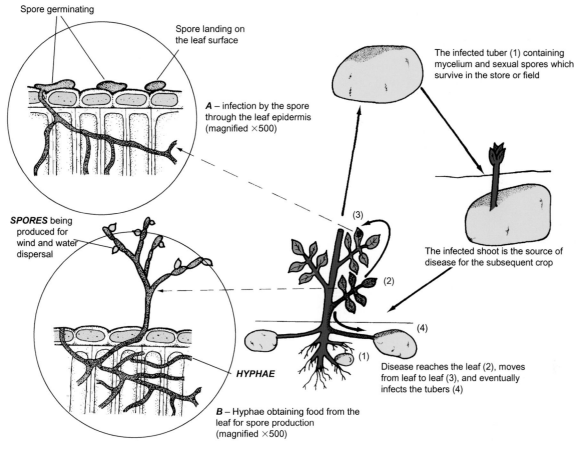

Spore germinating

Spore landing on
the leaf surface

A – infection by the spore
through the leaf epidermis
(magnified ×500)

The infected tuber (1) containing
mycelium and sexual spores which
survive in the store or field

SPORES being
produced for
wind and water
dispersal

The infected shoot is the source of
disease for the subsequent crop

(3)

(2)

(4)

(1)

HYPHAE

Disease reaches the leaf (2), moves
from leaf to leaf (3), and eventually
infects the tubers (4)

B – Hyphae obtaining food from the
leaf for spore production
(magnified ×500)

Figure 19.5 Potato blight life cycle

quickly. Later in the crop, badly infected plants may
have tuber infection as rainfall washes down spores
into the soil.

Control. The gardener should use clean seed bought
from a reputable source, choose resistant varieties and
apply a protective fungicide such as copper sulphate
(Bordeaux mixture) before damp conditions appear.
It is strongly recommended that a full cover of the
upper and lower surfaces of the leaves be achieved to
prevent even the smallest of initial infections, because
this fungus has such a potential for rapid increase.
There is, currently, no systemic fungicide (see p. 204)
available to gardeners for the control of potato blight.

Early potato cultivars usually complete tuber
production before serious blight attacks. Resistant
potato cultivars such as 'Sarpo Axona' prevent rapid
build-up of disease, although resistance may be
overcome by newly occurring fungal strains.

Predicting blight outbreaks

Nationally broadcast warnings are given on
the radio, based on temperature and humidity
measurements. UK government department

(Defra) scientists give a forecast to growers
when they have measured a 'Smith period'
(conditions most favourable for spore infection).
This period requires two consecutive days
where the minimum temperature is 10°C or
above and when each day has at least 11 hours
with a relative humidity greater than 90%. In
this way, protectant sprays of chemicals can be
applied before infection takes place.

Damping off (*Pythium and Phytophthora species*)

These two 'fungi' belong to the Zygomycota.

Damage. These two similar genera of 'fungi' cause
considerable losses to the delicate seedling stage. The
infection may occur below the soil surface, but most
commonly the emerging seedling plumule is infected
at the soil surface, causing it to topple (Figure 19.6).

Occasionally the roots of mature plants such as
cucumbers are infected, turn brown and soggy, and
the plants die.

Rose plants often have high levels of *Pythium* (along
with parasitic nematodes) around their roots as they

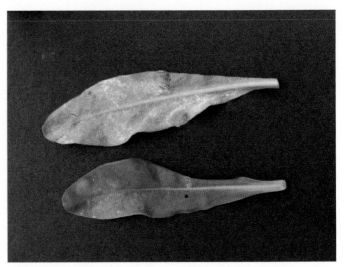

Figure 19.7 Cabbage downy mildew symptoms on a stock leaf. Note that the infection is on the lower leaf surface

Figure 19.6 Damping off on seedlings. Note the shrivelled appearance of the leaves on infected plants

get older. Although the mature plant is not seriously affected, it is a common experience that on removal of the plant and replacement with a young rose plant, there is a quite rapid decline in its vigour, called 'rose sickness'.

Life cycle and spread. Both *Pythium* and *Phytophthora* occur naturally in soils as saprophytes, but under damp conditions they produce the asexual spores that cause infection. These spores are spread by water. Sexual spores (oospores) are produced in infected roots (mostly in autumn) and may survive several months of dry or cold soil conditions.

Control. Prevention control is best achieved against these diseases by providing a disease-free growing medium. This may be produced by using fresh compost, or by partial sterilization of soil with heat. Seed producers often coat the crop seed with a protective seed dressing such as thiram to prevent early infection.

Water tanks with open tops, harbouring rotting leaves, are a common source of infected water and should be cleaned out regularly. Sand and capillary matting on benches in greenhouses should be regularly washed in hot water. The use of door mats soaked in a sterilant may prevent foot spread of the organisms from one greenhouse to another. Waterlogged soils

should be avoided, as these fungi increase most rapidly under these conditions. Copper formulation (known as 'Cheshunt mixture') may be used as a drench to slow down the increase of damping off. No systemic fungicide is currently available to gardeners for the control of 'damping-off' disease.

Downy mildew of cabbage and related plants (*Perenospora brassicae*)

This fungus is a member of the Zygomycota group of fungi. Downy mildews are not closely related to powdery mildews.

Damage. This serious disease causes a white bloom mainly on the under-surface of leaves (Figure 19.7) where the humid microclimate favours infection and spore production. Cruciferous plants such as stocks and wallflowers, cabbage and other brassicas, and occasionally weeds such as shepherd's purse are attacked by this 'fungus'. The disease is most damaging at a time when seedlings are germinating, particularly in spring when the young tissues of the host plant are susceptible and favourable damp conditions may combine to kill off a large proportion of the developing plants.

Life cycle and spread. Asexual spores (zoospores) are produced by hyphae present on the lower leaf surface, mainly in spring and summer, and are spread by wind currents. Thick-walled sexual spores (oospores) produced within the leaf tissues fall to the ground with the death of the leaf and survive the winter. Spring infections occur when rain splash carries the spores up from the soil to the lower leaf surface of seedlings and young plants.

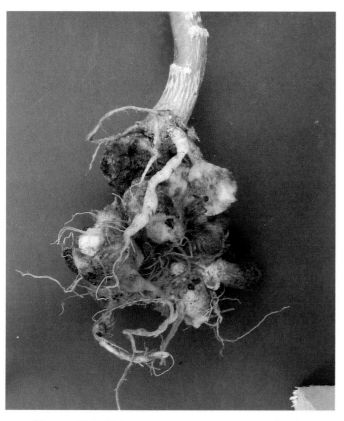

Figure 19.8 Club root symptoms on roots of cabbage

Control. It is not advisable to grow brassicas year after year in the same field, and particularly not to sow in spring next to overwintered crops.

The gardener can use a product containing the protectant fungicide copper oxychloride. As complete a cover of the spray as possible ensures that the 'fungus' is prevented from establishing itself.

Other crops such as lettuce and onions are attacked by equally important, but different species of downy mildews (*Bremia lactucae* and *Perenospora destructor*, respectively). No cross-infection is seen between crops such as cabbage, lettuce and onions, which belong to different plant families.

Club root (*Plasmodiophora brassicae*)

This fungus is classified into a quite separate group of fungi, the Plasmodiophorales.

Damage. It causes serious damage to most members of the Brassicaceae (Cruciferae) family, which includes edible crops such as cabbage, cauliflowers and Brussels sprouts; and ornamentals such as wallflowers, stocks and *Alyssum*. Infected plants show signs of wilting and yellowing of older leaves, and often severe stunting. On examination, the roots appear stubby and swollen (Figure 19.8), and may show a wet rot.

Life cycle and spread. The club root organism survives in the soil for more than five years as minute spores which germinate to infect the root hairs of susceptible plants. The fungus is unusual in forming a jelly-like mass (plasmodium), not hyphae, within the plant's root tissues. The plasmodium stimulates root cell division and causes cell enlargement, which produces swollen roots. The flow of food and nutrients in phloem and xylem is disturbed, with consequent poor growth of the plant. With plant maturity, the spores produced by the plasmodium within the root are released as the root rots. The disease is favoured by high soil moisture, high soil temperatures and acid soils.

Although this fungus does not spread much in undisturbed soils, it can easily be carried on infected plants, or on tools and wheels of garden machinery such as rotavators. In peat soil-growing areas, high winds may carry the disease a considerable distance.

Control. Several physical and cultural control measures may be used by the gardener. Rotation greatly helps by keeping cruciferous crops away from high spore concentrations in the soil. Liming of soil inhibits spore activity (see p. 176). Recently released cultivars of summer cabbage (such as 'Kilaxy'), cauliflowers (such as 'Clapton'), swedes (such as 'Invitation') and Brussels sprouts (such as 'Crispus F1') are claimed to have strong resistance to club root. Autumn-sown plants establish in soil temperatures less favourable to the disease and are normally less infected. Composting infected brassica plants should be avoided. Transplants grown in a club root-free bed are more able to establish when planted into vegetable plots. No chemical controls against this disease are at present available to gardeners.

Powdery mildew on strawberry (*Podosphaera aphanis*)

This belongs to the Ascomycota group of fungi. Powdery mildews should not be confused with downy mildews (see p. 255) that are an unrelated group of fungus-like organisms. Powdery mildews prefer hot dry conditions whilst downy mildews occur most commonly in cool damp conditions.

Damage. The first symptoms on strawberry are often purple spots seen on the upper leaf surface, and sometimes on the lower leaf surface. A slightly fluffy dry white infection (see Figure 19.9) is then seen gradually covering areas of the upper leaf surface. Darkened areas on the upper leaf may indicate infection on the lower leaf surface. Leaves eventually turn brown and leaf edges curl. The loss of photosynthesis in the leaves leads to reduced plant

19

Figure 19.9 Strawberry powdery mildew on leaves (source: James Lindsey's Ecology of Commanster)

growth and reduced fruit production. The mildew can infect strawberry flowers, causing a deep pink colouration. Affected fruit may appear distorted, and often look rather dull in colour with the seeds protruding out from the fruit surface. Infection of young fruit may cause small shrivelled strawberries.

Life cycle and spread. The fungus produces chains of asexual spores (conidia) from infected leaf surfaces. These spores are carried by wind to uninfected leaves. Above about 15°C, these spores are able to infect leaf tissue (mainly the upper surface) without there being a surface layer of water, although humid conditions and dampness do favour the disease. It can thus be seen that powdery mildews are well adapted for infection in summer conditions. Strawberry powdery mildew becomes more serious towards the end of summer as night humidity increases. It is also more serious in plastic-tunnel-grown strawberries, where temperatures and humidity are high.

Overwintering of the fungus is seen in two ways. During autumn, tiny (1 mm) black spherical spore cases (containing many spores) develop on the infected leaves and survive on dead leaf material. In mild winters, the fungus may survive within the green shoots of the dormant plant.

Spread is by means of summer spores. New planting material may introduce the disease. This species of powdery mildew (*P. aphanis*) is not harboured by any other garden species other than hops.

Cultural controls. Weak-looking plants may be removed in autumn as they will be the most likely ones to harbour the disease over winter. Organic matter added to soil helps maintain nutrient balance and soil moisture, and thus help maintain plant resistance to the disease.

Avoid overcrowding of plants that speeds up the disease spread. Remove any infected shoots when the disease is seen early in summer and avoid shaking this plant material that may release spores. Water in the morning and apply to the soil and not the leaves. Plants should not be allowed to dry out, as the weakened condition of the plants favours infection. Avoid over-fertilizing the plants with nitrogen fertilizers as this leads to increased susceptibility.

Resistance. At the present time, some mid-season cultivars such as 'Pegasus' and some late cultivars such as 'Florence' show a good resistance to this disease.

Chemical controls. Potassium bicarbonate, used as a spray by organic growers, helps provide resistance to the disease in the leaf cells. Sulphur may be applied as a fine dust. It may also be applied in a spray formulation, containing a fatty-acid ingredient that improves retention on the leaf. Myclobutanil is a systemic fungicide that may effectively reduce the disease.

Other powdery mildews

Apple mildew (*Podosphaera leucotricha*) survives the winter as mycelium within the buds, which often appear small and shrivelled on twigs that have a dried, silvery appearance. Fruit may develop a russeted surface. This fungus may affect other related species such as pears, quinces, medlars and ornamental *Malus*. Pruning of silvered twigs, and sprays with sulphur or myclobutanil may be used when mildew begins to spread through the tree.

Rose mildew (*Sphaerotheca pannosa*, Figure 19.10c), gooseberry mildew (*S. mors-uvae*) occurring mainly on the gooseberry fruit (see Figure 19.10b), and cucumber mildew (*S. fuliginea*) on cucumber and marrow leaves (Figure 19.10a) are three other important species.

It is important that gardeners realize that although hot dry summers may simultaneously lead to outbreaks of powdery mildew species in all the plant species mentioned above, cross-infection of the disease does not occur between strawberries, apple, rose, gooseberry and cucumber crops.

Rose black spot (*Diplocarpon rosae*)

This belongs to the Ascomycota group of fungi.

Damage. This common disease in garden and greenhouse roses is first seen as dark brown leaf spots which may be followed by general leaf yellowing and then leaf-drop (Figure 19.11). The infection of young shoots has a slow weakening effect on the whole plant.

Figure 19.10a (a) Cucumber powdery mildew on marrow leaves; (b) gooseberry powdery mildew infects the fruit; (c) rose powdery mildew

Figure 19.11 Rose black spot. Note the leaf-yellowing symptom that accompanies the black spot

Life cycle and spread. Asexual spores (produced within spore cases (see p. 253) embedded in the leaves) are released in wet and mainly warm weather conditions, and are then spread a few metres by rain drops or irrigation water before beginning the cycle of infection again. No overwintering sexual stage is seen in Britain and Ireland, and it is probable that asexual spores surviving in autumn-produced wood or in fallen leaves begin the infection process the following spring.

Control. Removal of fallen leaves is a very important aspect of control. Resistance is not common in rose cultivars. Gardeners can use a systemic fungicide

ingredient such as myclobutanil that reaches the internal leaf tissues and kills the fungus. Several sprays may be needed in each summer season.

Grey mould (*Botrytis cinerea*)

This is a fungus classified in the Deuteromycota group of fungi.

Damage. Grey mould is most commonly recognized by the fluffy, light grey fungal mass which follows its infection. In lettuce, the whole plant rots off at the base. The plant goes yellow and dies. In tomatoes, infection in damaged side shoots and light yellow spots (ghost spots) on the unripe and ripe fruit are found. Infected strawberry fruit may be covered by the fungus (Figure 19.12). In many flower crops (e.g. chrysanthemums), infected petals show purple spots which, in very damp conditions, lead to a mummified flower head. This disease may affect many crops.

Life cycle and spread. Grey mould normally requires wounded tissue for infection, which explains its importance in crops which are de-leafed, such as tomatoes, or disbudded such as chrysanthemums. Damp conditions are essential for its infection and spore production. The millions of spores are spread by wind to the next wounded surface. Black sclerotia (see p. 252), about 2 mm across, produced in badly infected plants, often act as the overwintering stage of the disease after falling to the ground, and are particularly infective in unsterilized soils on young seedlings and delicate plants such as lettuce.

Control. Preventive control may involve partial soil sterilization. Strict attention to greenhouse humidity control (particularly overnight) such as correct ventilation reduces the dew formation that is so important in the fungus infection. Cutting out of infected tissue is possible in sturdy stems such as

19

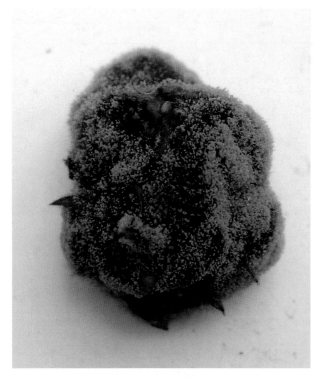

Figure 19.12 *Botrytis cinerea* infecting a strawberry fruit

tomatoes. Amateur gardeners at present have no effective chemical control against this disease.

Apple and pear canker (*Nectria galligena*)

This belongs to the Ascomycota group of fungi.

Damage. This fungus causes sunken areas in the bark of both young and old branches of ornamental Malus, apples or pears (Figure 19.13a). Poor shoot growth is seen, and the wood may fracture in high winds.

Life cycle and spread. The fungus enters through leaf scars in autumn or through pruning wounds during winter. Care is therefore necessary to prevent infection, particularly in susceptible apple cultivars such as the 'Cox's Orange Pippin' by avoiding pruning in mild, damp conditions. Spores produced in red spore cases (found embedded in the canker tissue) are spread by rain splash.

Control. Removal of cankered shoots may be necessary to prevent further infection, while in cankers of large branches, cutting out of brown infected tissue may allow continued use of the branch. Removed tissue should be burnt. The exposed wood area may be protected by a protective paste applied to the wound, especially in areas of the Britain and Ireland with high rainfall. Pruning knives should be cleaned after use on cankered tissue with

liquids (such as those containing citrus extracts). Amateur and professional horticulturists may apply a spray of copper (Bordeaux solution) at bud burst (spring) and leaf fall (autumn) to prevent entry of germinating spores.

Coral spot (*Nectria cinnabarina*)

This belongs to the Ascomycota group of fungi.

Damage. This disease is found on dead branches of trees and shrubs. Closer inspection shows a mass of pink pustules (about 1 mm across) sticking out from the dead wood (Figure 19.13b).

The disease is commonly seen on broadleaf species such as maple, horse chestnut, beech, hornbeam, lime, walnut, *Magnolia* and *Elaeagnus*. Cane fruit such as currant and gooseberry bushes are also quite susceptible. The conifer group of trees such as pine is not usually affected.

Life cycle and spread. This fungal species is a weak parasite, often living as a saprophyte on dead trees and on canes used to support climbing beans. The trees and shrubs mentioned above are infected by spores landing on dead wood, or through young tissue killed by frosts in late spring, or through lenticels (see p. 86). The fungus is then able to invade further into living wood. This infection is often made worse when the host plant has been subjected to stress such as drought, water logging or root disease. The pink pustules develop when the infected area of wood has died. The spores released from the pustules are spread by both wind and water splash.

Control. Pruning should be carried out in dry weather to reduce the chances of infection. Cuts are best located slightly away from junction points with the main branch or the stem, as wound healing is more rapid in this location. Infected stems should be cut out to leave healthy tissue and to minimize further spore production. Infected prunings should be taken away from the recently pruned area and burnt. Pruning paints (such as the one containing octane acid salts) are not normally recommended in dry weather, but may be of value when rain is forecast immediately after pruning, or where a particularly susceptible species has been pruned.

Rust fungi

This belong to the Basidiomycota group of fungi.

Rusts produce characteristic orange, brown or black raised leaf spots caused by the fungal tissue breaking through the leaf epidermis of the host (Figure 19.14).

Figure 19.13 (a) Apple and pear canker – note the swollen branch and exposed wood; (b) coral spot – note the swollen orange pustules

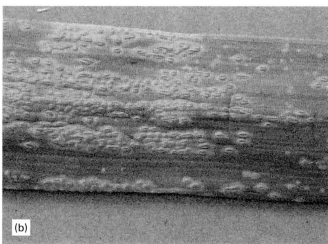

Figure 19.14 (a) Hollyhock rust; (a) leek rust

The rusts are a distinctive group of fungi often with complex life cycles involving up to five different spore-forms within the same fungal species.

Hollyhock rust (*Puccinia malvacearum*)

This is the most common disease on hollyhocks in UK. It is especially damaging in wet summers.

Damage. Leaves are usually the worst plant parts attacked. Starting on the lower leaves, there often appear numerous raised grey spots (see Figure 19.14a) on the **lower surface** of the leaf. Later in the summer, raised reddish-orange spots may be produced on the lower leaf surface. Flat yellow spots may show on the upper leaf surface indicating where the infection has taken place directly below. After severe levels of infection, the leaves often shrivel and die. **Stems** may show a cankerous appearance after infection. The **sepals** on the outer part of the

hollyhock flower can be affected in a similar way to the leaf. Continued infection in wet summers often leads to stunted plants.

Life cycle and spread. Two types of spore are involved in infection. Teliospores are relatively large and are able to survive the winter, and then, in damp conditions, infect young hollyhock shoots in late spring to begin the disease cycle. In mild winters, the fungus may live as **mycelium** (see p. 252) within the leaf and shoot tissues. In the warmer months, the infection cycle takes about 14 days from infection to mature leaf spot production; and at this time, the mass of hyphae and spores that have developed push through the lower epidermis to create the typical **raised spot** symptom. A new cycle of infection begins if the weather is damp. In the early summer period, the raised leaf spots are **grey** in colour from the numerous tiny **basidiospores** that are produced from the teliospores.

19

The much smaller size of these spores allows their **spread** by wind currents. Towards the end of the summer, spore production changes as teliospore production becomes more common, giving the leaf spots a richer reddish-orange colour. These much larger spores are carried more easily by rain droplets than by wind currents.

Hollyhock's related garden species in the family Malvaceae, *Abutilon, Hibiscus, Lavatera* and *Sidalcea* may be affected by this rust disease. The wild mallow (*Malva sylvestris*) may also be part of the rust cycle, especially if it is included as a plant member of a 'wildflower meadow' close to hollyhock plants.

Cultural control. This is not an easy disease to control since the tiny basidiospores are easily carried from garden to garden. Heavily infected plants should be removed and burnt. Gardeners should not use 'self-saved seed' as this may be infected by the fungus. Although hollyhocks are perennials, it is strongly advised that they be treated as biennials and removed after their flowering period. If rust disease levels are low, cut the stems off to ground level in early autumn and carefully dispose of all above-ground plant material. Avoid dense stands of hollyhock where high humidity favours infection. Avoid having other Malvaceous plants near to hollyhocks.

Chemical control. Several fungicides such as **difenoconazole** and **myclobutanil** are effective against the spores of this disease. They also are both systemic and thus potentially able to control the fungal mycelium inside the leaf and shoot tissues. However, the rapid increase of this rust once it becomes established, and the many spores travelling in from nearby gardens, often leads to inadequate chemical control. Regular applications may be needed **twice a week** to cope with newly arriving spores and newly expanding leaf tissue. The main advice, if fungicides are to be used, is that spraying should be started as **early** as possible in mid-May.

Honey fungus (*Armillaria mellea*)

This belongs to the Basidiomycota group of fungi.

Damage. The fungus primarily attacks trees and shrubs such as apple, lilac and privet. In spring the foliage wilts and turns yellow. Death of the plant may take a few weeks, or several years in large trees. Confirming symptoms are the toadstools (see Figure 19.1).

Life cycle and spread. The infection process involves rhizomorphs (sometimes referred to as 'bootlaces') that spread out underground from infected trees or stumps (see Figure 19.15) for a distance of 7 m and to a depth of 0.7 m. The infected stump may remain a

Figure 19.15 Honey fungus 'bootlaces' (rhizomorphs) that enable the fungus to move underground from a dead stump to infect other plants

serious source of infection for 20 years or more. The rhizomorphs are the main means of spread for this disease. The nutrients they are able to conduct provide the considerable energy required for the infection of the tough, woody shrub and tree roots. Mycelium then moves up the stem beneath the bark to a height of several metres and is visible (when the bark is pulled away) as white sheets, smelling of mushrooms. In autumn, clumps of light brown toadstools may be produced, often at the base of the stem. The millions of spores produced by the toadstools are not considered to be important in the infection process.

Honey fungus often establishes itself in newly planted trees and shrubs that have been planted too deeply. Too deep planting of shrubs and trees produces less vigorous plants that are more vulnerable to infection. Plant vigour is reduced because feeding roots which ideally should be growing near the surface of the soil have been located in the relatively infertile subsoil.

Control. This is difficult. Some genera of plants are less likely to be infected (see Table 19.1). Removal of the disease source (the infected stump) is strongly recommended. The use of a dead tree stump to hold a bird table is not a good idea. Specialist companies may be employed to quickly drill out the infected stump. Alternatively, with large stumps that are hard to remove, a surrounding trench is sometimes dug to a depth of 0.7 m (filled with coarse gravel) to prevent the progress of rhizomorphs.

Bacteria

These minute organisms (see Figure 19.2) measure about 0.001 mm and occur as single cells that divide

Table 19.1 Levels of resistance to honey fungus in garden shrubs and trees

Plant species	Latin name	Resistance level
Maple	*Acer spp.*	Susceptible
Box elder	*Acer negundo*	Very resistant
Birch	*Betula* spp.	Susceptible
Box	*Buxus sempervivans*	Resistant
Cedar	*Cedrus* spp.	Susceptible
Cypress	*Chamaecyparis lawsoniana*	Susceptible
	Cupressocyparis leylandii	Susceptible
Eleagnus	*Eleagnus* spp.	Resistant
Holly	*Ilex aquifolium*	Resistant
Privet	*Ligustrum* spp.	Susceptible
Lonicera	*Lonicera nitida*	Resistant
Mahonia	*Mahonia* spp.	Resistant
Apple	*Malus* spp.	Susceptible
Pine	*Pinus*	Susceptible
Cherry and plum	*Prunus* spp.	Susceptible
Laurel	*Prunus laurocerasus*	Resistant
Rhododendron	*Rhododendron*	Susceptible
Sumach	*Rhus typhina*	Resistant
Lilac	*Syringa* spp.	Susceptible
Tamarisk	*Tamarix* spp.	Resistant
Yew	*Taxus baccata*	Vert resistant

Figure 19.16 Fireblight on cockspur hawthorn (*Crataegus crus-galli*)

rapidly. They are important in the conversion of soil organic matter (see Chapter 13), but may, in a few parasitic species, cause serious damage or losses to horticultural plants. In hotter, tropical countries, bacteria are much more common as the cause of plant diseases.

Fireblight (*Erwinia amylovora*)

Damage. This disease, which first appeared in the Britain and Ireland in 1957, can cause serious damage to members of the Rosaceae family. Individual branches wilt, the leaves rapidly turning a chestnut brown (see Figure 19.16). When the disease reaches the main trunk, it spreads to other branches and may cause death of the tree within six weeks of first infection, the general appearance resembling a burnt tree (hence the name of the disease). Badly infected plants produce a bacterial slime on the outside of the branches in humid weather. On slicing through an infected stem, a brown stain will often be seen. Pears, hawthorn and *Cotoneaster* are commonly attacked, while apples and *Pyracantha* suffer less commonly.

Life cycle and spread. The bacterium is spread by bees as they pollinate, by harmful insects such as aphids and by small droplets of rain. Humid conditions and temperatures in excess of 18°C, which occur from June to September, favour the spread. Natural plant openings such as stomata and lenticels are common sites for infection. Flowers are the main point of entry in pears. The bacterial slime mentioned above is an important source of further infections. Fireblight, once notifiable nationally, must now be reported only in fruit-growing areas.

Control. The compulsory removal of the susceptible 'Laxton's Superb' pear cultivar in the 1960s eliminated a serious source of infection. Preventive measures such as removal of badly infected plants to prevent further infection, and removal of hawthorn hedges close to pear orchards, help in control. Careful pruning, 60 cm below the stained wood of early infection, may save a tree from the disease. Wounds should be sealed with protective paint, and pruning implements should be sterilized with liquids such as those containing citrus oils.

Bacterial canker of Prunus (*Pseudomonas mors-prunorum*)

Damage. This disease affects the plant genus *Prunus*, which includes ornamental species, and fruits such as

19

Figure 19.17 Bacterial canker of *Prunus*

plum, cherry, peach and apricot. Symptoms typically appear on the stem as a swollen area which often shows stem-cracking and exudes a light brown gum (see Figure 19.17). The angle between branches is the most common site for the disease. Severe infections girdling the stems cause death of tissues above the infection, and the resulting brown foliage can resemble the damage caused by fireblight. In May and June, leaves may become infected; dark brown leaf spots 2 mm across develop and the infected area may be blown out by heavy winds to give a 'shot-hole' effect.

Life cycle and spread. The bacteria present in the cankers are mainly carried by wind-blown rain droplets, infecting leaf scars and pruning wounds in autumn, and young developing leaves in summer.

Control. Preventive control involves the use of resistant rootstocks and scions (e.g. in plums). The careful cutting out of infected tissue followed by the application of a suitable paint and the use of autumn sprays of a copper compound (Bordeaux mixture) help to reduce this disease.

Soft rot (*Erwinia carotovora*)

This bacterium affects stored potatoes, carrots, bulbs and corms, where the bacterium's ability to dissolve the cell walls of the plant results in a mushy soft rot. High temperatures and humidity caused by poor ventilation promote infection through lenticels. A related strain of this bacterium causes blackened stems (black leg) on potato plants. Spread is mainly caused by infected planting material. In stores, the bacterium may be spread by insects or by liquid oozing from infected plant material. Preventive control measures are important. Plant material should be damaged as little as possible when harvesting, and diseased or damaged specimens should be removed before storage. Hot, humid conditions should be avoided in store. No curative measures are available.

Crown gall (*Agrobacterium tumifasciens*)

This bacterium affects apples, grapes, peaches, roses, *Euonymus* and many herbaceous plants. The disease is first seen just above ground level as a swollen, canker-like structure, often about 5 cm in size, growing out of the stem. It may occasionally cause serious damage, but usually is not a very important problem. The bacterium is able to survive well in soils, and infects the plant through small wounds in the roots.

Control of crown gall depends on cultural control methods such as disease-free propagating material, avoiding wounds at planting time and budding scions to rootstocks (rather than grafting) to avoid injuries near the soil level.

This bacterium is of special scientific interest in the area of plant breeding, having the ability to pass its genetic information to that of the plant cell. It does this by means of a small unit of DNA called a 'plasmid'. This plasmid ability of *A. tumifasciens* has been harnessed by plant breeders to transfer genetic information between unrelated plant species. Thus, it is the properties of this bacterium, more than any other organism, that have led to 'genetically modified' (GM) crops.

Viruses

Structure and biology of viruses

Viruses are extremely small; much smaller even than bacteria (see Figure 19.2). The light microscope is unable to focus in on them, but they can be seen as rods or spheres when seen under an electron microscope. The virus particle is composed of a DNA or RNA (see p. 77) core surrounded by a protective protein coat. On entering a plant cell, the virus takes over the organization of the cell nucleus to produce many more virus particles. Since the virus itself lacks any cytoplasm cell contents (see p. 77), it is often considered to be a non-living unit.

The virus's close association with the plant cell nucleus presents difficulties in the production of a curative virus control chemical that does not also kill the plant. No established commercial 'viricide' has yet been produced against plant viruses. Crops that are propagated vegetatively (such as potatoes) may continue indefinitely with the virus inside them if there is no strategy to remove the virus.

In recent years the broad area called 'virus diseases' has been closely investigated. Virus particles have, in most cases, been isolated as the cause of disease (e.g. cucumber mosaic). Other agents of disease to be discovered have been viroids (e.g. in chrysanthemum stunt disease), and these are even smaller than viruses.

Mycoplasmas (the cause of diseases such as aster yellows) are not related to viruses, but are a group of bacteria that induce symptoms similar to those produced by viruses.

Figure 19.18 Cucumber mosaic virus

Spread. A number of organisms (collectively called 'vectors') spread viruses from plant to plant and then transmit the viruses into the plant. Peach-potato aphid is reported as transmitting over 200 types of virus (e.g. cucumber mosaic) to different plant species. The aphid stylet (see p. 232) injects salivary juices containing virus into the parenchyma and phloem tissues, enabling the virus to travel to other parts of the plant. 'Persistent virus transmission' is seen in some vector/virus combinations such as peach-potato aphid with potato virus X, and *Xiphinema* dagger nematode transmits arabis mosaic. This persistence occurs because the virus is able to survive and increase within the vector's body for several weeks. However, in many vector/virus combinations such as plum pox, the virus survives only briefly as a contaminant on the insect's stylet, with the result that the vector does not spread the virus very far amongst a population of plants. Other examples of vector/virus combinations include bean weevils/broad bean stain virus and *Olpidium* soil fungus/big vein agent on lettuce. Important methods of spread also involve vegetative material (e.g. chrysanthemum stunt viroid

and plum pox), infected seed (e.g. bean common mosaic virus), seed testa (e.g. tomato mosaic virus) and mechanical transmission by hand (e.g. tomato mosaic virus).

Symptoms. The presence of a damaging virus in a plant is recognizable to gardeners only by means of its symptoms. (Commercial: growers may need to consult a virology specialist, whose identification techniques include electron microscopy, transmission tests on sensitive plants such as *Chenopodium* species and serological reactions using specific antiserum samples.)

Leaf mosaic, a yellow mottling, is the most common symptom (e.g. cucumber mosaic virus; see Figure 19.18). Other symptoms include leaf distortion into feathery shapes (cucumber mosaic virus), flower colour streaks (e.g. tulip break virus), fruit blemishing (tomato mosaic and plum pox), internal discoloration of tubers (tobacco rattle virus, causing 'spraing' in potatoes) and stunting of plants (chrysanthemum stunt viroid). Symptoms similar to those described above may be caused by misused herbicide sprays, genetic 'sports', poor soil fertility and structure (see deficiency symptoms, p. 267) and mite damage (e.g. on raspberry leaves).

In the following descriptions of seven important viruses, Latin names of genus and species are not included, since no consistent classification is yet accepted.

Cucumber mosaic

Damage. Several strains of virus cause this disease. In addition to cucumber, the following may also be affected: spinach, celery, tomato, *Pelargonium* and

19

Petunia. On cucumbers, a mottling of young leaves occurs (see Figure 19.18) followed by a twisting and curling of the whole foliage, and fruit may show yellow sunken areas. On the shrub *Daphne odora*, a yellowing and slight mottle is commonly seen on infected foliage, while *Euonymus* leaves produces bright yellow leaf spots. Infected tomato leaves are reduced in size (fern-leaf symptom).

Life cycle. The virus may be spread by infected hands, but more commonly an aphid (e.g. peach-potato aphid) is involved. Many crops such as lettuce, maize, *Pelargonium* and privet, and weeds such as fat hen and teasel, may act as a reservoir for the virus.

Control. Since there are no curative methods for control, care must be taken to use resistance and cultural methods. Examples of resistant cultivars available to the gardener are: 'Defender' courgette, 'Bonica' aubergine and 'Crispy Salad' cucumber. Choice of uninfected stock is vital in vegetatively propagated plants such as *Pelargonium*. Careful control of aphid vectors may be important where susceptible crops such as lettuce and cucumbers are grown in succession, and when crops are next to other susceptible species. Removal of infected weeds, particularly from greenhouses, may help prevent widespread infection.

Tomato mosaic (tobacco mosaic)

Damage. This disease may cause serious losses in the Solanaceae family of plants that includes tomatoes. Infected seedlings have a stunted, spiky appearance. On more mature plants leaves have a pale green mottled appearance, or sometimes a bright yellow ('aucuba') symptom. The stem may show brown streaks in summer when growing conditions are poor, a condition often resulting in death of the plant. Fruit yield and quality may be lowered, the green fruit appearing yellow/bronze and the ripe fruit hard, making the crop inedible (Figure 19.19).

Life cycle and spread. The virus is able to enter plant leaves through microscopic wounds such as broken leaf hairs, or via the broken bases of de-leafed side shoots. The period from plant infection to symptom expression is about 15 days. The virus may survive within the seed coat (testa) or endosperm of the tomato seed. It is very easily spread on fingers of gardeners as it is present in large numbers in the leaf hairs of infected plants.

Control. Heat treatment of dry seed at 70°C for four days by seed merchants helps to remove initial infection in susceptible cultivars. Peat growing bag and nutrient-film methods enable the grower to avoid

Figure 19.19 Tomato mosaic symptoms on tomato fruit

soil infection on roots of the previous crop. Hands and tools should be washed in soapy water after working with infected plants. Clothing may harbour the virus.

Cultivars and rootstocks containing several factors for resistance are commonly grown. 'Shirley', 'Sonatine', 'Cherry Wonder' and 'Estrella' are a few of many that at present claimed have resistance. The virus' ability to produce new strains always brings the possibility of a variety's resistance breaking down.

Leaf roll virus of potato

Potatoes are one of the gardener's favourite crops. The occurrence of this important virus disease in a crop that is vegetatively propagated (see p. 133) presents problems for control.

Damage. The leaves of the plant show an upward rolling (see Figure 19.20) caused by an accumulation of starch in the pallisade mesophyll (see p. 215). These leaves often are light green in colour; and may turn red on the upper side, and purple on the lower side. The leaves may also cause a rattling sound when shaken against each other. Leaf roll virus does not show any symptoms in the tubers, but can cause a serious reduction in potato yield, especially when the gardener re-uses tubers from the previous year's crop, and when aphid numbers are high.

Life cycle and spread. The main vector of leaf roll is the peach-potato aphid (*Myzus persicae*, see also p. 232). The aphid, while feeding on the sugars present in phloem tissues of an infected potato, will take up the virus, especially if it feeds for a few hours. The virus may multiply in the aphid to such an extent that the insect remains an active vector for

Figure 19.20 Leaf roll virus on potatoes. The leaves are thickened and roll upwards

the whole growing period of the crop. The aphid is likely to move from plant to plant and may be blown considerable distances by wind currents. A high spring incidence of aphids usually results in increased levels of leaf roll.

Control. A range of resistance and cultural controls are available. A cultivar such as 'Valor' has resistance to leaf roll (also to 'spraing', a nematode-transmitted virus that causes brown staining in tubers, and to some strains of potato cyst nematode). Do not grow your potato crop from last year's tubers, but buy certified potato 'seed' (see also p. 194). Remove any plants showing leaf roll symptoms. Some gardeners spray their potato crop in June with an insecticide such as deltamethrin against the aphid vector, but the virus may already be in the plants before spraying takes place.

Reversion disease on blackcurrants

Damage. This virus disease, caused by blackcurrant reversion virus, can seriously reduce blackcurrant yields. Flower buds on infected bushes are almost hairless, and appear brighter in colour than healthy buds. Infected leaves often have fewer main veins than healthy ones (Figure 19.21). After several years of infection, the bush may cease to produce fruit. The virus is spread by the blackcurrant gall mite, and reversion infected plants are particularly susceptible to attack by this pest. Removal and burning of infected plants is an important form of control. Use of certified plant material, raised in areas away from infection and vectors, is strongly recommended. Control of the mite vector in spring and early summer is described on p. 245.

Figure 19.21 Reversion disease on blackcurrant. Note that the infected leaf (bottom) has fewer main veins and leaf lobes than the healthy leaf (top)

Tulip break

The petals of infected tulips produce irregular coloured streaks and may appear distorted. Leaves may become light green and the plants stunted after several years' infection.

The virus is spread mechanically by knives, while three aphid vectors are known: the bulb aphid in stores, the melon aphid in greenhouses, and the peach-potato aphid outdoors and in greenhouses. Preventive control must be used against this disease. Removal of infected plants in the garden prevents a source of virus for aphid transmission.

Plum pox

This virus disease, also called 'Sharka', has increased in importance in Britain and Ireland since 1970, after its introduction from mainland Europe. Plums, damsons, peaches, blackthorn and ornamental plum are affected, while cherries and flowering cherries are immune. Leaf symptoms of faint inter-veinal yellow blotches can best be seen on leaves from the centre of the infected tree. The most reliable symptoms, however, are found on fruit, where sunken dark blotches are seen. The virus is spread by several species of aphids. The speed of spread is quite slow because the virus is not able to live and multiply in the aphid. Preventive control is the only option. Certified planting stock should be used. Aphid-controlling insecticides should be applied in late spring, summer and autumn. Suspected infected trees should be reported.

Arabis mosaic

This virus infects a wide range of horticultural crops. On strawberries, yellow spots or mottling are produced on the leaves, and certain cultivars become severely stunted. On ornamental plants (e.g. *Daphne odorata*), yellow rings and lines are seen on infected leaves, and the plants may slowly die back, particularly when this virus is associated with cucumber mosaic inside the plant. Several weeds (e.g. chickweed and grass species) may harbour this disease. The virus is spread by a common soil-inhabiting nematode, *Xiphinema diversicaudatum*, which may retain the virus in its body for several months. Certified virus-free soft fruit planting material can be bought.

Plant disorders

There are several symptoms that show on plant leaves, stems and flowers that are not caused by pests or diseases. The main causes are: frost, low temperature, high temperature, shade, drought, water-logging (see p. 151), high humidity (such as 'rose balling' and oedema), incorrect soil pH, nutrient deficiencies and excess fertilizer (see p. 170), fasciation (see also p. 271) and reversion.

A plant disorder is a condition in the plant resulting from a non-living (abiotic) factor such as an environmental factor, a nutrient or water being present at the incorrect level, or a genetic disorder.

Frost

Plants differ in their tolerance (hardiness) to low temperatures. Low temperatures slow down the plant's growth. Frost often causes the above-ground parts of sensitive plants to blacken (see Figure 19.22) and then collapse into a mess of dead tissue (after ice has formed inside the plant and fractured all the cells). *Pelargonium* species left to grow outside in autumn will die when the first frost occurs. At the beginning of the growing season, potatoes planted too early in spring may be killed by late frosts in June (see also p. 21).

A list of the hardiness of garden plants is summarized on the companion website: www.routledge.com/cw/adams.

Figure 19.22 Frost-damaged plant, showing browning of the petals

Low temperature

This usually cause slow plant growth, resulting from a reduced rate of photosynthesis. Plant species vary in their temperature preferences. *Cyclamen coum* grows well under cool conditions. Snapdragons (*Antirrhinum* spp.) grow better when night temperatures are no higher than 12°C, while most bedding plants prefer a higher night temperature.

High temperature

The effects of high temperature may differ with each species. As the species approaches its upper level of tolerance, photosynthesis slows down due to enzymes becoming less efficient. Plants such as sweetcorn originating from the tropics have different photosynthesis systems to help cope with this problem (see p. 112). The direct effects of high temperature are usually to cause leaf margins to dry off before the whole leaf dies (see Figure 19.23).

Shade

House plant species are sometimes placed in parts of the house unsuitable for their ideal growth. For example, a poinsettia, needing high light, will grow poorly in the back corner of a room. Plants outdoors may suffer from the same oversight. *Pelargonium* used as bedding plants should be given full sunlight and will develop a pale foliage colour if placed in a shady spot. *Impatiens*, however, is able to withstand considerable shade and maintain its rich, dark green foliage.

Drought

The plant needs sufficient water to carry nutrients around, to be present as an ingredient for making sugar, to transpire from the leaf in order to keep a desirable leaf temperature, and to maintain turgidity

Figure 19.23 (a) Healthy *Salvia* plant; (b) high temperature effect on a *Salvia* plant. The leaves are wilting, and edges of leaves are going dark brown

in some plant tissues. In some plant species, leaves change from shiny to dull as a first signal of water stress, and also may change from bright green to a grey green. New leaves wilt, but in species such as holly and conifers only the very youngest leaves wilt. Flowers may fade quickly and fall prematurely. Older leaves often turn brown, dry and fall off. Digging a few centimetres into the soil may indicate the need for watering with shallow-rooted perennials and annual border plants. Shrubs with deep roots rarely need watering, although transplanted older shrubs may show summer water stress for a number of years (see also Chapters 10 and 12).

Water-logging

Overwatering (see Figure 19.24) replaces the air spaces in soil and growing composts with water, thus excluding the oxygen necessary for root respiration that is needed to supply energy for root growth and nutrient uptake. Overwatering symptoms may include the following: the whole plant may wilt, the lower leaves turn yellow and drop. New foliage may have brown spots. The whole plant may become stunted, and stems and roots become brown and decayed.

Figure 19.24 Waterlogging effect on Begonia. The plant has collapsed, and leaves are yellowing and browning

Oedema

The condition called oedema is seen as raised corky spots on the under surface of leaves. Species such as *Pelargoniums* (Figure 19.25), *Rhododendron*, *Begonia*, pansies, violets and some fleshy-leaved plants such as *Peperomia* are affected. Orchids can show oedema on their petals.

19

Figure 19.25 Oedema on *Pelargonium* leaf. Note the raised, corky spots

This effect is most often seen in greenhouses, where overwatering, lack of heating or poor ventilation can affect plant growth. Oedema occurs when the roots' ability to supply water exceeds the leaves' ability to release the water by transpiration. Conditions favouring oedema occur most commonly in late winter and early spring, especially during extended periods of cool, cloudy weather. Warm, moist soil or compost occurring alongside cool, moist air brings on the condition most severely. The symptoms are commonly seen in unheated greenhouses. The problem can be greatly reduced by glasshouse heating and automatic venting.

Rose balling

This is a condition in which the outer petals of the rose flower die and become stiff. The inner petals are thus prevented from emerging to produce a normal bloom. The flower bud remains globular in shape (see Figure 19.26). Rose balling is caused by damp, cool conditions, and is most often seen in places that are shaded for part of the day. The wet outer petals are not able to dry out before being scorched by the direct sun. The outer petals are turned first into a soft dead tissue that quickly dries to form a quite tough layer, preventing any petals inside from emerging. Rose cultivars that have many slender petals ('double-petalled') are particularly prone to this problem. The withered blooms eventually drop off the plant.

Affected buds should be cut off. Care should be taken not to increase the problem by splashing water onto buds when watering roses. Grey mould (see p. 258)

Figure 19.26 Rose balling. Note the papery, dead petals on the outside of the rose flower

may grow on damaged buds and cause stem die-back. In such circumstances, shoots should be pruned back to the healthy part of the stem.

Soil pH

Garden species of plants have originally developed in different habitats and different parts of the world (see Chapter 2). Most species grow best at pH 6 and 7 but may have a tolerance to more acid or more alkaline conditions (see p. 175). Outside their limits of tolerance, roots may be subjected to two problems: an inability to absorb a balanced level of nutrients and an inappropriate balance of soil micro-organisms. Above-ground, growth is stunted and main veins may become blackened (see Figure 19.27).

Lime-induced chlorosis

This is a quite common condition that illustrates how an unsuitably high pH may affect nutrient uptake and plant growth (see Chapter 14, p. 171). Here, the leaves of the plant show an unhealthy yellow appearance, especially between the veins (interveinal chlorosis, see Figure 19.28). Young leaves are particularly affected. Raspberries, top fruit, *Hydrangea*, *Rhododendron*, *Camellias* and *Skimmia* are examples of plants prone to this problem.

Figure 19.27 Effect of a low pH ericaceous compost on *Salvia* spp. Note the stunted growth and blackened main leaf veins

Figure 19.28 Lime-induced chlorosis on raspberry. Note the pale areas between the main veins

Nutrient deficiencies

Each nutrient (the most common being nitrogen, phosphorus, potassium, calcium and magnesium) is required in the correct amounts to enable the plant to carry out its chemical processes. When nutrients are present in too low amounts, deficiencies begin to appear, usually showing as leaf symptoms (see pp. 169–171). Care should be taken to provide regular applications of a suitable fertilizer, especially during the summer months and in situations where the roots are restricted (as in pots). Two common gardening problems should be noted.

Figure 19.29 Blossom end rot

Figure 19.30 Bitter pit in apple fruit. Numerous small brown spots appear on the fruit surface in late summer

Blossom end rot (Figure 19.29) occurs on tomatoes and peppers. It produces a symptom of a black, concave lesion on the fruit that looks at first sight like a fungal disease. It is caused by an imbalance between potassium and calcium in the soil or compost. It occurs most often when the soil or compost is allowed to dry out while the fruits are swelling. It is seen more often in greenhouse container-grown plants than in plants growing in the open garden or greenhouse borders. It is most common when plants are grown in grow bags, where they have a small, shallow root run that dries out easily. Although there is no cure for blossom end rot once the symptoms begin to appear, the obvious recommendation is that fruiting crops should never be allowed to have dry roots.

Bitter pit in apples (Figure 19.30). Here the fruit develop many small, dark brown, sunken pits. The tissues below are stained to depth of about

19

Figure 19.31 (a) Fasciation on *Sedum* species – note the flattened stem in the affected (lower) plant; (b) fasciated flower

2 mm. Cultivars such as 'Bramley's Seedling' and 'Egremont Russet' are most susceptible. Young overbearing trees show the worst effects. The disorder is thought to be caused by low boron and calcium levels in the fruit, influenced by irregular water supply in the tree. Five recommendations are given to help prevent this problem: ensure a steady water supply to the tree during dry spells; mulch around trees to help moisture retention; summer prune young, vigorous trees, especially when they are holding too many fruit; use occasional foliar sprays of calcium nitrate plus detergent in the evening during summer; drench the soil with a dilute borax solution at the rate of about 15 g of borax per 10 square metres to the soil.

Excess fertilizer

When soluble fertilizers are present at too high levels, roots are scorched and are unable to provide nutrients for the other parts of the plant, often resulting in the death of the plant. This condition is sometimes described as high conductivity (see osmosis p. 121). Careful consideration of the appropriate frequency and amounts of fertilizer will prevent this embarrassing situation.

Fasciation

This is a condition in which the plant stem grows abnormally because of a disturbance in the stem apex (see p. 82). Fasciated stems are often wider and flatter than normal stems (see Figure 19.31). This disturbance may be caused by aphids sucking into the tender meristem, or by a mechanical disturbance. Sometimes, fasciation results in an abnormal growth of flowers around a central flower. Gardeners have used the genetic tendency towards fasciation shown in plants such as the fasciated willow *Salix udensa* 'Sekka', and have perpetuated it as a desirable type by

Figure 19.32 Reversion from the variegated to the green leaf state in *Euonymus fortunei*

means of propagation. Cockscomb (*Celosia argentea* var. *cristata*), however, is a garden annual of tropical origin, that carries the genetic tendency to fasciation from generation to generation through its seed.

Reversion

Many plant species have been selected and propagated by botanists and gardeners for

Table 19.2 Some symptoms of diseases and physiological disorders

Symptom	Cause	Other cause
Leaf spot	Fungus, e.g. apple scab	Bacterial canker (Prunus)
Raised leaf spots	Rusts	Oedema (croky spots)
White covering on leaf	Powdery mildew – upper	Spraying hard water
	Downy mildew – lower	
Leaf yellowing	Low nitrogen levels	Root disease
Brown edge to the leaf	Low potassium	
Leaves curl and go brown	Underwatering	
Dry, crumbling leaves	Plants overheated	Too much fertilizer
Yellow leaf veins	Low magnesium/iron	
Dark coloured leaves	Low phosphorus	
Lower leaves yellow	Wilt fungus	Overwatering
Yellow/green leaf mottle	Virus mosaic	Mutation/chimaera
Fruit spots	Fungus	Bitter pit (apple)
Sunken fruit lesions	Blossom end rot (tomatoes)	
Bud or leaf drop	Sudden change of temperature	
Stems elongated	Too little light	
Whole plant wilts	Severe underwatering	Wilt disease, vine weevil grubs
Fluffy mould	*Botrytis (grey)*	*Penicillium (blue)*
Brown stem lesions	Tomato mosaic virus	
Swollen woody stems	Fungal/bacterial canker	
Oozing from woody stems	Bacterial canker/fireblight	
Brown roots	Root rot	Overfertilizing

Many examples of variegations are genetically chimaeras that may revert, by producing shoots and leaves with the original green leaf colour.

Pruning out the green shoots that appear may require constant attention and patience.

Symptoms of disease and plant disorders

Table 19.2 presents a summary of the most important symptoms to help the reader 'home-in' on disease problems and physiological disorders.

Further reading

British Crop Protection Council (2014) *UK Pesticide Guide*. BCPC.

Brown, L.V. (2008) *Applied Principles of Horticulture*. 3rd edn. Butterworth Heinemann.

Buczacki, S. and Harris, K. (2005) *Pests, Diseases and Disorders of Garden Plants*. Collins.

Costello, L.R. (2003) *Abiotic Disorders of Landscape Plants* University of California, Agriculture and Natural Resources

Greenwood, P. and Halstead, A. (2009) *Pests and Diseases*. Dorling Kindersley.

Hessayon, D.G. (2009) *Pest and Weed Expert* Transworld Publishers.

RHS (2013) Fungicides for home gardeners. Online: www.rhs.org.uk/media/pdfs/advice/fungicides. RHS Advisory Service.

19

their interesting leaf appearance, especially leaf variegation (see p. 90). Common examples of these are *Euonymus fortunei* 'Emerald and Gold' (see Figure 19.32) and Norway maple (*Acer platanoides* 'Drummondii'). Variegation is also seen in varieties and cultivars of species such as *Hedera, Cornus, Daphne, Ilex* and *Eleagnus*.

Please visit the companion website for further information:
www.routledge.com/cw/adams

Glossary

active transport	movement of a substance into a cell across the cell membrane against a concentration gradient. It requires energy and substances are taken up selectively
adult growth	reproductive growth
adventitious buds	buds which do not arise from a stem
adventitious roots	roots which do not derive from the radicle of the embryo
aerobic respiration	process by which sugars are broken down in the presence of oxygen to yield energy, the end products being carbon dioxide and water
air filled porosity (AFP)	percentage of air in a growing medium immediately after it has stopped draining having been saturated with water
alternate host	a shrub/weed/garden plant that harbours a pest or disease
anaerobic respiration	respiration in the absence of oxygen, with the release of energy, carbon dioxide and ethanol
androecium	the male part of the flower consisting of the stamens
angiosperms	seed bearing flowering plants
annual	a plant that completes its life cycle within a growing season
apical meristem	area of cell division at the root and shoot tips responsible for increase in length
asexual reproduction	the formation of new individuals without fusion of gametes
autotroph	see producer
Available Water Content (AWC)	the water held in the soil between field capacity and the permanent wilting point
base dressing	fertilizer that is applied to the soil and worked into the seedbed, or incorporated in composts, before sowing/planting
binomial	a two part name made up of a generic epithet and a specific epithet
biodiversity	encompasses the genetic variation within a species, the total number of species present and the variety of habitats they live in
biological control	the use of a natural enemy to reduce the damage caused by a pest or disease
biomass	the weight or volume of living plant and animal material in an ecosystem
biome	a group of communities with their own type of climate, vegetation and animal life. It covers a large geographical area
bitter pit	small spots on apple fruits resulting from insufficient calcium and boron in the fruit
blossom end rot	black lesions produced on tomato fruit, caused by insufficient watering
buffering capacity	the ability of the soil to retain nutrients including lime against loss by leaching
bulb	an underground modified shoot which mainly consists of food storage leaves (scale leaves)
calcicoles	plants that are adapted to grow on calcareous soils
calcifuges	plants that are adapted to grow on acid soils below pH 5.5

calyx	collective name for all the sepals
capillarity	the movement of water against gravity within thin tubes
carnivore	an organism that feeds on animal tissue
carpel	the female reproductive unit made up of a stigma, style and ovary. Individual carpels are often fused together
centre of origin	a geographical area where plant diversity is high
clay	soil particle less than 0.002mm in diameter
clay soil	soil with more than 35% clay particles
climax community	the final community in the successional sequence
community	a group of populations in a given area or habitat
compatible pollen	pollen which germinates and grows enabling delivery of male gametes
companion planting	an association of plant species that derives benefit of some kind from each other
compost	a growing medium used for growing plants in containers
composting	the decomposition of organic matter in a pile before it is applied to soils
compound fertilizers	those that supply two or more of the nutrients nitrogen, phosphorus and potassium
consumers	organisms which feed on other living organisms
corm	a swollen and compressed underground stem which acts as a food storage organ
corolla	collective name for all the petals
cross pollination	pollen transfer between flowers on different plants
cultivar	a variation within a species which has usually arisen, and has been maintained, in cultivation
cultivated plant	a plant that does not exist in the wild anywhere
cultural control	a procedure, or manipulation of the growing environment, that results in weed, pest, or disease control
deciduous	a plant that sheds all its leaves at once, often at the end of the growing season
decomposer	an organism that breaks down dead organic matter (see saprophyte)
dehiscent fruit	a fruit that splits open to release the seeds
detritivore	an organism that feeds on decomposed organic matter
development	the changes in structure, form and behaviour that take place through a plant's life cycle
differentiation	the changes that take place in a cell, tissue or organ enabling it to perform a specific function
diffusion	the movement of a substance from a high concentration to a lower concentration
dioecious	plants with male and female flowers on different plants
dormancy	the condition when viable seed fails to germinate even when all germination requirements are met
drainage	the removal of gravitational (excess) water
ecoystem	a community of living organisms together with their non-living environment functioning together as a unit
ephemerals	a plant that has several life cycles in a growing season
epigeal germination	germination in which the cotyledon/s emerge above the ground, initially enclosed in the testa
erosion	the movement of rock fragments or soil
essential minerals	inorganic substances necessary for the plant to grow and develop. Often referred to as 'nutrients'
etiolation	abnormal growth occuring in low light levels
evergreen	a plant retaining leaves in all seasons

false fruit	a fruit containing structures derived from flower parts other than the ovary (a pseudocarp)
fasciation	an abnormal flattened-stem or flower growth caused by a disturbance in the stem apex
fertigation	describes irrigation when it is used to deliver nutrients as well as water
fertilisation	the fusion of a male sex cell (the male gamete) from a pollen grain with a female sex cell (the female gamete) in the ovule to produce an embryo
fertilizer	a concentrated source of plant nutrients that are added to growing media
fibrous root	many roots growing from the base of the stem with no dominant root
Field Capacity (FC)	the amount of water the soil can hold against the force of gravity
foliar feed	describes the application of nutrients in a spray to the foliage
formulation	the read-to-use pesticide product containing active ingredient and other necessary ingredients
friable	the consistency of the soil when it is easily cultivated i.e. readily forms crumbs
fruit	the structure formed from the ovary wall usually after fertilisation
fungicide, contact	a fungicide that acts against fungi on the leaf surface
fungicide, systemic	a fungicide that travels through vascular tissues
gamete	sex cell
genus	a group of individuals within a family which have characteristics in common
germination of seeds	the emergence of the young root or radicle through the testa, usually at the micropyle
gravitational water	the water that can be removed from the soil by the force of gravity
growth	the increase in size (dry weight) of cells, organs or the whole plant
gymnosperms	seed bearing, non-flowering plants
gynoecium	the female part of the flower consisting of the carpels
habitat	an area occupied by a community of organisms
half-hardy	a plant able to survive temperatures between 1°C to −5°C
hardiness	a plant's ability to survive low temperatures
hardy	a plant able to survive below −5°C
hazard	something with the potential to do harm
heated glasshouse (hardiness)	a plant requiring temperatures above 5°C to survive
hemiparasite	a plant which is parasitic on other plants and also photosynthesises
herbaceous perennial	a perennial that is non-woody and generally loses its stems and foliage at the end of the growing season
herbicide, contact	a herbicide that acts at the point of contact
herbicide, residual	a herbicide that remains chemically stable and active over a period of months
herbicide, selective	a herbicide that kills only certain chosen weeds
herbicide, translocated	a herbicide that travels through pholem and xylem to reach all plant organs
herbivore	an organism that feeds on plant tissue
hermaphrodite flower	a flower with both male and female parts
heterotroph	see consumers
hydroponics	the cultivation of plants in nutrient solution without soils
hypogeal germination	germination in which the cotyledon/s remain below the ground inside the testa
incompatible pollen	pollen which fails to germinate or grow
indehiscent fruit	a fruit that does not split open to release the seeds
interspecific competition	competition for resources between individuals of different species
intraspecific competition	competition for resources between individuals of the same species
juvenile growth	non-reproductive growth

larva	immature stage of some insect
lateral meristem	area of cell division within the stem and root responsible for increase in width
lateral root	roots branching from a taproot
lime requirement	the quantity of calcium carbonate required to raise the soil pH to 6.5
liquid feed	describes a fertilizer solution watered on to soils or composts
manure	a source bulky organic matter comprising animal faeces and bedding
molluscicide	a pesticide effective against slugs and snails
monoculture	where only one species of plant is present
monoecious	plants with separate male and female flowers on the same plant
mulches	materials applied to the surface of the soil to suppress weeds, modify soil temperatures, reduce water loss, protect the soil surface and/or reduce erosion
mutualism	a relationship between two organisms which is beneficial to both
mycelium	a mass of fungal tissue
native plant	present at the end of the last Ice Age when Britain and Ireland separated from the rest of Europe
naturalized plant	introduced by humans and now reproduces in the wild
niche	the position or role of each species within a habitat
nymph	immature stage of some insect
oedema	a dry, corky spotting on leaves and petals of some greenhouse plants caused by excess root water-pressure
organ	a collection of plant tissues carrying out a particular function
osmosis	the movement of water from a high water (low solute) concentration to a low water (high solute concentration) across a selectively permeable membrane
parasite	an extreme form of predation where the predator often lives within the host/prey
parent material	the rock from which a soil is made
perennating organ	a storage organ which enables a plant to survive unfavourable seasons e.g. bulb, corm, tuber
perennial	a plant living through several growing seasons
Permanent Wilting Point (PWP)	the water content of the soil when a wilted plant does not recover overnight
pesticide	a chemical product used to control weeds, pests , and diseases
phloem	the tissue which transports sugars made in photosynthesis from the leaves to other plant organs
photosynthesis	the process in the chloroplasts by which green plants trap light energy from the sun, convert it into chemical energy and use it to produce food in the form of carbohydrates such as sugars and starch
phyllosphere	a community of bacteria and fungi that live on the leaf surface
physical control	a material, mechanical or hand control where the weed, pest or disease is directly blocked or destroyed
pioneer community	the first community in the successional sequence
plant disorder	a plant condition caused by a non-living factor (such as low temperature)
pollination	the transfer of pollen from stamen to stigma of a flower or flowers
population	a group of individuals of one species which interbreed
predation	a relationship between two organisms which is beneficial to one partner (the predator) and harmful to the other (the prey)
prickle	a sharp pointed outgrowth of the epidermis modified for defence
primary root	see taproot

producer	an organism which manufactures its own food from simple molecules e.g. green plants through the process of photosynthesis
propagule	any part of a plant that can be used to create a new plant
pupa	resting stage of some insects such as moths
quiescent seed	a viable seed which does not germinate because the environmental requirements (water, oxygen, a suitable temperature) are not present
response periods	when the plant/crop benefits from the addition of water
reversion	a return to a green-leaf condition in a plant selected for its variegated foliage
rhizome	a horizontal, generally underground stem which may act as a food storage organ
rhizomorph	'bootlaces' produced by some toadstool fungi, that are able to infect plants underground
risk	can be measured by the chances of something happening and the level of consequence if it did
rose balling	rose flower buds fail to open when outer petals are scorched and killed
rotation	planting particular crops in different plots each year
runner	a stem that runs along the ground and produces new plantlets at the nodes e.g. *Ranunculus repens* or at the tip e.g. *Fragaria* (strawberry)
sand	a soil particle between 0.06 and 2.0mm in diameter
saprophyte	an organism that lives on dead plant material (*see* decomposers and detritivores)
saturation point of soils	when water has filled all the soil pores (i.e. no air in the pores); waterlogging
scarification	physically damaging the seed coat to break dormancy
sclerotium	a small dark-coloured clump of fungal tissue that enables some fungi to survive for long periods
secondary root	see lateral root
seed	the structure that develops from the ovule after fertilization
self-pollination	pollen transfer between flowers on the same plant or within the same flower
semi-evergreen	a plant that retains some of its leaves through the year but may shed most leaves under severe weather conditions
sexual reproduction	the formation of new individuals through fusion of male and female gametes (sex cells)
shrub	a multistemmed woody perennial plant having side branches emerging from near ground level
silt	a soil particle between 0.002 and 0.06 mm in diameter
soil horizons	specific layers in the soil that can be seen by digging a soil pit
Soil Moisture Deficit (SMD)	the amount of water required to return the soil to field capacity (FC)
soil structure	the arrangement of particles in the soil
soil texture	the relative proportion of the sand, silt and clay particles in the soil
species	a group of individuals within a genus which have characteristics in common and are able to breed among themselves
spine	a sharp pointed leaf modified for defence and to reduce water loss
spore-case	a small fungal structure that contains spores
stolon	a long arching branch which roots at its tip to produce a new plant e.g. in *Rubus fruticosus*
stones	soil particles larger than 2 mm in diameter
straights	fertiizers that supply only one of the major nutrients
stratification	moist storage of seeds in cold or warm temperatures to break dormancy
subsoil	the layer (horizon) below the cultivated layer (topsoil) and lighter in colour because of its low organic matter level
succession	a sequence of changes in the composition of plants and animals in an area over time.

succulent fruit	a soft, fleshy fruit
sustainable	an action or process which provides the best for the environment and people, socially and economically, both new and indefinitely into the future
symptoms	a visible feature that is characteristic of a disease-causing organism or agent
taproot	a single large root, derived from the radicle
tender	a plant able to survive temperatures 1–5°C. Tolerant of low temperatures but will not survive frost.
tepal	a fused petal and sepal found in monocotyledonous flowers
thorn	a sharp pointed branch or shoot modified for defence
tilth	the crumb structure of the seedbed
tissue	a collection of cells carrying out a specific function
top dressing	a fertilizer applied to the surface of soils
topsoil	an upper layer (horizon) of soil (above the subsoil and below the litter layer) normally moved during cultivation. It is typically 10–40 cm deep and darkened by the decomposed organic matter it contains.
transpiration	the evaporation of water vapour from the leaves and other plant surfaces
transpiration pull	the main mechanism by which water moves in the xylem due to transpiration
tree	a large woody perennial unbranched for some distance above ground, on a single stem
tropism	a directional growth response to an environmental stimulus
tuber	a swollen underground food storage organ which may be formed from a root or stem
variety	a general, non-botanical, term for plants which vary from the species
vascular plants	plants with conducting tissue i.e. xylem and phloem
vector	an organism (such as an aphid) that transfers viruses or pollen from plant to plant
vegetative propagation	involves asexual reproduction and results in a clone
viable seed	seed that has the potential for germination when the required external conditions are supplied
water-holding capacity (WHC)	the amount of water held at field capacity
weathering	the breakdown of rocks (or soil particles)
weed, annual	a weed with one life cycle in a growing season
weed, ephemeral	a weed with several life cycles in a growing season
weed, perennial	a weed that lives through several growing seasons
woody perennial	a perennial that maintains a live woody framework of stems at the end of the growing season
xylem	the tissue which transports water and dissolved mineral nutrients from the roots, up the stem to the leaves and other plant organs
zygote	cell resulting from the fusion of male and female gametes

Index

Page numbers in *italic* indicate illustrations, **bold** indicates tables

Index